Natural Language Processing on Oracle Cloud Infrastructure

Building Transformer-Based NLP Solutions Using Oracle AI and Hugging Face

Hicham Assoudi

Apress®

Natural Language Processing on Oracle Cloud Infrastructure: Building Transformer-Based NLP Solutions Using Oracle AI and Hugging Face

Hicham Assoudi
Montreal, QC, Canada

ISBN-13 (pbk): 979-8-8688-1072-5
https://doi.org/10.1007/979-8-8688-1073-2

ISBN-13 (electronic): 979-8-8688-1073-2

Copyright © 2024 by Hicham Assoudi

This work is subject to copyright. All rights are reserved by the Publisher, whether the whole or part of the material is concerned, specifically the rights of translation, reprinting, reuse of illustrations, recitation, broadcasting, reproduction on microfilms or in any other physical way, and transmission or information storage and retrieval, electronic adaptation, computer software, or by similar or dissimilar methodology now known or hereafter developed.

Trademarked names, logos, and images may appear in this book. Rather than use a trademark symbol with every occurrence of a trademarked name, logo, or image we use the names, logos, and images only in an editorial fashion and to the benefit of the trademark owner, with no intention of infringement of the trademark.

The use in this publication of trade names, trademarks, service marks, and similar terms, even if they are not identified as such, is not to be taken as an expression of opinion as to whether or not they are subject to proprietary rights.

While the advice and information in this book are believed to be true and accurate at the date of publication, neither the authors nor the editors nor the publisher can accept any legal responsibility for any errors or omissions that may be made. The publisher makes no warranty, express or implied, with respect to the material contained herein.

    Managing Director, Apress Media LLC: Welmoed Spahr
    Acquisitions Editor: Celestin Suresh John
    Development Editor: Laura Berendson
    Coordinating Editor: Gryffin Winkler

Cover designed by eStudioCalamar

Cover Image by sfkjrgk from Pixabay

Distributed to the book trade worldwide by Apress Media, LLC, 1 New York Plaza, New York, NY 10004, U.S.A. Phone 1-800-SPRINGER, fax (201) 348-4505, e-mail orders-ny@springer-sbm.com, or visit www.springeronline.com. Apress Media, LLC is a California LLC and the sole member (owner) is Springer Science + Business Media Finance Inc (SSBM Finance Inc). SSBM Finance Inc is a **Delaware** corporation.

For information on translations, please e-mail booktranslations@springernature.com; for reprint, paperback, or audio rights, please e-mail bookpermissions@springernature.com.

Apress titles may be purchased in bulk for academic, corporate, or promotional use. eBook versions and licenses are also available for most titles. For more information, reference our Print and eBook Bulk Sales web page at http://www.apress.com/bulk-sales.

Any source code or other supplementary material referenced by the author in this book is available to readers on GitHub (https://github.com/Apress). For more detailed information, please visit https://www.apress.com/gp/services/source-code.

If disposing of this product, please recycle the paper

To my family,

I can't thank you enough for your endless support and patience. You stood by me through all the late nights, weekends, and even during vacations when I had to keep working. To my wife, Imane, your understanding meant everything, and to my little one, Salim, your patience when I couldn't always be there is something I deeply appreciate. This book is as much yours as it is mine.

Table of Contents

About the Author xi

About the Technical Reviewers xiii

Acknowledgments xvii

Introduction xix

Part I: Foundations and Case Study Introduction 1

Chapter 1: NLP Essentials 3

Introduction to Natural Language Processing 3
    NLP Tasks 6
    NLP Key Concepts 8
    Common Challenges 14

Transformers for NLP 15
    Transformer Architecture 15
    Transformer Taxonomy 20
    Transfer Learning 23
    Hugging Face Ecosystem 25

Strategic Considerations for NLP Adoption 28
    Models 29
    Data 32
    Team 32

Summary 33

References 34

TABLE OF CONTENTS

Chapter 2: Oracle Cloud for NLP 35

Introduction to Oracle Cloud Infrastructure (OCI) 35

    History 35

    Core Concepts and Terminology 37

Oracle's AI Overview 52

    AI Strategy 52

    AI Stack 53

    OCI AI Services 54

    OCI ML Services 56

    AI Infrastructure... 57

OCI for NLP... 58

    OCI Language 59

    OCI Data Science 62

    OCI Data Labeling 64

    AI Samples... 66

    High-Level Flow for Building NLP Models Using OCI 67

Summary... 69

References 70

Chapter 3: Healthcare NLP Case Study 73

MedTALN Inc. Case Study 73

    Company Background 73

    Healthcare NLP.. 76

    Healthcare NER Initiative 79

Healthcare NER Inception 84

    Scope and Requirements 85

    Assembling the Team 88

Healthcare NER Elaboration 89

    Architectural Design 90

Solution Blueprint 97

High-Level Architecture	97
High-Level Approach	99
Project Preparation	101
Summary	104
Reference	105

Part II: Case Study Implementation 107

Chapter 4: Tenancy Preparation 109

Getting Started	109
Cost-Saving Strategies	109
OCI Tenancy Preparation	110
Compartment Creation	111
Network Configuration	114
Storage	119
Identity and Security	124
Data Science Environment Setup	139
Project	139
Notebook Sessions	141
Summary	170

Chapter 5: Dataset Preparation 171

Preliminaries	171
Labeled Datasets	172
Cost Saving	174
Dataset Life Cycle	179
Framing the Problem (Step 1)	181
Dataset Selection (Step 2)	182
Training Dataset Preparation	191
Dataset Collection and Wrangling (Steps 3 and 4)	193
Dataset Labeling (Step 5)	215

TABLE OF CONTENTS

    Dataset Creation (Step 6) 234

    Additional Notes 239

  Summary 246

  References 247

Chapter 6: Model Fine-Tuning 249

  Preliminaries 249

    Language Models (LMs) 249

    Healthcare-Specific Pretrained Language Models 256

    Cost-Saving Strategies for the Training Phase 259

    Transfer Learning–Based Fine-Tuning Workflow 260

  Pretrained Model Selection 262

    Framing the Problem (Step 1) 263

    MLM Model Selection from Hugging Face (Step 2) 265

  Healthcare NER Model Fine-Tuning 275

    Training Dataset Creation Notebook 276

    Training Notebook 282

  Healthcare NER Model Evaluation 299

    Evaluation Notebook 300

  Summary 317

  References 318

Part III: Case Study Deployment and Wrap-Up 321

Chapter 7: Model Deployment and Monitoring 323

  Model Inference Preliminaries 324

    Understanding Inference vs. Training 324

    Preparing the Environment 328

  Deployment Process 334

    Oracle Data Science Model Catalog 335

    Oracle Data Science Model Deployment 336

    Oracle ADS HuggingFacePipelineModel 338

Deployment Process Notebook 340
    Initializing the ADS Class "HuggingFacePipelineModel" 341
    Authenticate 341
    Save the Model to the Model Catalog... 347
    Deploy and Invoke 353
Monitoring and Maintenance 358
    Logs 359
    Metrics 361
Summary 362
References 363

Chapter 8: MLOps and Conclusion 365

MLOps with OCI Data Science 365
    OCI Data Science Pipelines 365
    Pipeline Example 366
    Pipeline Creation Step-by-Step 367
Journey Through NLP: From Theory to Practice 380
    Healthcare NER Model Life Cycle Summary 380
Responsible AI 391
Summary 394
Reference 395

Index 397

About the Author

Hicham Assoudi is an accomplished IT professional and AI expert with over 30 years of experience, including more than 25 years of specializing in Oracle technologies. He holds a PhD in Computer Science and is also an OCI Certified Architect. Hicham has held key roles such as Technology Manager at Oracle and IT Architect at IBM, offering highly specialized technical consulting to major corporations across Canada, the United States, and Europe.

Hicham's journey into AI began over a decade ago during his doctoral studies, where he initially focused on intelligent agents and general machine learning. He later specialized in Natural Language Processing (NLP) from the early days of Transformer models, positioning him at the forefront of NLP innovation. As the founder of typica.ai, an AI startup, Hicham applies cutting-edge research to practical NLP solutions, empowering organizations to leverage NLP for significant business impact.

In addition to his industry contributions, Hicham maintains strong ties to academia as an External Research Associate at the AI Lab of UQAM University in Montreal, bridging the gap between academic research and real-world applications.

About the Technical Reviewers

Karanbir Singh is an accomplished engineering leader with over seven years of experience leading AI/ML engineering, distributed systems, and microservices projects across diverse industries, including fintech and automotive. Currently working as a Senior Software Engineer at Salesforce, he focuses on backend technologies as well as AI. His career has been marked by a commitment to building high-performing teams, driving technological innovation, and delivering impactful solutions that enhance business outcomes.

At TrueML, as an Engineering Manager, he managed a critical team to develop and deploy machine learning models in production. He successfully expanded and led engineering teams, significantly improving feature development velocity and client engagement through strategic collaboration and mentorship. His leadership directly contributed to increased revenue, client retention, and substantial cost savings through innovative internal solutions. His role involved not only steering technical projects but also shaping the company's roadmap in partnership with data science, product management, and platform teams.

Previously, at Lucid Motors and Poynt, he developed critical components and integrations that advanced product capabilities and strengthened industry partnerships. His technical expertise spans across AI/ML, cloud computing, and software architecture, and he is adept at utilizing cutting-edge technologies and methodologies to drive results.

Karanbir holds a master's degree in Computer Software Engineering from San Jose State University and has been recognized for his innovative contributions, including winning the Silicon Valley Innovation Challenge. He is passionate about mentoring and coaching emerging talent and thrives in environments where he can leverage his skills to solve complex problems and advance technological initiatives.

ABOUT THE TECHNICAL REVIEWERS

Prashanth Josyula is a dynamic force in the tech world whose journey is marked by an unyielding passion for innovation and an extraordinary depth of expertise in both technical literature and software engineering. As a Principal Member of Technical Staff (PMTS) at Salesforce, Prashanth doesn't just meet expectations—he consistently exceeds them, pushing the boundaries of what's possible in technology.

With over 16 years of robust experience in the IT industry, Prashanth has mastered a multitude of programming languages and technologies, establishing himself as a true polyglot programmer. His proficiency spans a cross Java, Python, Scala, Kotlin, JavaScript, TypeScript, shell scripting, SQL, and an array of open source solutions. Since beginning his professional journey in 2008, he has delved into various domains, each time leaving a mark of excellence.

In the realm of Java/JavaEE and Spring, Prashanth has been instrumental in designing and building resilient, scalable backend systems that power critical applications across industries. His deep understanding of these technologies ensures robust and high-performance solutions tailored to meet complex business needs.

Prashanth's expertise in UI technologies is equally impressive. He has crafted intuitive, responsive user interfaces using frameworks like ExtJS, JQuery, DOJO, Angular, and React. His commitment to creating seamless user experiences shines through in every project, bridging the gap between complex backend processes and user-friendly front-end interfaces.

Venturing into big data, Prashanth has leveraged platforms like Hadoop, Spark, Hive, Oozie, and Pig to transform massive datasets into valuable insights, driving strategic decisions and innovations. His ability to harness the power of big data showcases his analytical mindset and his knack for tackling large-scale data challenges.

In the field of microservices and infrastructure, Prashanth has been a pioneer, in engineering robust and scalable solutions with cutting-edge tools like Kubernetes, Helm, Terraform, and Spinnaker. His contributions to open source projects reflect his commitment to collaborative innovation and continuous improvement.

Moreover, Prashanth is at the forefront of AI and machine learning, exploring and advancing the capabilities of these transformative technologies. His work in this area is characterized by a fearless approach to experimentation and a relentless pursuit of knowledge.

Each day for Prashanth is an exciting adventure, filled with opportunities to learn, innovate, and lead. His career is a testament to his dedication to advancing technology, not just for the sake of progress but to truly make a difference. With his unparalleled skills and a visionary mindset, Prashanth continues to inspire peers and push the envelope of technological possibility.

Ankur Goel is a seasoned Principal Solutions Engineer at Confluent Inc. and brings more than 18 years of versatile expertise, primarily in the digital native and fintech industries. With key roles as cloud architect, technical lead, and solutions architect, Ankur is a certified professional in AWS, Kafka, Oracle Exadata, and Hadoop, showcasing a profound understanding of distributed technology and cloud services.

In his current role, Ankur serves as a global advisor for key and strategic accounts, specializing in cloud platform adoption, event-driven architectures, and real-time stream processing. Noteworthy is his impactful contribution to establishing robust event-driven architectures for major digital native clients in the United States and leading complex database implementations for prominent banks and telecom giants in India.

Ankur's unwavering commitment to customer satisfaction, evident in his prior role as the primary database architect for multiple Yahoo websites, continues to resonate in his recent collaborations across industries. Beyond his professional pursuits, Ankur enjoys culinary endeavors and sports, adding depth and vibrancy to his multifaceted life.

Acknowledgments

Writing this book has been a rewarding journey, and I would like to express my gratitude to a few individuals who contributed to its completion.

First, I would like to thank the technical reviewers for their encouraging feedback and insightful comments. Your thoughtful reviews motivated me and ensured the content was as clear and accurate as possible.

I would also like to acknowledge the team at Apress for their support throughout the publishing process. Your professionalism and guidance made the experience smooth and enjoyable.

Lastly, I'm grateful to everyone who supported me throughout this journey, directly or indirectly. Your encouragement made a significant difference.

Introduction

Welcome to *Natural Language Processing on Oracle Cloud Infrastructure*. This book serves as a comprehensive guide to creating real-world NLP solutions on Oracle Cloud Infrastructure (OCI). The motivation behind this book stemmed from recognizing the need for a dedicated, all-encompassing guide to constructing NLP solutions on OCI. While existing resources are available, they are often dispersed and challenging to consolidate into a single source that systematically guides the entire NLP implementation process on OCI. This book aims to bridge that gap.

By combining OCI's robust infrastructure with cutting-edge NLP technologies, we will explore how to tackle practical challenges efficiently and deliver effective and cost-effective NLP models. Whether you are new to NLP or looking to leverage OCI for your current projects, this guide equips you with the insights and tools necessary for success.

The book is structured around the typical NLP project life cycle, starting with foundational concepts and progressing to advanced implementation. In Part 1, we begin with the essentials. Chapter 1 provides a comprehensive overview of NLP, guiding you through its evolution and key developments. Chapter 2 introduces the OCI ecosystem, focusing on the AI infrastructure and services best suited for building and scaling NLP models.

Chapter 3 presents the case study of MedTALN Inc., a fictional Canadian healthcare analytics company that serves as the backdrop for this book. MedTALN Inc. faces the unique challenge of developing a domain-specific Named Entity Recognition (NER) model for healthcare that supports French—an essential requirement in their Canadian context. This case study not only addresses the complexities of building healthcare-specific NLP solutions but also the added challenge of supporting multiple languages, including French. Through MedTALN's journey, we'll walk step-by-step through the implementation of this NER solution on OCI, demonstrating how to design, build, and deploy custom NLP models tailored to specific industry and language needs.

In Part 2, we shift focus to hands-on implementation. Chapter 4 guides you through setting up your OCI environment, covering configurations proven effective in production. Chapter 5 explains how to create a robust training dataset, starting with a prelabeled dataset from Hugging Face and enriching it using OCI's Data Labeling

INTRODUCTION

Service. In Chapter 6, we explore the process of fine-tuning pretrained language models for healthcare applications that support French, leveraging GPU-based OCI Data Science Notebooks and models sourced from Hugging Face.

Part 3 brings the project to completion, focusing on deployment and operationalization. Chapter 7 provides a detailed, step-by-step guide to deploying the NER model using OCI Data Science Model Deployment, streamlining the process for real-world applications. Finally, Chapter 8 addresses how to implement MLOps using OCI Data Science Pipelines, reflecting on key lessons learned and concluding with discussions on cost-effectiveness and responsible AI in NLP implementations.

The book primarily focuses on a fictional case study of building an NER model for French in the healthcare sector. However, its main goal is to guide you through the process of developing an NLP model from start to finish on OCI, regardless of the specific task, domain, or language. The techniques, strategies, and methods discussed are adaptable to other NLP tasks, such as sentiment analysis, and can be applied across industries like legal. Additionally, these approaches are applicable to a range of languages, from Spanish to non-Latin languages like Arabic.

As we embark on this journey, my goal is to provide not only the technical expertise but also the confidence to approach NLP projects with clarity and purpose. By consolidating best practices and practical insights, this book aims to be the resource I wished for when I first started working with NLP on OCI.

Let's dive in and start building impactful, scalable, and responsible NLP solutions on Oracle Cloud Infrastructure (OCI) together!

PART I

Foundations and Case Study Introduction

Part 1 aims to equip readers with a solid foundation in NLP, familiarize them with OCI's capabilities for building NLP solutions, and introduce the case study that will be developed in detail throughout the book.

Chapter 1, "NLP Essentials," introduces Natural Language Processing (NLP), offering a comprehensive exploration of its core principles, historical evolution, and the strategic dimensions of NLP implementations. The aim is to give readers a thorough understanding of NLP, emphasizing key concepts necessary for developing practical NLP-based applications.

Chapter 2, "Oracle Cloud for NLP," focuses on Oracle Cloud Infrastructure (OCI) and its capabilities for building advanced NLP solutions. This chapter delves into OCI's functionalities pertinent to artificial intelligence, specifically NLP. By the end of this chapter, readers will understand how OCI supports and facilitates the implementation of NLP solutions.

Chapter 3, "Healthcare NLP Case Study," lays the groundwork for implementing this book's practical example. From understanding the business drivers that led to this project to grasping the implementation blueprint, we provide a comprehensive view that equips readers with the understanding needed to build this practical example from scratch on OCI, as explored and explained in the book's upcoming chapters.

Collectively, these initial chapters not only ground readers in essential theoretical knowledge but also prepare them for the hands-on, applied exploration that follows, bridging the gap between understanding and implementation.

CHAPTER 1

NLP Essentials

In the digital era, there has been a significant increase in text data, ranging from social media content to corporate documents. To effectively utilize this data, a thorough grasp of Natural Language Processing (NLP) is essential. This chapter examines the evolution of NLP, from its initial stages to its contemporary advanced methods. We will discuss the core tasks and inherent challenges of NLP. Lastly, we will consider the key strategic choices that businesses encounter when integrating NLP solutions.

Introduction to Natural Language Processing

Natural Language Processing (NLP) represents a critical convergence of computer science and linguistics. Situated within the broad spectrum of artificial intelligence (AI), NLP aims to enhance machine comprehension of human language. This domain fuels a range of modern applications, from semantic search and AI-driven conversational tools to document translation and summarization. Moreover, advancements in NLP have paved the way for the generation of text, expanding the horizon of possibilities.

In today's digital era, the omnipresence of NLP, as showcased in the burgeoning popularity of systems like ChatGPT, might be perceived as a given. Yet, this wasn't always the case. The genesis of NLP can be traced back to the mid-20th century, a period synonymous with Alan Turing's groundbreaking Turing test, originally conceived as the "imitation game." Turing's audacious proposition centered around machines simulating humanlike intelligence, challenging our conventional wisdom on machine cognition and sowing the seeds for future machine–human interactions.

The initial forays into NLP were, understandably, primitive. Early techniques resembled a mosaic of basic rules and dictionary lookups, striving for complex tasks while often only achieving rudimentary ones. One can recall the 1954 Georgetown experiment, a milestone in machine translation, wherein a handful of Russian sentences were translated into English. It was an endeavor viewed as groundbreaking at its time, though history would prove that the layers of complexity within machine translation ran far deeper than initially estimated.

Enter Eliza, a creation from the esteemed corridors of the Massachusetts Institute of Technology (MIT), considered by many as one of the earliest chatbots. Crafted by Joseph Weizenbaum in the mid-1960s, Eliza simulated conversation by emitting strategically vague responses, hinting at an artificial emotional intelligence.

As the pages of NLP's history turned through the 1970s and 1980s, hand-coded rule systems were the norm until machine learning heralded a new dawn in the late 1980s. These novel systems leveraged statistical inference, fashioning models rooted in real-world data, thereby reducing the need for intricate hand-crafted rules. The turn of the millennium only accelerated this shift, with burgeoning data access paving the way for increasingly successful statistical methods.

The 2010s witnessed a revival of an old acquaintance: neural networks. Their ability to produce unparalleled results across AI disciplines, not least in NLP, coincided with the proliferation of online data and the advent of modern GPU capabilities. Word embeddings emerged as a sophisticated tool, mapping words into multidimensional spaces, bringing semantically similar terms closer and revolutionizing word representation.

While recurrent neural networks (RNN) and long short-term memory nets (LSTM) offered great promise in modeling sequences, they weren't without challenges. Enter BERT in 2018, a transformative deep learning model by Google, rooted in the "Transformer" architecture. Offering bidirectional understanding and remarkable training efficiency, Transformers, as exemplified by BERT, have redefined the benchmarks in NLP tasks, at times even eclipsing human performance.

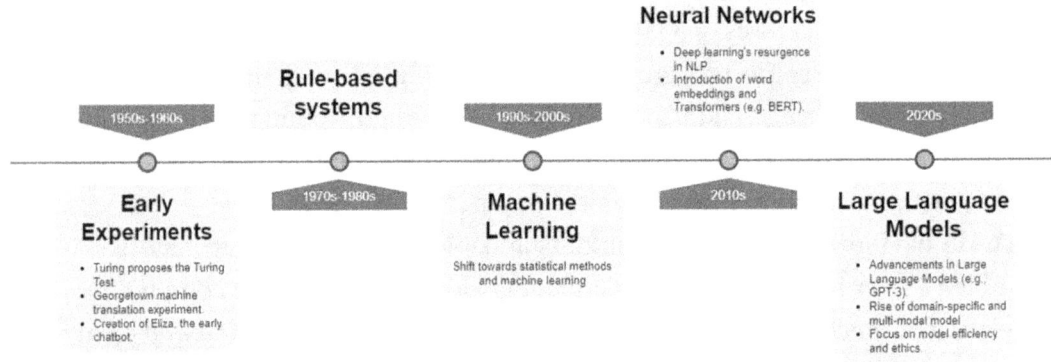

Figure 1-1. *NLP history timeline*

Natural Language Processing (NLP) has undergone an outstanding transformation over the years. Initially relying on rudimentary rule-based systems, NLP was hampered by limited adaptability and a narrow understanding of human language nuances. However, several pivotal advancements catalyzed its ongoing maturity. The exponential increase in available data, coupled with advances in computational power, particularly through GPUs, led to the development of more intricate machine learning models. The democratization of machine learning frameworks and the extensive research collaborations further fueled this evolution. Among these advancements, the emergence of large language models (LLMs)[1] stands out. Pioneers in this domain, such as OpenAI's ChatGPT, Facebook's LLaMA, Mistral AI's Mistral 7B, and Cohere LLM, epitomize the zenith of current NLP capabilities, showcasing how far the field has come and hinting at the potential that lies ahead.

Here's a chronological outline highlighting the key milestones in the continuous advancement of NLP's maturity (see Figure 1-1):

- Rule-Based Systems: NLP's inception was closely tied to rule-based systems. These systems leaned heavily on predefined linguistic rules. While foundational, their inability to handle nuanced linguistic variations was evident.

- Statistical and Machine Learning-Based Paradigms: As we transitioned away from hard-coded rules, the latter part of the 20th century saw a surge in statistical models. Machine learning models such as Naive Bayes, hidden Markov models, and conditional random fields came into play. Although they were adaptive to a degree, they often demanded extensive labeled datasets and occasionally overlooked intricate linguistic structures.

- Neural Network Paradigms

- Basic Neural Networks: With the dawn of deep learning, neural networks, specifically RNNs and CNNs, started playing a pivotal role in NLP, enabling the modeling of more complex linguistic relationships.

- Neural Network-Based Embeddings with Unsupervised Learning: Using neural networks and leveraging the power of unsupervised learning, significant strides were made in training on vast amounts of unlabeled data. Embeddings like Word2Vec laid the foundation for this era.

---

[1] The term "large" in LLMs refers to their size in terms of parameters.

- Transformers and LLMs: The Transformer architecture was a watershed moment in NLP. Models like BERT utilized vast amounts of data to train, and then, these pretrained models were refined for specific tasks—a practice termed as transfer learning. This era witnessed a rapid increase in model sizes, from BERT's 340 million parameters to massive models like OpenAI's ChatGPT-3 with 175 billion parameters.

Reflecting on this trajectory, the progression from rule-based systems to sophisticated LLMs underscores the ongoing evolution and promise of NLP.

In summary, the progression from foundational methods to advanced techniques has marked NLP's growth. With ongoing research and technological advancements, the field continues to make strides toward more efficient and nuanced language processing systems.

NLP Tasks

Modern Natural Language Processing (NLP) models are proficient in an expansive range of NLP tasks. These tasks serve as the foundation for numerous applications and are often integral components in intricate NLP architectures. Let's systematically explore these cornerstone NLP tasks:

- Sentiment Analysis: A widely embraced NLP task, sentiment analysis discerns the emotional tone of a given text, categorizing it as positive or negative. Its prevalent application is in the automated segmentation of customer feedback. Typically, deep learning architectures address this as a binary or multiclass classification challenge.

- Named Entity Recognition (NER): NER focuses on pinpointing named entities within a textual corpus. These entities can encompass individuals, geographical locations, institutional names, or temporal references. Recognized entities often aid downstream processes, such as enhancing the precision of search engine outcomes or flagging sensitive personal data for redaction.

- Machine Translation: The objective here is the transformation of a text snippet from its source language to a target language. In recent times, deep learning architectures have ushered in considerable advancements in the accuracy and fluency of translations.

- Text Summarization: This involves distilling lengthy content into a concise rendition. There are two primary approaches:

- Extractive Summarization: Selecting pivotal segments of the original content for representation.

- Abstractive Summarization: Generating a novel summary, ensuring syntactical correctness and coherence. Notably, training models for abstractive summarization is more intricate compared to its extractive counterpart.

- Question Answering: Often referred to as machine comprehension, this task necessitates generating answers from a reference text in response to posed questions. While deep learning has made notable strides here, challenges persist, especially when the reference text lacks a direct answer.

- Topic Modeling: This task is geared toward discerning dominant themes in a document and quantifying their prominence. Such an analysis is invaluable for understanding large-scale textual databases, ranging from corporate communication to historical archives or academic literature.

- Speech Recognition: Venturing into the auditory domain, speech recognition transcribes spoken content into textual format. While NLP predominantly focuses on textual content, the intertwined nature of language processing brings speech recognition within its ambit.

- Text to Speech (TTS): The converse of the previous task, TTS endeavors to vocalize written text. The process is twofold: Initially, the text undergoes a phonetic transformation via NLP, followed by a conversion into audio signals leveraging digital signal processing techniques.

- Language Generation: Central to NLP, this task pertains to the synthesis of context-aware and grammatically consistent text based on specific cues or prompts. The emergence of sophisticated models, epitomized by ChatGPT, has significantly elevated the caliber and versatility of automated language generation.

In summation, while this list encapsulates some pivotal NLP tasks, the domain's breadth extends further. Given the rapid innovations in deep learning, one can anticipate even more refined and powerful NLP capabilities in the forthcoming era.

NLP Key Concepts

Understanding the intricate process of developing modern NLP models goes beyond merely learning popular terminologies and jargon prevalent in NLP literature. It requires a deep comprehension of these concepts, their interrelationships, and their practical applications. This section provides an overview of some foundational elements, aiming to simplify and clarify the complex process of building NLP models.

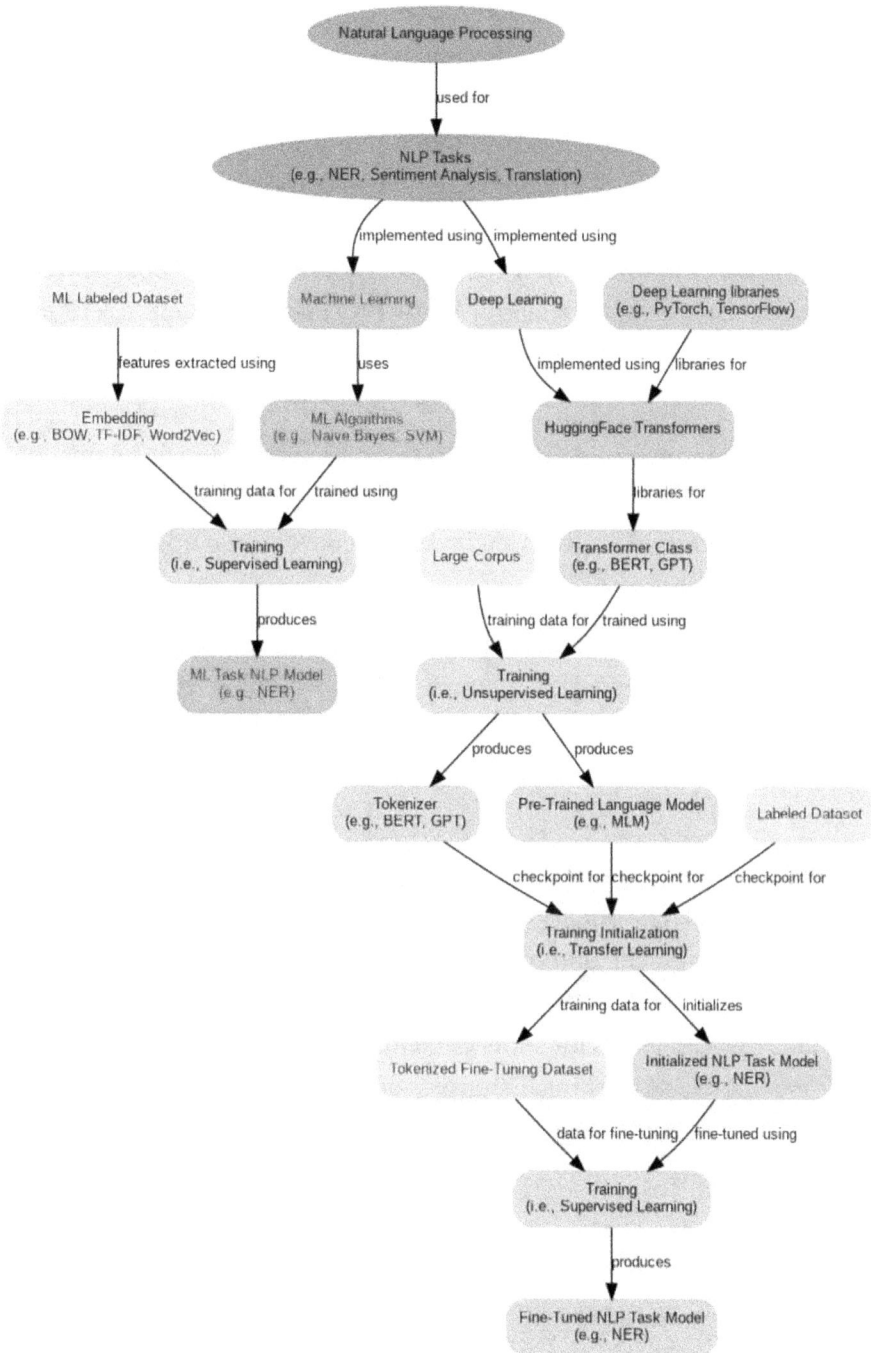

Figure 1-2. *Overview of Natural Language Processing (NLP) approaches and workflow*

CHAPTER 1   NLP ESSENTIALS

Figure 1-2 is a diagram designed to introduce the complex process of developing modern NLP models. It depicts some fundamental concepts and their relationships in a clear and structured manner, illustrating the top-down flow between these concepts. This visual representation simplifies the understanding of NLP model development approaches and workflow tasks, enhancing your comprehension of the essential techniques and methods that drive the creation of effective NLP models.

The following descriptions provide an overview of the key concepts depicted in the diagram. Each concept plays a crucial role in the process of developing NLP models, illustrating how different components interact to create effective NLP solutions.

- Machine Learning (ML): A subset of artificial intelligence that involves training models on data to make predictions or decisions without being explicitly programmed. Training ML-based NLP models involves algorithms such as Naive Bayes and Support Vector Machines (SVM).

- Deep Learning (DL): A subset of machine learning that uses neural networks with many layers (deep neural networks) to learn from data. Training DL-based NLP models leverages architectures such as Transformers, which are implemented using libraries like PyTorch and TensorFlow.

- Datasets: Datasets are collections of text used to train and evaluate NLP models. High-quality datasets are crucial for building effective models. They can be domain-specific (e.g., medical, legal) or general-purpose (e.g., Wikipedia, news articles). Datasets can be annotated with labels for supervised learning or remain unlabeled for unsupervised learning. Examples of popular NLP datasets include the Penn Treebank, IMDB reviews, and the Common Crawl.

- Tokenization: Tokenization, a crucial preprocessing step within datasets, divides text into smaller units known as tokens, typically words or subwords. This process is essential for converting raw text into a structured format that NLP models can use. Different tokenization techniques exist, such as word tokenization, subword tokenization (e.g., Byte Pair Encoding), and character tokenization. Effective tokenization helps ensure that datasets are appropriately prepared for training and evaluating NLP models.

- Language Models: Language models are a foundational component in NLP, enabling machines to understand and generate text. These models provide the basis for text prediction and generation, allowing machines to comprehend and produce human language in a meaningful way. They predict the probability of a sequence of words. It is fundamental to many NLP tasks, as it helps generate coherent text, translate languages, and understand context. Language models are trained on big datasets to learn the human language's underlying structures. Examples include the Transformer-based language models such as BERT, GPT, and T5.

- Large Language Models (LLMs): Large language models (LLMs), such as the renowned ChatGPT models (e.g., ChatGPT-3.5, ChatGPT-4, and ChatGPT-4o), represent a special class of Transformer-based language models. These models are trained on massive datasets using extensive GPU processing, often involving clusters of thousands of GPUs. Their vast training datasets and computational power enable them to perform exceptionally well in understanding and generating human language.

- Embeddings: Embeddings are numerical representations of words or phrases in a continuous vector space. They capture semantic meanings and relationships between words, allowing models to perform better on various NLP tasks. Word embeddings like Word2Vec, GloVe, and contextual embeddings from models like BERT and GPT transform text into dense vectors that capture semantic relationships between words in multidimensional space.

- Transformers: Transformers are a deep learning–oriented architecture that has revolutionized NLP. The Transformer model leverages self-attention mechanisms to process entire sentences simultaneously rather than sequentially. This allows for more efficient handling of long-range dependencies and has led to significant improvements in performance across various NLP tasks.

- Training: Training a model in NLP involves exposing a model to large amounts of text data to learn language patterns and structures. Training can be supervised (using labeled data) or unsupervised (using unlabeled data). This process typically includes data preparation, model training, and evaluation. Evaluation involves validating the model's performance on a separate dataset that was not used during training, ensuring that the model generalizes well to new, unseen data and preventing overfitting.

- Pretraining: Pretraining involves creating a pretrained language model by training on massive unlabeled datasets using unsupervised learning (usually publicly available data on the Internet).

- Fine-Tuning: Fine-tuning is a training process that adapts a pretrained model to a specific task or domain. This involves further training the model on a smaller, task-specific dataset (supervised learning). Fine-tuning allows models to leverage general language understanding from large-scale pretraining while specializing in particular applications, such as Named Entity Recognition, sentiment analysis, translation, or question answering.

As depicted in Figure 1-2, there are two primary approaches for building models in NLP: the ML-based approach and the DL-based approach. Each approach has distinct methodologies and applications, significantly influencing how NLP models are trained to learn patterns from textual data. Understanding these differences helps approach each NLP project's specific requirements and constraints with the appropriate set of tools and approaches.

- Machine Learning-Based Models for NLP: This approach involves algorithms like SVM and Naive Bayes alongside feature extraction methods such as Bag of Words and TF-IDF. Machine learning in NLP typically involves algorithms like Support Vector Machines (SVM), Naive Bayes, and decision trees, combined with manual feature extraction techniques such as Bag of Words or Term Frequency-Inverse Document Frequency (TF-IDF). This approach relies on transforming raw text into numerical features that these algorithms can process. Feature extraction methods like Bag of Words convert text into fixed-length vectors based on word counts or occurrences, while TF-IDF adjusts these counts by the importance of words across documents. Once the text is converted into numerical features, traditional ML algorithms are applied to classify, cluster, or make predictions based on these features. This approach works well with smaller datasets and requires less computational power compared to deep learning. However, it often struggles with capturing complex linguistic patterns and long-range dependencies in text.

- Deep Learning-Based Models for NLP (with Transformers): This approach includes advanced neural network models like BERT, GPT, and other Transformer-based architectures that leverage self-attention mechanisms to process text data. Deep learning with Transformers has brought about a paradigm shift in NLP. It harnesses the power of neural networks that autonomously learn features and complex patterns directly from raw text data. This approach is exemplified by models such as BERT (Bidirectional Encoder Representations from Transformers) and GPT (Generative Pretrained Transformer). Transformers use self-attention mechanisms to process entire sentences simultaneously, enabling them to capture long-range dependencies and contextual relationships between words more effectively than traditional ML methods. While deep learning models typically require large datasets and significant computational resources to train, they offer superior performance on various NLP tasks, from language translation to sentiment analysis and text generation.

The major differences between machine learning (ML) and deep learning (DL) approaches in NLP lie in their methodologies, algorithms, training data requirements, and computational requirements. ML-based NLP focuses on manual feature extraction and simpler algorithms, making it suitable for projects with smaller datasets and limited computational power. However, it may struggle with capturing complex linguistic patterns and long-range dependencies in text. In contrast, DL-based NLP, particularly with Transformer models, leverages advanced neural networks that automatically learn from raw text. This approach requires larger datasets and significant computational resources but achieves state-of-the-art performance on diverse NLP tasks by effectively handling long-range dependencies and contextual relationships. Understanding these differences allows you to choose the appropriate approach based on each NLP project's specific requirements and constraints, ensuring optimal performance and resource utilization.

While this overview of essential NLP notions might not be exhaustive and may not be sufficient to grasp every concept or technique in the NLP field, it can serve as a good starting point. The realm of NLP is vast, but throughout this book, we will have the opportunity to dig into specific concepts and see how they are applied in real-world projects.

Common Challenges

Natural Language Processing, while a groundbreaking field, isn't devoid of intricacies and hurdles. Understanding its challenges is essential for harnessing its true potential. Here's a breakdown of the significant challenges faced by NLP systems:

- Ambiguity: The inherent nature of language brings about ambiguities. A single word can have varied meanings, requiring NLP systems to decipher the intended one, akin to solving a complex computational puzzle.

- Polysemy and Homonymy: Linguistic phenomena like polysemy and homonymy present unique challenges. Words like "bark" (referring to a dog's sound) and "bark" (referring to a tree's outer layer) necessitate NLP systems to derive meaning based on context.

- Contextual Understanding: The meaning of words often hinges on surrounding content. Hence, an NLP system must understand the entirety of the input rather than isolated words, ensuring accurate semantic interpretation.

- Data Limitations: The efficacy of an NLP model correlates with the volume and diversity of the data it's trained on. Sparse data, especially for certain languages or niche topics, can constrain the system's learning capabilities.

- Multilingualism: With thousands of languages worldwide, each with its unique grammar and semantics, NLP systems face the arduous task of understanding and processing multiple languages seamlessly.

- Domain Adaptation: Specialized jargon, such as those in the medical or legal fields, demands specialized understanding. NLP systems must be adept at domain-specific nuances to process such information accurately.

- Ethical and Bias Considerations: An essential aspect of NLP development is ensuring systems are free from biases. Addressing and eliminating potential prejudices in NLP models is paramount for ethical and balanced outputs.

Indeed, the advent and proliferation of large language models (LLMs) in the NLP landscape have brought to the forefront a unique set of challenges. Model hallucinations, or the generation of information not present in the input, present issues of reliability and accuracy. Furthermore, the significant computational costs associated with training and deploying LLMs underscore environmental and economic concerns. Biases embedded in training data pose risks of perpetuating stereotypes and reinforcing societal inequalities. Additionally, the extensive capabilities of LLMs also spark a myriad of ethical quandaries, emphasizing the imperative for responsible use and governance in NLP applications.

Transformers for NLP

Transformers have revolutionized the field of Natural Language Processing (NLP) by enabling models to process and understand language with unprecedented accuracy and efficiency. In this section, we provide a broad overview of Transformers, including their architecture, functionality, and guidelines for choosing the appropriate architecture for specific NLP tasks.

Our objective is not to delve into the detailed inner workings of Transformers. Instead, we aim to give you a conceptual understanding of Transformers' high-level architecture and basic functionality. A deeper dive is required to fully grasp these complex concepts, which is beyond the scope of this book. Excellent references are available for those who want a more detailed exploration and in-depth explanations of Transformers. Notable among these are the seminal paper "Attention Is All You Need" (Vaswani, et al., 2017) and "Chapter 3: Transformer Anatomy" in the book *Natural Language Processing with Transformers, Revised Edition* (Tunstall, Werra, & Wolf, 2022).

Transformer Architecture

The Transformer architecture offers significant advantages over traditional models. Its parallelization capability allows it to process sequences much faster, as it does not require sequential processing like recurrent networks. Moreover, its ability to handle

CHAPTER 1   NLP ESSENTIALS

long-range dependencies and capture complex relationships within the text has set new standards in NLP, enabling models to achieve state-of-the-art results in various tasks such as language translation, text generation, and question answering.

Transformers, a specific class of deep learning models, have brought about a paradigm shift in text processing. Since their introduction, Transformers have demonstrated their supremacy in the NLP field across various tasks. This breakthrough has enabled models to understand and generate language with unprecedented accuracy and efficiency.

Unlike their predecessors, such as RNNs (recurrent neural networks) and CNNs (convolutional neural networks), Transformers employ self-attention mechanisms. This unique feature enables them to grasp the significance of different words in a sentence, irrespective of their position, thereby enhancing their ability to capture contextual relationships in text.

The research paper "Attention Is All You Need" (Vaswani, et al., 2017) made a groundbreaking contribution to the NLP field by introducing the Transformer architecture. This paper proposed a new network architecture based solely on attention mechanisms, completely dispensing with recurrence and convolutions. The authors argue that the Transformer model is the first to rely entirely on self-attention to compute representations of its input and output without using sequence-aligned RNNs or convolutional networks.

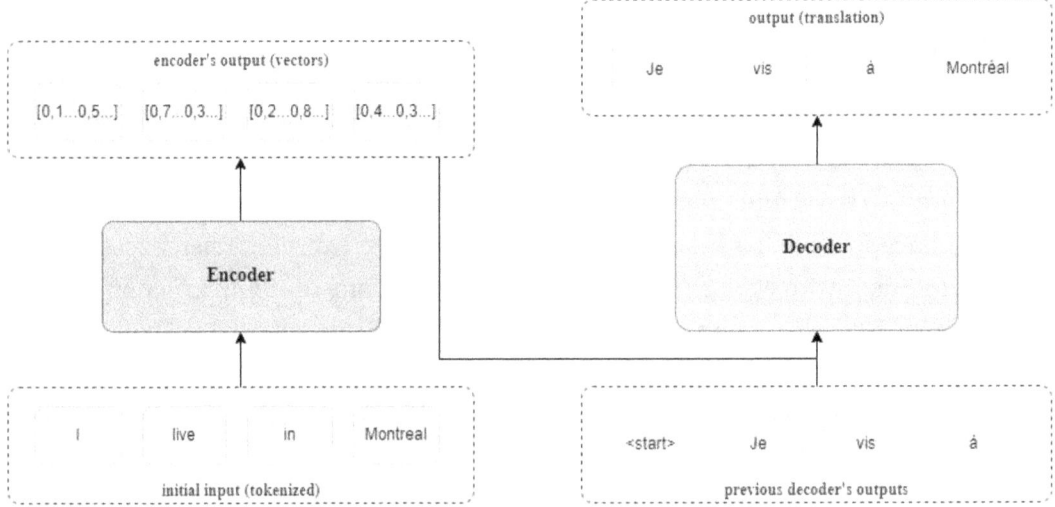

Figure 1-3. *Encoder–decoder high-level architecture (simplified)*

At the core of the Transformer architecture lies the encoder–decoder structure as illustrated in Figure 1-3.

CHAPTER 1 NLP ESSENTIALS

The encoder's primary role is to process the input sequence. It converts the input into numerical representations that capture the semantics of the text. As depicted in Figure 1-4, the encoder comprises a stack of several layers (e.g., six layers), each containing a self-attention mechanism and a feed-forward network. These layers help the encoder to focus on different parts of the input sequence dynamically, ensuring that the model effectively captures the semantics of the text.

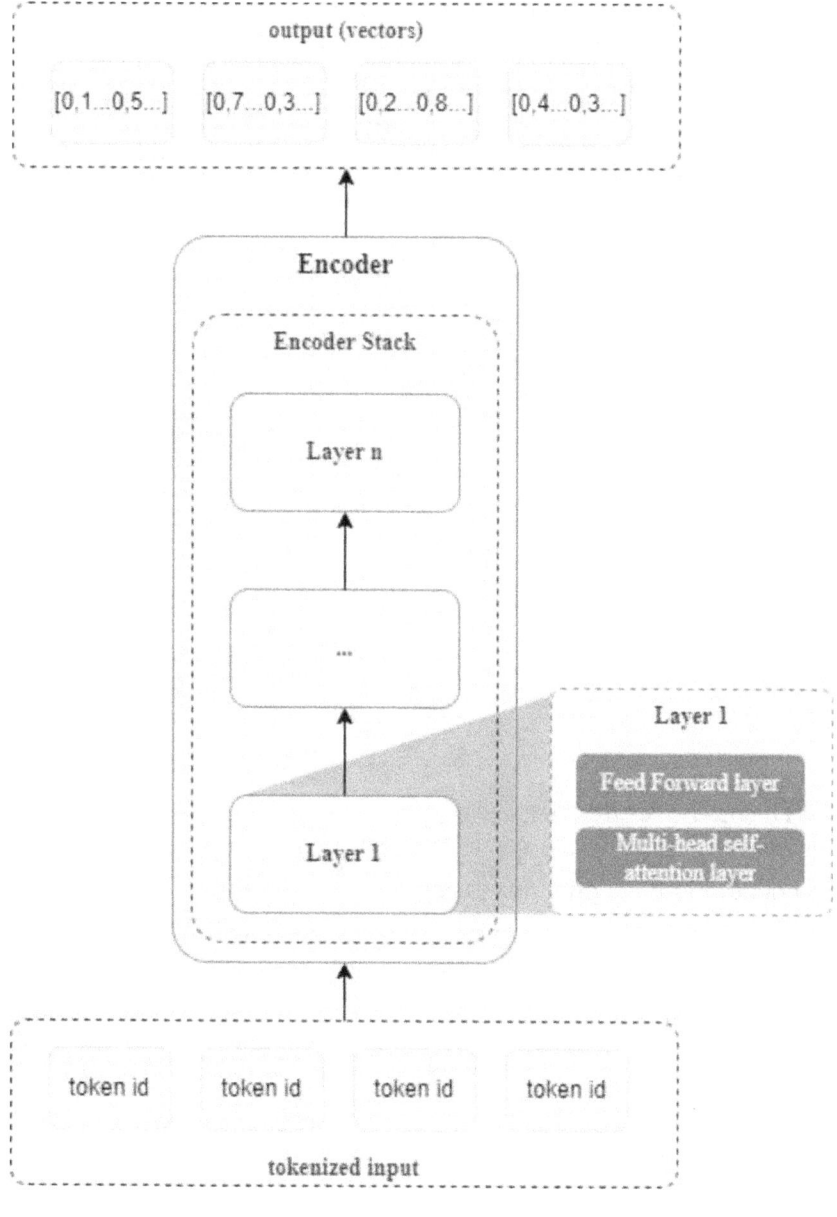

Figure 1-4. *Encoder architecture*

CHAPTER 1 NLP ESSENTIALS

The decoder, on the other hand, takes the vector representations generated by the encoder and use them to generate the output sequence. Like the encoder, the decoder consists of multiple layers, each with a self-attention mechanism and a feed-forward network as shown in Figure 1-5.

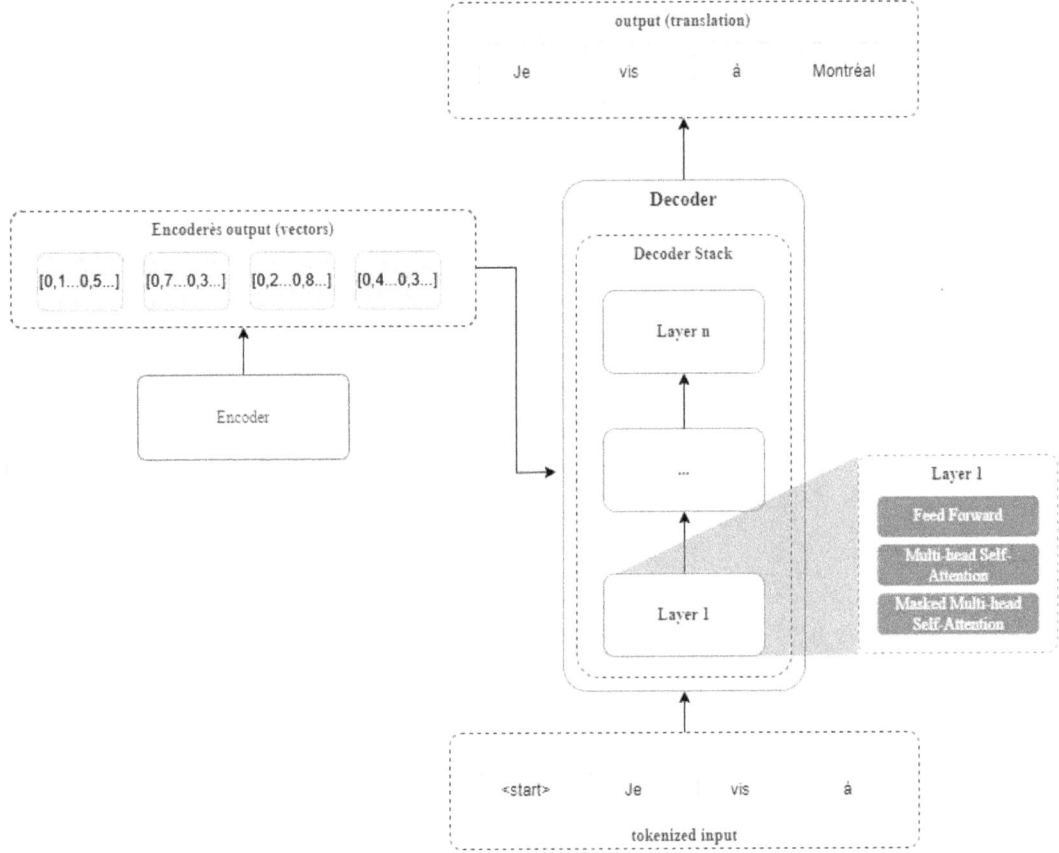

Figure 1-5. *Decoder architecture*

Both the encoder and decoder layers contain a self-attention mechanism, which is central to the Transformer's performance. This mechanism allows the model to weigh the importance of different words in the input sequence, regardless of their positions. It enables the model to capture dependencies and relationships between words within the text, which is essential for understanding context and meaning. Additionally, the self-attention mechanism allows for parallelization during training by processing all words in the sequence simultaneously, significantly speeding up the process and providing scalability, as it can efficiently handle long sequences by focusing on relevant parts of the input.

Since Transformers do not inherently understand the order of words in a sequence, positional encodings are added to provide information about the position of each token. This ensures that the model retains the order of words, which is essential for understanding the syntax and structure of the text.

Input and output embeddings are another crucial aspect of the Transformer architecture. Before being processed by the encoder, input tokens are converted into continuous vector representations, known as embeddings. Similarly, output tokens are embedded before being processed by the decoder. These embeddings provide dense representations of the tokens, capturing their meanings in a high-dimensional space.

Let's explain how the Transformer model works in simple terms by going through a toy example for a translation task. Suppose we need to translate this English sentence to French: "I live in Montreal."

The input to the encoder of a Transformer model is a sequence of tokens, which can be words, subwords, or characters, depending on the tokenization method used. For example, "I live in Montreal" would be tokenized into ["I", "live", "in", "Montreal"].

The encoder's output is a sequence of vector representations, one for each input token. These representations capture the contextual information of each token based on its position and surrounding tokens in the input sequence. The encoder processes the input sequence in a bidirectional manner, considering both the left and right context of each token.

The input to the decoder is typically the target sequence (the desired output) shifted by one position to the right, with a special start token added at the beginning. For example, if the target sequence is "Je vis à Montréal," the input to the decoder would be ["<start>", "Je", "vis", "à", "Montréal"].

The decoder output is a sequence of probability distributions over the vocabulary, one for each position in the target sequence. Given the previous tokens and the encoder's output, these probability distributions represent the model's prediction of the next token in the sequence. The decoder generates the output sequence token by token, using the encoder's output and its previous predictions as input.

The data flow in a Transformer model can be summarized as follows:

1. The input sequence is fed into the encoder, producing a sequence of vector representations (encoder output).

2. The encoder output and the shifted target sequence (decoder input) are fed into the decoder.

3. The decoder generates the output sequence token by token, using the encoder output and its previous predictions as input.

4. The final output is the sequence of tokens generated by the decoder, representing the model's prediction for the target sequence.

The Transformer model uses self-attention mechanisms in both the encoder and decoder, which allow it to capture long-range dependencies and contextual information more effectively than traditional sequence-to-sequence models like recurrent neural networks (RNNs).

Transformer Taxonomy

Understanding the taxonomy of Transformer models helps categorize and differentiate their applications and functionalities. Figure 1-6 provides a summarized overview of the main Transformer families with the well-known models for each category.

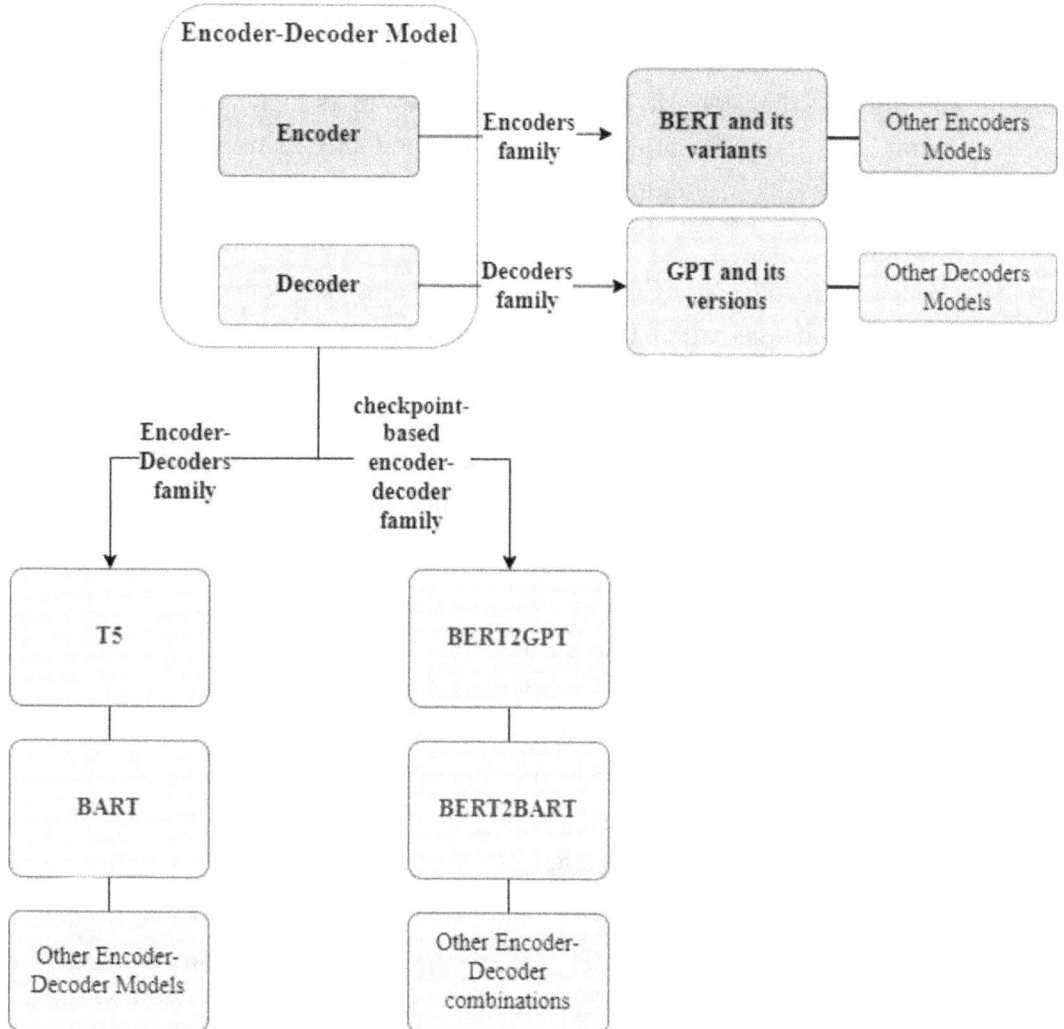

Figure 1-6. *Transformer taxonomy*

Transformers can be divided into four categories, each with specific purposes, designs, and training methods. The four categories are

- Encoder-Only Models: These models are designed for the understanding tasks. They process the entire input sequence simultaneously and are well-suited for tasks like sentence classification, Named Entity Recognition, and question answering. BERT (Bidirectional Encoder Representations from Transformers) is a prime example.

- Decoder-Only Models: These models are used for generation tasks where the output sequence is generated sequentially based on the preceding context. They are effective for tasks such as text completion, text generation, and language modeling. Examples include the GPT (Generative Pretrained Transformer) series by OpenAI.

- Encoder–Decoder Models: These models are designed for sequence-to-sequence tasks, where the input sequence is transformed into an output sequence. They are highly effective for tasks like machine translation, text summarization, and question answering. Notable examples include T5 (Text-to-Text Transfer Transformer) and BART (Bidirectional and Auto-Regressive Transformers).

- Checkpoint-Based Encoder–Decoder Models: These models combine pretrained checkpoints for both the encoder and the decoder. They can be initialized from a pretrained encoder checkpoint and a pretrained decoder checkpoint. Any pretrained auto-encoding model, such as BERT, can serve as the encoder. Additionally, the decoder can be pretrained auto-encoding models (e.g., BERT), pretrained causal language models (e.g., GPT-2), or the pretrained decoder part of sequence-to-sequence models (e.g., the decoder of BART).

To differentiate between encoder–decoder models like T5 or BART and encoder–decoder models based on loading checkpoints such as BERT2GPT, one can focus on their design and training methods. Pretrained encoder–decoder models like T5 and BART are designed and trained end-to-end as integrated architectures. These models are purpose-built for tasks such as translation, summarization, and text generation, ensuring that the encoder and decoder work seamlessly together because they are trained from scratch as a single unit.

In contrast, checkpoint-based encoder–decoder models, like BERT2GPT, involve a hybrid design where pretrained encoders and decoders from different models are combined. This approach uses pretrained checkpoints that may have been trained on separate tasks, allowing for the strengths of different models to be leveraged, which can enhance performance on complex tasks.

CHAPTER 1 NLP ESSENTIALS

As shown in Figure 1-7, one could imagine using a BERT checkpoint to initialize the encoder for better input understanding and choosing a GPT-2 model as the decoder for superior text generation. This flexibility allows you to combine strengths from different models to achieve better results. For example, in translation, you might use a pretrained English encoder with a French decoder to translate text from English to French. In summarization tasks, you could use an encoder trained on long-form text and a decoder optimized for generating concise summaries. For multimodal tasks, combining an image encoder with a text decoder can generate descriptions of images. Domain-specific applications also benefit from this flexibility, such as using encoders and decoders fine-tuned on medical or legal texts for specialized document translation or summarization in those fields.

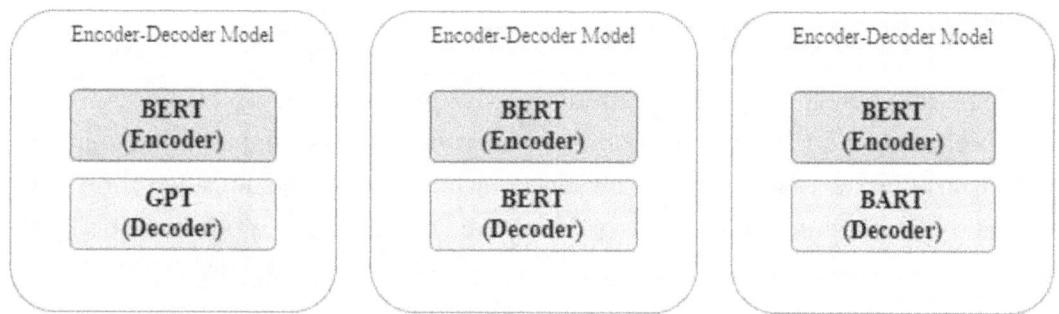

Figure 1-7. Examples of hybrid seq-to-seq encoder–decoder models

The effectiveness of initializing sequence-to-sequence models with pretrained checkpoints for sequence generation tasks was demonstrated in the paper "Leveraging Pre-trained Checkpoints for Sequence Generation Tasks" (Rothe, Narayan, & Severyn, 2020). In this study the authors conducted an extensive empirical evaluation, initializing both the encoder and decoder with publicly available pretrained model checkpoint (e.g., BERT, GPT-2, and RoBERTa). The results showcased new state-of-the-art performance in tasks such as machine translation and text summarization.

Transfer Learning

Transfer learning, a fundamental concept in modern NLP, involves using pretrained models on large datasets to improve performance on specific tasks with limited data. By initializing models with pretrained checkpoints, such as using BERT for the encoder and GPT-2 for the decoder, we significantly improve training effectiveness and efficiency.

23

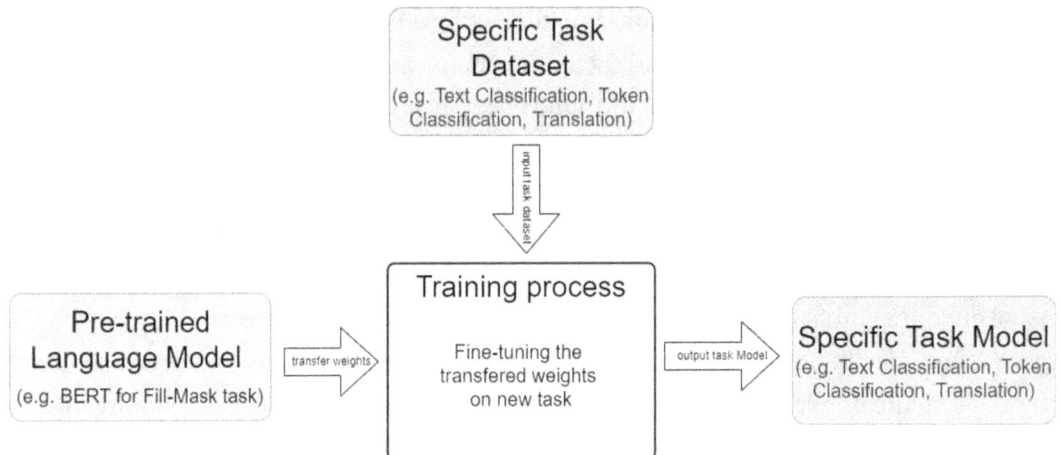

Figure 1-8. *Transfer learning–based training process*

Figure 1-8 shows how transfer learning speeds up the training of NLP models by using pretrained models that have been trained on general tasks and applying them to specific tasks like text classification, token classification, table question answering, and translation. This is done by fine-tuning a pretrained model, which was initially trained using unsupervised learning on large datasets, with a task-specific dataset through a supervised learning approach.

For example, when fine-tuning a pretrained model like BERT for Named Entity Recognition (NER), the model's initial training weights come from a pretraining task called Masked Language Modeling (MLM). During the BERT pretraining phase, MLM helps the model learn to predict missing words in sentences, enhancing its understanding of context and language patterns. The model undergoes additional training on a smaller, NER-specific dataset in the fine-tuning phase. This allows the model to adapt its prelearned language features to solve the new NER task.

When it comes to the training phase of NLP models, transfer learning differs significantly from the conventional model life cycle. The main distinction lies in the use of pretrained models, which eliminates the need for extensive training from scratch in the transfer learning approach. This results in significant resource efficiency, as the fine-tuned model requires less computational resources and time for training. Furthermore, pretrained models bring prior knowledge, resulting in better initial performance and quicker convergence during fine-tuning.

CHAPTER 1 NLP ESSENTIALS

It's critical to select the appropriate pretrained model for the task at hand because it significantly influences the performance and accuracy of the final fine-tuned model.

As illustrated in Figure 1-9, the transfer learning life cycle emphasizes the utilization of pretrained models to enhance performance and efficiency across various NLP tasks.

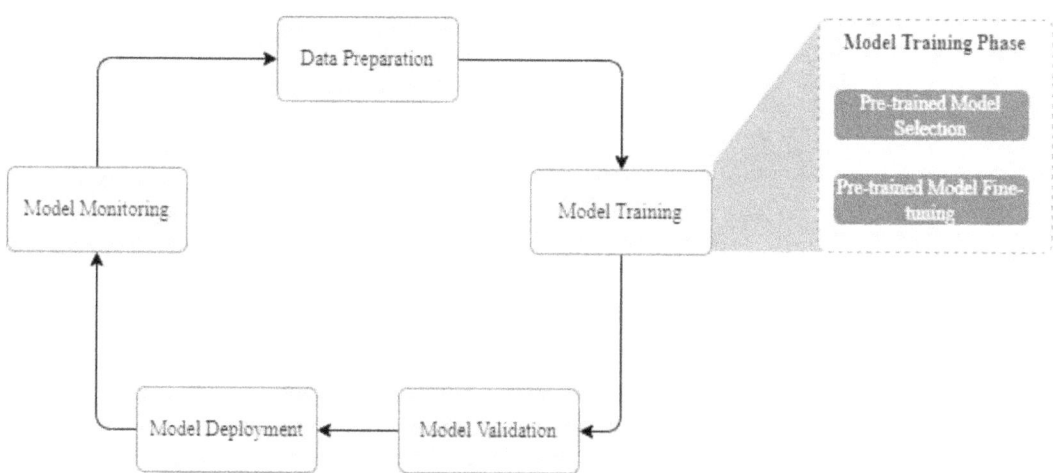

Figure 1-9. *NLP model life cycle in the context of transfer learning*

The transfer learning approach not only enhances the performance and precision of the fine-tuned models but also offers additional benefits, including the reduction of the required training data and minimization of training time (GPU compute resources), and thus decreases NLP training costs and the environmental training footprint.

Hugging Face Ecosystem

As we delve deeper into the benefits and processes of Transformers and transfer learning, it's important to recognize the tools and platforms that facilitate these advancements. One such ecosystem is Hugging Face, a leading platform that provides comprehensive support for the development of Transformer-based models. Hugging Face offers an extensive library of pretrained models, along with user-friendly interfaces and resources to streamline the transfer learning process. This ecosystem simplifies the deployment of advanced NLP models and accelerates the development cycle, making state-of-the-art NLP accessible to a wider audience. The subsequent paragraphs will explore the Hugging Face ecosystem, its key components, and its impact on the field of NLP.

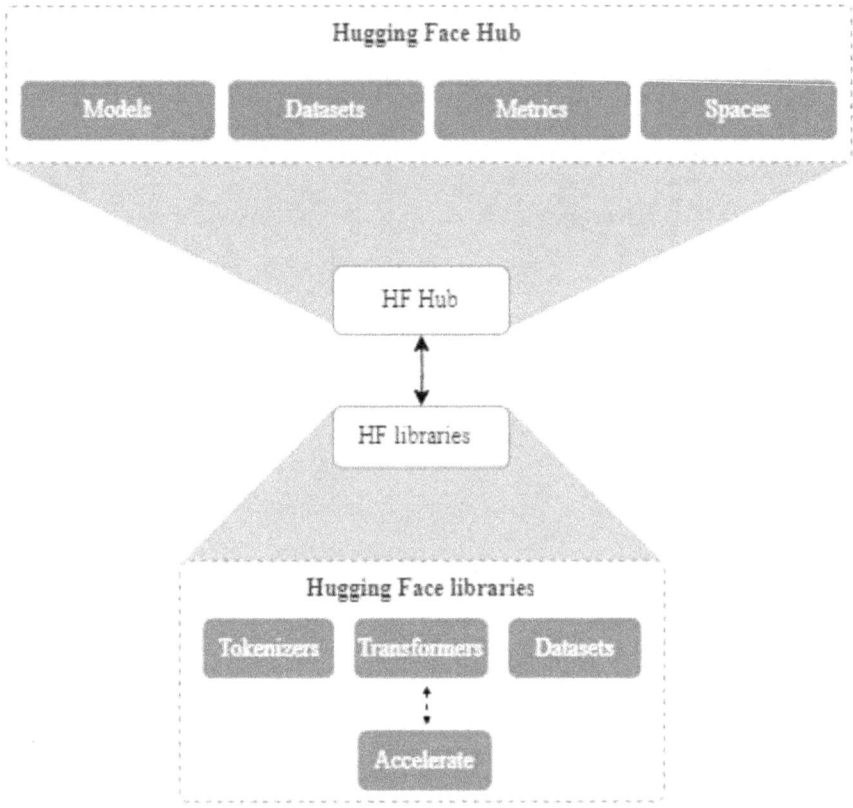

Figure 1-10. Hugging Face ecosystem

As illustrated in Figure 1-10, the Hugging Face ecosystem (Hugging Face, 2022) is divided into two main components: the Hugging Face Hub and the Hugging Face libraries. The Hugging Face Hub serves as a central repository for pretrained models, datasets, metrics, and documentation. It provides the resources needed for initializing and fine-tuning Transformer-based models. The Hugging Face libraries, which include tools like the Transformers, Datasets, and Tokenizers libraries, offer the essential tools and frameworks to utilize these resources effectively. Together, these components support the entire development life cycle of Transformer-based models, from initial training and fine-tuning to deployment and evaluation, ensuring seamless integration and efficient development processes.

The Hugging Face Hub (Hugging Face) is an online platform that provides a virtual space for hosting a vast repository of machine learning models, datasets, and demo applications. With over 500k models, 130k datasets, and 160k demo apps,[2] this platform promotes knowledge sharing and collaboration among machine learning experts. Its overarching goal is to facilitate the exploration, experimentation, collaboration, and development of cutting-edge technologies in machine learning. As a central hub for machine learning resources, the Hugging Face Hub offers an unparalleled opportunity for practitioners to stay current with field advancements and exchange ideas, insights, and best practices. It includes a repository of pretrained models that can be used for various NLP tasks, a collection of datasets available for training and fine-tuning models, tools and benchmarks for evaluating model performance (i.e., Metrics), as well tools for building and sharing interactive machine learning demo applications (i.e., Spaces).

The Hugging Face libraries (Hugging Face) form the backbone of the ecosystem, providing essential tools for working with Transformer models. The Transformers library offers a unified API to access pretrained Transformer models for various NLP tasks, including text classification, translation, summarization, and question answering. It supports a range of models such as BERT, GPT-2, GPT-3, RoBERTa, T5, and others, enabling users to leverage cutting-edge technologies with ease. The Datasets library provides a vast repository of datasets tailored for machine learning and NLP, simplifying the process of loading, preprocessing, and managing large datasets. This library supports diverse formats and storage solutions, making it a flexible tool for handling data. The Tokenizers library excels in providing fast, efficient, and customizable tokenization, including implementations for various tokenization techniques required for different Transformer models. Additionally, Hugging Face's inference API simplifies the deployment of NLP models in real-world applications by providing a straightforward way to integrate powerful NLP models into applications through a simple API call.

A vibrant community and extensive documentation are vital components of the Hugging Face ecosystem. The platform provides comprehensive guides, tutorials, and support forums, helping users of all skill levels to maximize the potential of the available tools and libraries. The community-driven approach ensures continuous improvement and support, fostering an environment where users can share knowledge, seek assistance, and collaborate on projects. This support infrastructure is crucial for maintaining the ecosystem's relevance and usability.

---

[2] At the time of writing.

The Hugging Face ecosystem has significantly lowered the barrier to entry in NLP, making advanced AI accessible to a broader audience. Its tools are widely adopted in both academia and industry, facilitating rapid prototyping and deployment of NLP solutions. By democratizing access to state-of-the-art tools and models, Hugging Face enables researchers, developers, and businesses to harness the power of NLP effortlessly.

Strategic Considerations for NLP Adoption

Natural Language Processing (NLP) has the potential to revolutionize how businesses interact with data, engage customers, and derive insights from vast amounts of unstructured information. However, adopting NLP presents a strategic dilemma for decision-makers. A prime example of this dilemma is whether an organization should invest in ready-to-use cloud NLP models or develop custom NLP models.

As organizations consider adopting NLP technologies, they must make several strategic decisions to ensure the success of their initiatives (see Figure 1-11). This section explores key strategic considerations in three main areas: model development and deployment, data acquisition and preparation, and team building.

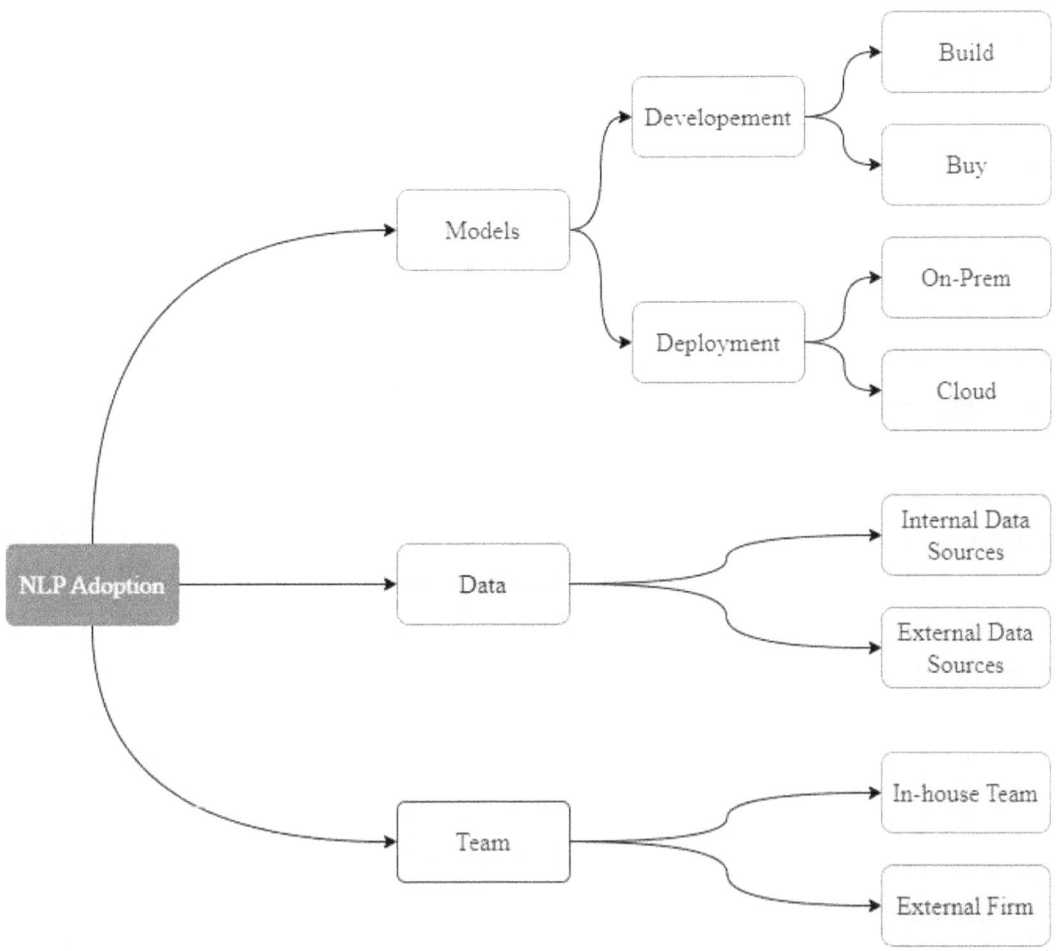

Figure 1-11. Strategic decisions for NLP adoption

Models

When it comes to NLP model development, organizations face the choice of building in-house, purchasing pretrained models, or using cloud NLP services. Building NLP models in-house provides a unique advantage in terms of control and customization. This approach allows organizations to tailor their models to their specific needs, incorporating unique features and domain-specific knowledge that off-the-shelf solutions may lack. It's particularly beneficial for businesses with proprietary data or specialized requirements, ensuring that the models are perfectly aligned with their operational contexts. Developing custom models means all data remains in-house, ensuring tighter control over sensitive or proprietary information.

Additionally, organizations can innovate and potentially develop unique NLP capabilities, leading to a competitive advantage and even potential new revenue streams if they decide to commercialize their solutions. However, this approach requires a significant investment in skilled personnel, computational resources, and time. The field of NLP is competitive, and attracting and retaining top NLP talents (i.e., deep learning experts) can be challenging and expensive. It also necessitates ongoing updates and refinement to stay aligned with state-of-the-art advancements in NLP, ensuring that the models not only remain relevant and perform optimally but also leverage the latest techniques and innovations to maintain a competitive edge.

On the other hand, purchasing pretrained models or using cloud-based NLP services can significantly reduce the time and resources needed to deploy NLP capabilities. NLP cloud services offer a compelling alternative. They present a cost-efficient option, eliminating the expenses of talent acquisition, infrastructure setup, and continual model updates. Additionally, these services promise rapid deployment with ready-to-use, pretrained models, ensuring businesses can swiftly integrate advanced NLP functionalities. Scalability, a crucial factor for growing businesses, is seamlessly addressed by these platforms, which can effortlessly adapt to fluctuating workloads. Pretrained models and cloud services can be deployed quickly, enabling organizations to leverage NLP capabilities without the lengthy development process. This approach is often more cost-effective, as it minimizes the need for extensive in-house resources and infrastructure.

Furthermore, vendors typically provide ongoing support and updates, ensuring the models remain state of the art. Additionally, availing cloud services grants businesses indirect access to the invaluable expertise of top-tier AI researchers and engineers. This ensures that enterprises remain at the forefront of NLP advancements and allows them to concentrate on their core competencies, relegating the intricacies of NLP to those who specialize in it. In the fast-paced world of AI and NLP, staying updated is vital, and cloud-based solutions ensure consistent access to the latest breakthroughs without ceaseless in-house R&D investments. However, off-the-shelf models may only partially align with specific business needs and may require significant adaptation. Additionally, using external services can pose risks related to data privacy and compliance, particularly if sensitive data is involved.

CHAPTER 1 NLP ESSENTIALS

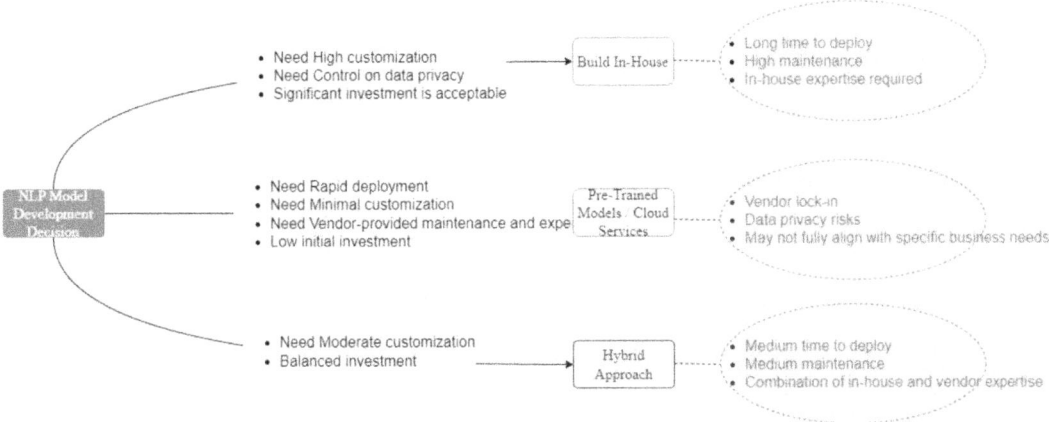

Figure 1-12. *NLP model development decision tree*

As depicted in Figure 1-12, the decision between adopting ready-to-use NLP models or building in-house custom models is not black and white. It's a strategic choice that hinges on the organization's immediate needs, long-term vision, available resources, and risk appetite. A hybrid approach, where some functionalities are outsourced while others are developed in-house, might also be worth considering.

Deploying NLP models on-premises offers enhanced control over data security and compliance. It is suitable for organizations handling sensitive data or operating in regulated industries. However, it requires substantial infrastructure, IT support, and ongoing maintenance and updates. Cloud deployment provides scalability, flexibility, and ease of management. It allows organizations to leverage powerful computing resources without significant upfront investment in infrastructure. Cloud providers often offer built-in tools for monitoring, scaling, and maintaining NLP models, making this option attractive for many organizations. However, concerns about data security and compliance may arise depending on the provider and the specific use case. The deployment location of NLP models, whether on-premises or in the cloud, is a pivotal decision for any organization. Both options offer distinct benefits and come with their sets of challenges. The organization's specific needs, regulatory environment, budget constraints, and strategic vision should influence the choice. Hybrid solutions, which leverage both on-premises and cloud resources, might also provide a balanced approach, offering the best of both worlds. As with all strategic decisions, CIOs and lead architects must consider the broader implications.

CHAPTER 1   NLP ESSENTIALS

Data

Data is the backbone of any NLP initiative, and strategic decisions around data collection, management, and utilization are critical. Using internal data sources ensures that the data is highly relevant and specific to the organization's operations. This includes customer interactions, internal documents, transaction records, and more. The primary advantage of internal data is its direct relevance to the business context and specific use cases, which can enhance the accuracy and effectiveness of NLP models. Moreover, organizations have full control over data quality, preprocessing, and integration with existing systems, ensuring that the data is consistent and reliable. Keeping data internal also reduces exposure risks and helps maintain privacy.

However, relying solely on internal data may have limitations. Internal data can be fragmented across different departments and systems, requiring significant effort to consolidate and preprocess. There may also be limitations in the volume and variety of data available internally, which can restrict the model's ability to generalize and perform well on diverse inputs. To address these limitations, organizations can supplement internal data with external data sources. External data can enhance the capabilities of NLP models by providing additional context and diversity. This includes publicly available datasets, third-party data providers, and social media data. While external data can offer broader insights and improve model generalization, it also comes with challenges. Ensuring compliance with data privacy regulations when using external data can be complex, and external data sources may vary in quality and relevance, requiring rigorous validation and preprocessing.

Team

Given their technical complexity and potential for competitive advantage, Natural Language Processing (NLP) initiatives prompt a critical decision for business leaders: Should the organization invest in building a dedicated in-house NLP team, or is it more strategic to delegate this task to specialized external firms?

Building an in-house NLP team ensures that expertise is readily available and aligned with the organization's goals and culture. It allows for greater customization and faster iteration on projects. This approach requires investment in hiring and training skilled professionals and ongoing management and development of the team. An in-house team typically includes data scientists, machine learning engineers, and domain experts who work closely with the organization to ensure that NLP initiatives are directly relevant to business needs. This alignment with the organization's goals and culture ensures the team can iterate quickly and adapt models as requirements evolve.

Additionally, having an in-house team helps retain expertise within the organization, building long-term capability. However, building and maintaining an in-house team requires significant investment in hiring, training, and managing skilled personnel. This can be challenging and costly, particularly in competitive job markets.

Partnering with an external firm can provide access to specialized expertise and resources that may not be available in-house. It can be a cost-effective way to implement NLP solutions, especially for organizations with limited internal capabilities. External firms bring extensive knowledge and experience in NLP, ensuring high-quality implementation and offering flexibility in scaling resources based on project needs. This approach can accelerate the implementation process, as external partners can leverage their existing frameworks and methodologies. However, relying on external firms may lead to challenges in aligning external efforts with internal goals and maintaining long-term knowledge within the organization. There are also concerns regarding intellectual property (IP) when outsourcing. Ensuring the proprietary nature of technologies and data can be challenging, potentially risking the leakage of competitive advantages. Furthermore, overreliance on external firms can create dependency, making it difficult to bring NLP development back in-house or switch vendors in the future. This dependency can limit organizational flexibility and control over NLP projects, potentially hindering long-term strategic goals and innovation in Natural Language Processing.

The decision to invest in an in-house NLP team or delegate to an external firm depends heavily on the organization's strategic goals, budgetary constraints, project timelines, and desired level of control. While an in-house team provides deep integration and control, an external firm offers immediate expertise and potential cost savings. Leaders should weigh these factors carefully, considering both immediate needs and long-term business objectives, to make an informed decision that best serves the organization's interests.

Summary

This chapter provides the theoretical framework for understanding Natural Language Processing (NLP). It delves into NLP's core principles, historical evolution, and strategic dimensions of NLP implementations to familiarize you with key concepts and theoretical groundwork necessary to understand subsequent chapters.

The introductory chapter covers essential concepts in NLP, focusing on Transformers and introducing the Hugging Face ecosystem for working with Transformer models. It also discusses strategic decisions decision-makers may face when adopting NLP technologies, such as choosing between prebuilt SaaS solutions and developing custom models.

While not comprehensive, this overview is a starting point for readers to understand fundamental NLP notions before delving into specific concepts and their real-world applications in subsequent chapters.

The next chapter will introduce the OCI ecosystem, emphasizing its artificial intelligence and machine learning offerings to provide a comprehensive understanding of harnessing the platform's potential for developing NLP solutions.

References

Hugging Face. (2022). *The Hugging Face Course, 2022.* Retrieved from Hugging Face: `https://huggingface.co/course`

Hugging Face. (n.d.). *Documentations.* Retrieved from Hugging Face: `https://huggingface.co/docs`

Rothe, S., Narayan, S., & Severyn, A. (2020). Leveraging Pre-trained Checkpoints for Sequence Generation Tasks. *Transactions of the Association for Computational Linguistics*

Tunstall, L., Werra, L. v., & Wolf, T. (2022). *Natural Language Processing with Transformers, Revised Edition.* O'Reilly Media, Inc.

Vaswani, A., Shazeer, N., Parmar, N., Uszkoreit, J., Jones, L., Gomez, A. N., ... Polosukhin, I. (2017). Attention Is All You Need. *CoRR*

CHAPTER 2

Oracle Cloud for NLP

Oracle Cloud Infrastructure (OCI) provides a robust platform for the development, deployment, and scaling of Natural Language Processing (NLP) applications. Its comprehensive suite of managed services abstracts the complexities of infrastructure management, allowing data scientists and developers to concentrate on core NLP tasks.

This chapter explores the OCI ecosystem, focusing on its artificial intelligence and machine learning offerings. It provides an overview of OCI's capabilities for developing practical Natural Language Processing (NLP) solutions.

Introduction to Oracle Cloud Infrastructure (OCI)

Oracle has become a major player in the cloud computing market, offering diverse services such as computing, storage, networking, databases, and AI/ML capabilities. In this section, we will explore the core concepts and components that form the foundation of OCI after providing a brief overview of its evolution.

History

Oracle Cloud Infrastructure (OCI) has undergone significant evolution since its inception, reflecting Oracle's commitment to becoming a major player in the cloud computing market. Oracle entered the cloud market relatively late compared to competitors like Amazon and Microsoft. The company's initial cloud offerings, launched in the early 2010s, were primarily focused on Software as a Service (SaaS) products, leveraging Oracle's strong position in enterprise software. However, these early efforts did not include a comprehensive Infrastructure as a Service (IaaS) platform.

Recognizing the growing importance of IaaS in the cloud ecosystem, Oracle began developing its own cloud infrastructure platform. The first generation of OCI, sometimes referred to as "OCI Classic," was introduced in 2016 under the name "Oracle Bare Metal

Cloud Services." This initial offering provided basic compute, storage, and networking services but lacked the scalability and advanced features of more established cloud platforms.

A turning point came in 2018 with the rebranding of Oracle Bare Metal Cloud Services to Oracle Cloud Infrastructure (OCI) and introducing OCI's second generation, labelled "Generation 2 Cloud." This marked a significant leap forward in Oracle's cloud capabilities. The new iteration of OCI was built from the ground up, focusing on high performance, security, and enterprise-grade reliability. It introduced a new network architecture designed to minimize latency and improve security, as well as bare metal compute instances for high-performance workloads (Oracle, 2018).

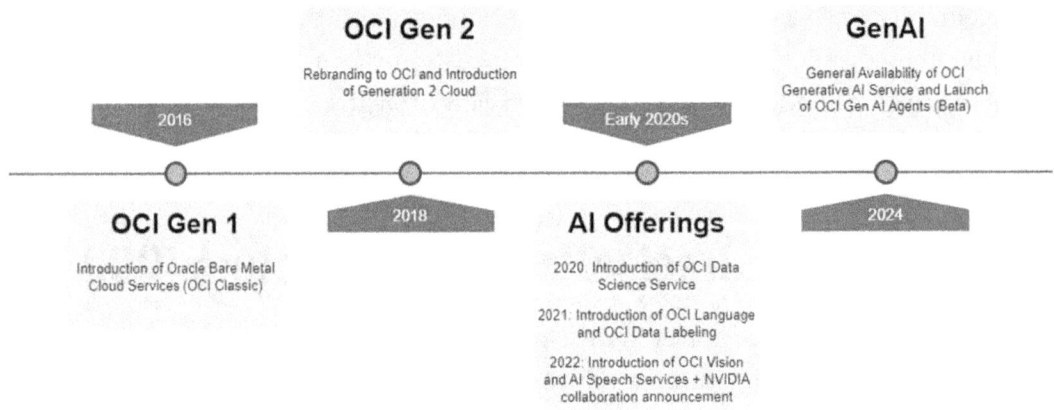

Figure 2-1. *Key AI milestones in OCI's evolution*

Since introducing OCI's second generation, Oracle has continuously expanded and refined OCI's capabilities. The company has aggressively invested in global data center expansion, rapidly increasing its cloud regions worldwide. This expansion has been crucial in addressing data sovereignty concerns and reducing latency for global customers.

Oracle has also focused on making OCI more accessible and flexible. This includes introducing always-free tier services, improving interoperability with other cloud providers, and developing partnerships to expand OCI's reach. Notable among these is the partnership with Microsoft, announced in 2019, enabling interconnection between OCI and Azure (Oracle, 2019).

In recent years, Oracle has strongly emphasized adding advanced services to OCI, particularly in areas like artificial intelligence, machine learning, and autonomous systems. Oracle expanded its AI offerings significantly in the early 2020s, with a strong focus on enterprise-grade AI capabilities. For example, the company introduced the OCI Data Science Service in 2020, OCI Language and OCI Data Labeling in 2021, and OCI Vision and AI Speech Services in 2022.

In 2022, Oracle announced a multiyear partnership with NVIDIA to enhance AI adoption. The collaboration aims to bring the full NVIDIA accelerated computing stack—from GPUs to systems to software—to Oracle Cloud Infrastructure (OCI), thus helping customers solve business challenges with accelerated computing and AI (Oracle, 2022).

The last important AI-related OCI development is the Generative AI Service, which became generally available in early 2024. It provides access to prebuilt large language models (LLMs) from partners like Meta and Cohere. It allows enterprises to connect LLMs with their proprietary data sources, enhancing the relevance and accuracy of AI-generated outputs (Oracle, 2023).

OCI has consistently focused on performance, security, and cost-effectiveness throughout its evolution, often positioning itself as a more efficient alternative to other major cloud providers. While it entered the market later than some competitors, OCI has rapidly evolved into a comprehensive and competitive cloud platform, particularly appealing to enterprises with existing Oracle investments and those requiring high-performance cloud infrastructure.

Core Concepts and Terminology

Understanding the fundamental concepts of Oracle Cloud Infrastructure (OCI) is crucial for developing robust cloud applications, particularly our practical Natural Language Processing (NLP) solution.[1] This foundational knowledge will serve as the cornerstone for our exploration throughout this book, enabling you to effectively leverage OCI's capabilities in the context of NLP.

Below are the core OCI concepts and services:

- Regions and Availability Domains: OCI is a geographically distributed cloud platform with multiple regions around the world. Each region contains several isolated availability domains, providing high availability and fault tolerance for your cloud resources. This geographical distribution will be a key consideration as we design the NLP solution to meet performance and data sovereignty requirements.

---

[1] This NLP application will be introduced as a comprehensive case study in the following chapter, demonstrating the practical implementation of OCI concepts in a real-world scenario.

- Virtual Cloud Networks (VCNs): At the heart of OCI's networking capabilities are Virtual Cloud Networks (VCNs). VCNs allow you to provision logically isolated, customizable virtual networks within the OCI environment, enabling secure communication between your cloud resources. The NLP solution will leverage VCNs to ensure a secure and scalable network architecture.

- Compute Instances: OCI offers various compute instance types, including bare metal, virtual machine (VM), and GPU-accelerated options. These instances can be dynamically scaled to meet the demands of your workloads. As we develop the NLP solution, we will evaluate the appropriate compute resources required for model training and inference tasks.

- Storage Services: OCI provides storage services, including Block Volumes, Object Storage, and File Storage. These options cater to diverse data storage and retrieval requirements within your cloud-based applications. The NLP solution will leverage these storage services to effectively manage the data required for training our NLP solution's models.

- Identity and Access Management (IAM): The OCI IAM service lets you centrally manage user access, permissions, and security policies across your cloud resources. This ensures controlled and auditable access to critical infrastructure and data. As we build the NLP solution, we will implement robust IAM practices to ensure the application's security and compliance.

Subsequent section will provide detailed explanations of these core OCI concepts and services, equipping you with the necessary knowledge to build sophisticated NLP applications on the OCI platform.[2]

---

[2] For those interested in a deeper understanding of OCI, the official OCI documentation provides comprehensive information on the platform's features, capabilities, and services (the official Oracle Cloud Infrastructure documentation can be found at https://docs.oracle.com/en-us/iaas/Content/home.htm). Additionally, numerous training resources and books are available to explore the OCI ecosystem in greater detail.

Regions and Realms

Oracle Cloud Infrastructure is physically hosted in regions and availability domains. A region is a localized geographic area, and an availability domain is one or more data centers located within a region. Oracle cloud regions are globally distributed data centers that provide secure, high-performance, local environments (refer to Figure 2-2). These regions allow businesses to move, build, and run all workloads in the cloud from infrastructure to applications while meeting regional data regulations. For more information about each offering, see Regions and Availability Domains and Dedicated Regions (Oracle, 2023).

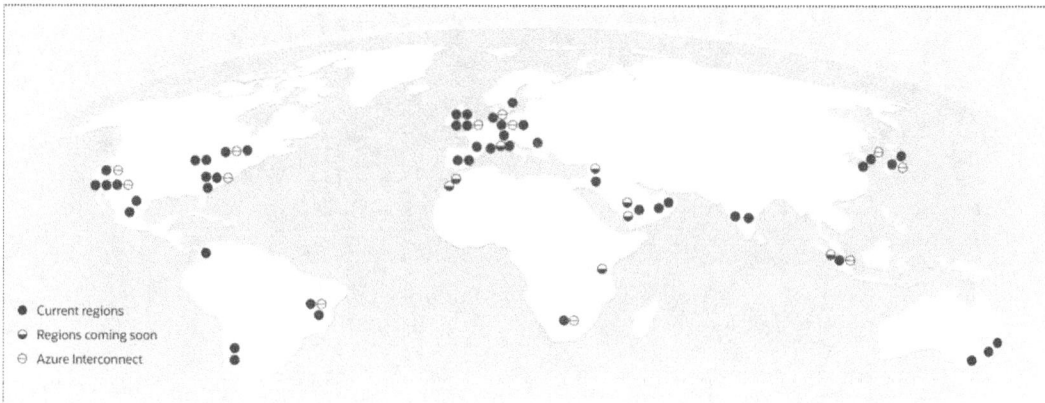

Figure 2-2. *OCI regions worldwide*

A realm is a logical collection of regions. Realms are isolated from each other and do not share any data. Your tenancy exists in a single realm and has access to the regions that belong to that realm. OCI currently offers realms for commercial regions, government regions, and dedicated regions (Oracle, 2023).

Oracle is expanding its global presence by continually adding more regions to its public cloud network. Oracle is continuing to expand its global presence by adding more regions to its public cloud network. This expansion demonstrates Oracle's dedication to providing enterprise cloud services to local and regional organizations across Africa and worldwide, as shown in Figure 2-1. For example, at the time of writing this book, Oracle recently announced its plans to open two public cloud regions in Morocco, positioning Oracle as the first *hyperscaler to open public cloud regions in North Africa* (Oracle, 2024).

Tenancy/Compartment

When you sign up or subscribe to Oracle Cloud services, Oracle creates a tenancy for you. You can think of the tenancy as your account, but it is also a secure and isolated partition within Oracle Cloud Infrastructure where you can create, organize, and administer your cloud resources. When you sign up, your tenancy is created in your home region, but you can subscribe your tenancy to as many regions as you need. Large organizations can have multiple tenancies (Oracle, 2023).

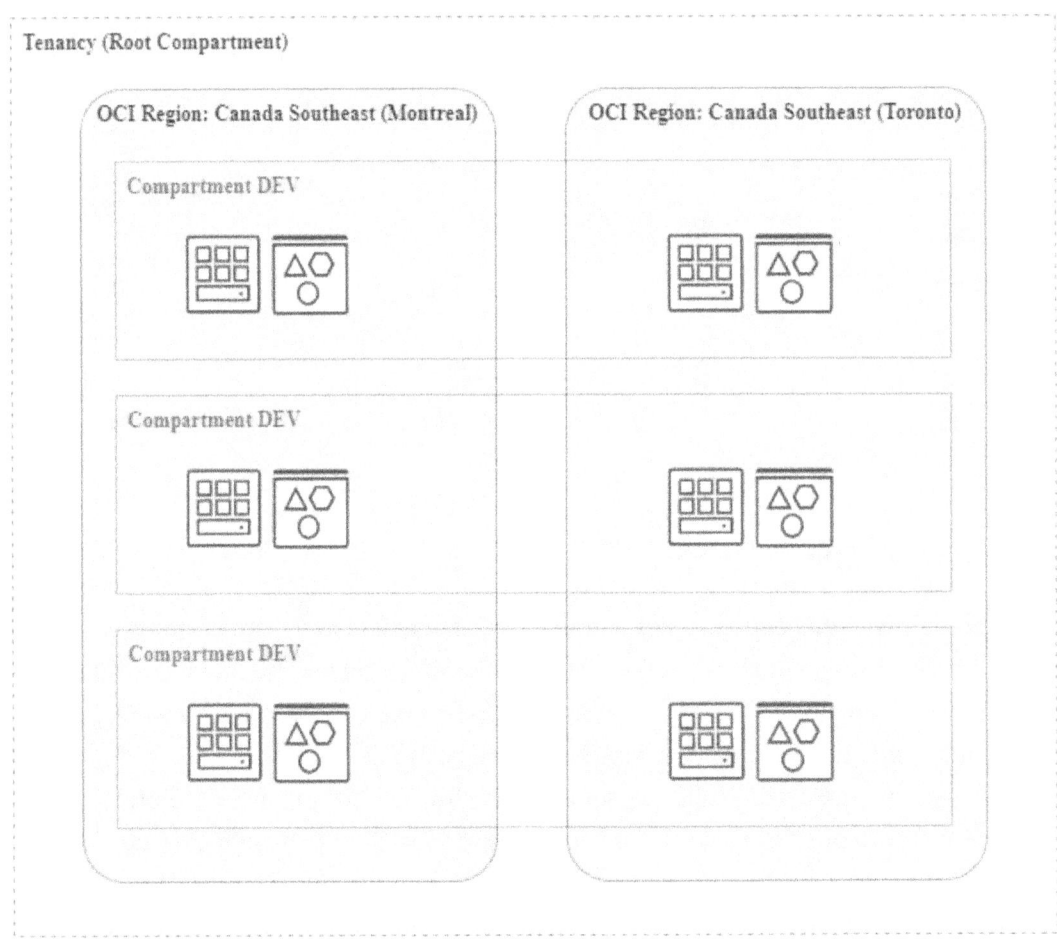

Figure 2-3. OCI tenancy, regions, and compartments

As shown in Figure 2-3, compartments allow you to organize and control access to your cloud resources. A compartment is a collection of related resources (such as instances, Virtual Cloud Networks, block volumes) that can be accessed only by certain groups that have been given permission by an administrator. A compartment should be thought of as a logical group and not a physical container. When you begin working with resources in the console, the compartment acts as a filter for what you are viewing.

When you sign up for Oracle Cloud Infrastructure, Oracle creates your tenancy, which is the root compartment that holds all your cloud resources. You then create additional compartments within the tenancy (root compartment) and corresponding policies to control access to the resources in each compartment. When you create a cloud resource such as an instance, block volume, or cloud network, you must specify to which compartment you want the resource to belong (Oracle, 2023).

In Oracle Cloud Infrastructure (OCI), compartments serve as organizational units that group related resources, such as network and storage, to improve governance and isolation. Each OCI tenancy includes a default root compartment, named after the tenancy, which is the parent compartment for all others within the tenancy. Best practices recommend avoiding the use of the root compartment for user-created OCI resources.

OCI accommodates a hierarchical compartment structure, allowing up to six nested levels. This feature enables nuanced segregation of resources, facilitating the management of different phases of a project or the operational needs of various departments. Each cloud resource must be linked to a designated compartment to maintain this structured organization.

Compartments are tenancy-wide, across regions. When you create a compartment, it is available in every region that your tenancy is subscribed to. You can get a cross-region view of your resources in a specific compartment with the tenancy explorer.

Core OCI Resources

OCI provides a complete set of cloud services, each offering various resource types to meet diverse needs. Here are some key OCI resources grouped by their primary use.

OCI Networking

In this section, we discuss OCI Virtual Cloud Network (VCN)[3] and its resources, such as security lists, network security groups, route tables, and connectivity to the outside world, such as the Internet or to an on-premises environment.

---

[3] For detailed information, refer to the Networking Overview found at https://docs.oracle.com/en-us/iaas/Content/Network/Concepts/overview.htm

CHAPTER 2   ORACLE CLOUD FOR NLP

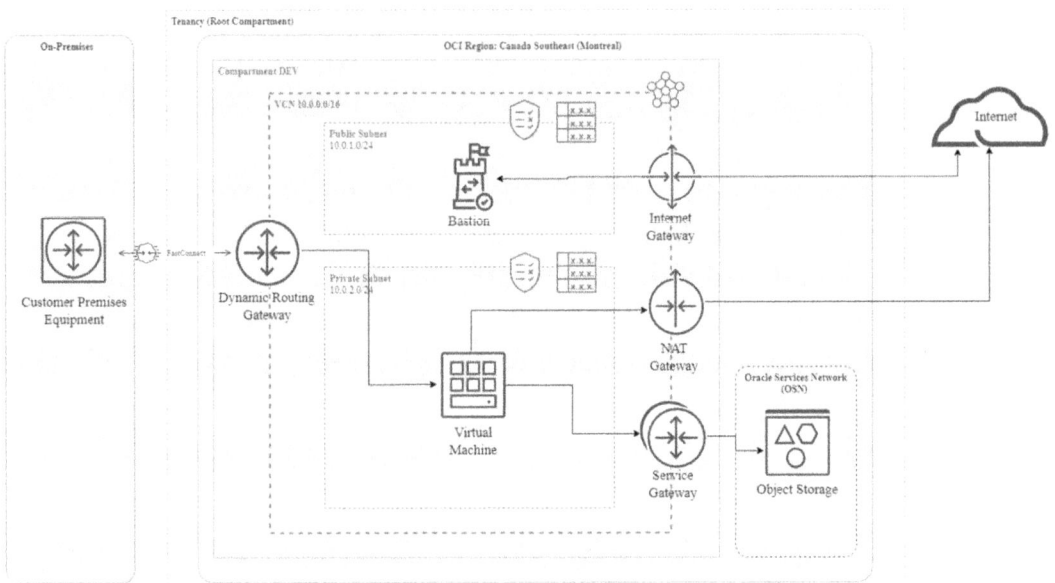

Figure 2-4. *OCI VCN and subnets*

As depicted in Figure 2-4, an OCI Virtual Cloud Network (VCN) resides in a single region. Oracle has many regions worldwide. When you create a VCN, it is specific to one region. Notably, within a VCN, we can have multiple IP CIDR ranges, up to five. The fundamental constituents of a VCN encompass the following:

- Subnets: Subnets divide a VCN and are necessary for deploying resources such as compute instances, database nodes, or load balancers. Subnets can be public or private and regional or specific to an availability domain (AD). Public subnets can optionally have public IPs, while private subnets cannot. Regional subnets span all ADs within a region, allowing resource distribution across ADs.

- Security Lists: Each subnet includes a firewall called a security list, applying to all resources within the subnet. Security lists allow or deny traffic based on rules. By default, three permissions are included: port 22 for SSH and two ICMP rules for fragmentation and troubleshooting. You can modify or remove these rules as needed. Security lists can be stateful or stateless. Stateful rules automatically allow return traffic, while stateless rules require explicit rules for both directions.

- Network Security Groups: Network security groups (NSGs) provide more granular control than security lists, allowing you to create exceptions for specific resources within a subnet. For example, if only certain web servers need port 80 and 443 open, you can create an NSG for those servers rather than applying the rule to the entire subnet. NSGs can be mixed with security lists, and you can use up to five NSGs per vNIC. Oracle recommends using NSGs over security lists for finer control, though they can be more complex.

- Gateways: The gateways used for connectivity between the VCN and external networks, such as Oracle Services Network (OSN), on-prem networks, or the Internet, are as follows:

- Internet Gateway: Allows bidirectional traffic between the VCN and the Internet for public subnets.

- NAT Gateway: Enables outbound traffic from private subnets to the Internet without allowing inbound traffic.

- Service Gateway: Provides access to Oracle Cloud Infrastructure services like object storage without traversing the Internet, staying within OCI for efficiency.

- Local Peering Gateway: Connects VCNs within the same region.

- Dynamic Routing Gateway (DRG): Connects VCNs across different regions or to on-premises networks.

- Route Tables: Route tables enable communication outside the VCN, such as to the Internet, on-premises networks, or peered VCNs. Local routing within the VCN handles traffic between subnets automatically.

We covered Virtual Cloud Networks, subnets, security lists, network security groups, route tables, and various gateways (Internet, NAT, service, local peering, and dynamic routing).

OCI Compute

The OCI Compute Service is a fundamental component of the Oracle Cloud. It allows us to set up compute instances, which are either physical or virtual computers in the cloud, as exemplified in Figure 2-5. Compute instances provide the necessary computational power for running software applications in the cloud. These instances are configured using images, which define a preinstalled software stack, and shapes, which specify the virtual hardware profile. These compute instances can be utilized directly by OCI users or by other OCI services. Each compute instance contains CPU, memory, and storage.

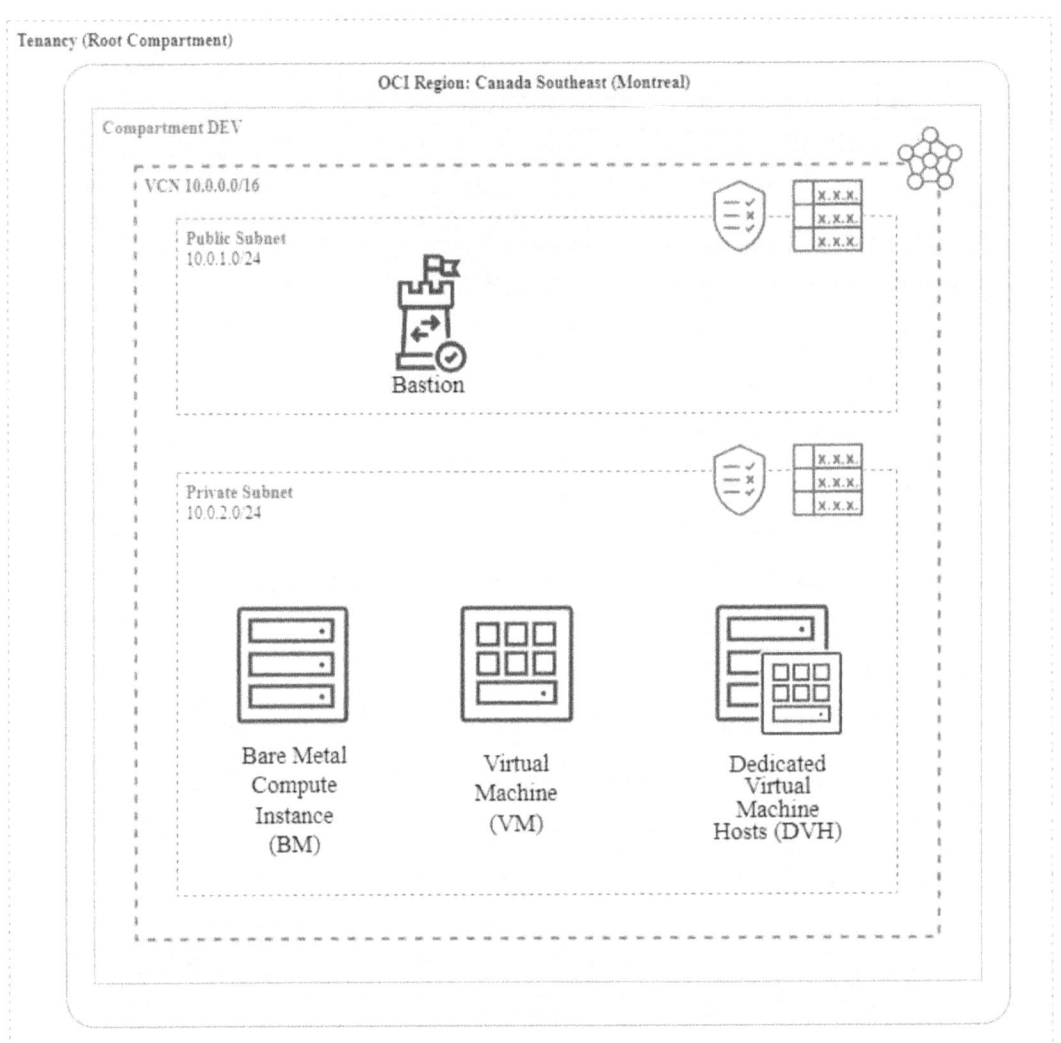

Figure 2-5. *OCI compute instances*

Before we discuss the Compute Service on OCI, it's important to understand how CPUs are measured and used. In most cloud environments, compute resources are measured using virtual CPUs (vCPUs), which represent a thread in a multithreaded processor core and are abstracted from the physical hardware. This abstraction often means that end users are unaware of the actual hardware being used or how many compute cores the cloud provider has allocated.

On the other hand, Oracle Cloud Infrastructure (OCI) uses Oracle Compute Units (OCPUs) as the measurement for compute resources. An OCPU represents an entire physical CPU core with hyper-threading enabled, effectively equating to two vCPUs or two threads. This difference is important for understanding how to scale your infrastructure and manage software licensing effectively.

With a clear understanding of how OCI measures compute resources using OCPUs, we can now explore the types of compute instances available. OCI Compute Service offers two primary options for compute instances:

- Bare Metal (BM): Provides direct hardware access with all the security capabilities, elasticity, and scalability of Oracle Cloud Infrastructure. Typical use cases include performance-intensive workloads, nonvirtualizable workloads, or workloads requiring a specific hypervisor

- Virtual Machines (VMs): Independent computing environments that run on top of physical bare metal hardware, isolated from each other. Users do not need to manage virtualization host maintenance.

OCI offers the option to customize VM instances for specialized workloads:

- Burstable Instances: VM instances with a baseline level of CPU performance that can handle occasional usage spikes, depending on available capacity

Currently, OCI offers AMD, Intel, and Ampere processors. The processor choice is part of the instance creation process and is included in the instance's shape. A shape defines the resources allocated to an instance and can be either fixed or flexible:

- Fixed Shapes: Predefined CPU and memory allocations, including special shapes like dense I/O shapes with local NVMe disks, GPU shapes with NVIDIA graphics processors, and HPC shapes with high-frequency processor cores and cluster networking support

- Flexible Shapes: Allow CPU and memory adjustments at creation or during the instance's lifespan. Available only for virtual machines.

Every compute instance, whether bare metal or a virtual machine, is created using an image. An image is a template for the boot volume, defining partitions and containing the OS and additional software. Four types of images can be used in OCI:

- Platform Images: Oracle-provided standard images, including multiple Linux distributions and Windows images

- Custom Images: Created from configured instances to replicate exact copies

- Marketplace Images: Oracle products and third-party images ready for deployment

- Bring Your Own Image (BYOI): Allows migrating on-premises virtual machines to OCI, supporting cloud migration projects and infrastructure flexibility

When launching compute instances, you can choose the type of host capacity:

- On-Demand Capacity: Resources are allocated immediately upon instance creation and released upon termination, with billing based on usage.

- Capacity Reservation: Ensures resources are available when needed by creating a reservation ahead of time. Capacity reservations can be created, changed, or deleted anytime and charged at a discounted rate when not used.

- Preemptible Capacity: Uses excess compute capacity on a first-come, first-served basis without guaranteed ownership, ideal for short-lived, fault-tolerant workloads. These instances are 50% cheaper than on-demand instances but can be terminated if capacity is needed elsewhere.

- Dedicated Capacity: Dedicated Virtual Machine Hosts (DVH) allow running compute VM instances on dedicated single-tenant servers, meeting compliance and regulatory requirements for isolation and supporting node-based or host-based software licensing.

OCI Compute Service offers scaling capabilities:

- Vertical Scaling: Flexible-shaped instances can adjust OCPU and memory counts, while fixed-shaped instances can change to different shapes with more or fewer resources, requiring a reboot.

CHAPTER 2 ORACLE CLOUD FOR NLP

- Horizontal Scaling (Autoscaling): Instance pools can scale based on metric- or schedule-based policies, increasing or decreasing the number of instances as needed.

OCI Compute Services provides a comprehensive and adaptable solution for running diverse workloads. From performance-intensive applications on bare metal instances to scalable virtual machines, OCI offers various capacity options including on-demand, preemptible, reserved, and dedicated instances to meet different performance, budget, and compliance requirements.

OCI Storage

This section briefly overviews OCI's main categories of storage services: Block Volumes, File Storage, Object Storage, and local NVMe as illustrated in Figure 2-6.

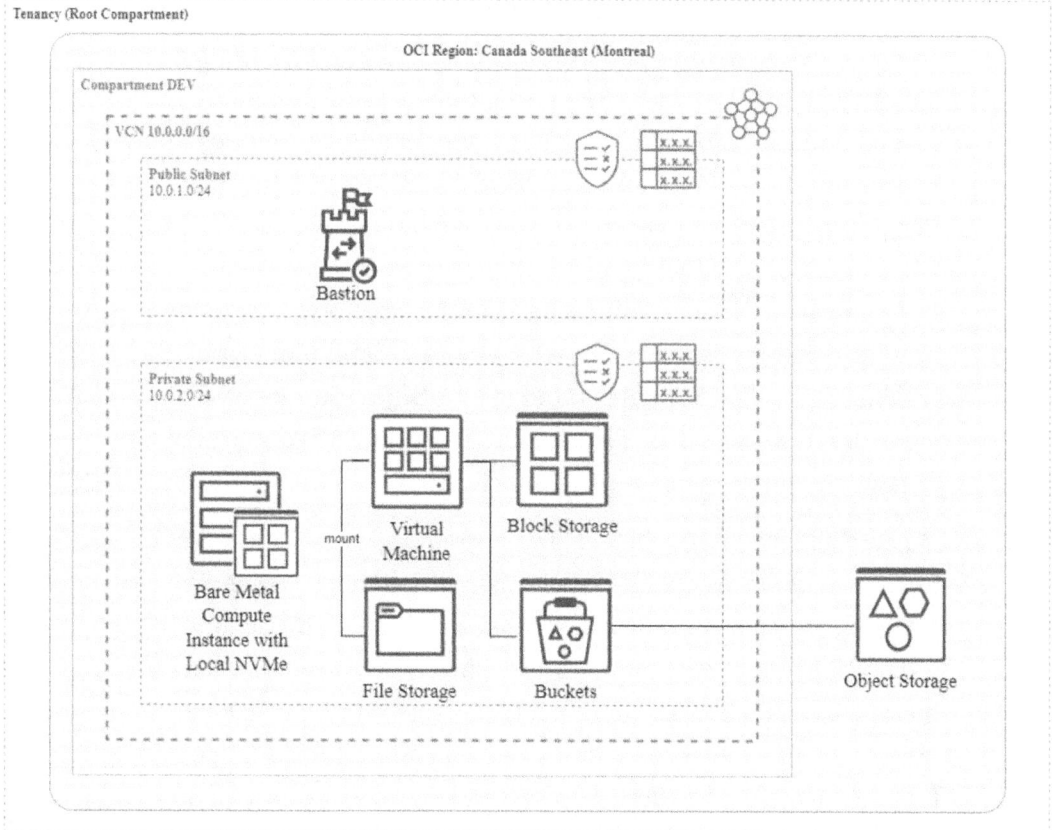

Figure 2-6. OCI Storage Services

- Block Storage Service: The Block Storage Service lets you dynamically provision and manage block storage volumes. Unlike local NVMe storage, these are accessed through a secure and high-speed network, similar to a SAN (storage area network). Volumes can be block volumes or boot volumes, with block volumes used for data and boot volumes for booting instances.

- Block Volume: Block volumes are elastic, allowing you to set the size according to your needs, from 1 gigabyte up to 32 terabytes per volume in 1 gigabyte increments. The size of the volume determines performance, and costs are defined by the space allocated and the performance level set. NVMe solid-state drives ensure outstanding performance, running on a high-performance network with submillisecond latencies. Each instance can connect up to 32 volumes, which appear to the OS as regular block devices. Volumes can be attached to multiple instances for sharing content and are encrypted at rest and in transit. Durability is ensured through several replicas across the availability domain. Block volumes can be resized online or offline but only to increase the size. Performance tiers include Basic, Balanced (default), Higher Performance, and Ultra High Performance, with IOPS ranging from 90 to 225 IOPS per gigabyte. OCI Block Volumes allows the creation of complete point-in-time snapshot copies, which can be incremental or full backups. These backups are stored in OCI Object Storage after they are encrypted and can be restored as new volumes in any availability domain within the same region (Oracle, 2024).

- Object Storage Service: Object Storage is an Internet-scale, high-performance storage platform for unstructured data. Use cases include content repositories, large datasets from pharmaceutical trials, application logs, and backups. Objects are stored in a flat hierarchy within buckets, and metadata is also stored with each object. Object Storage offers three tiers: Standard (Hot Storage) for frequent access, Infrequent Access (Cool Storage) for infrequent access with lower costs, and Archive Storage (Cold Storage) for seldom accessed data with a 90-day retention requirement. Auto-tiering can move objects between tiers based on access patterns, optimizing costs. Object life cycle policies allow for transitioning data between tiers based on time-based rules, facilitating cost management by moving objects to lower-cost tiers or deleting them after a set period (Oracle, 2024).

- File Storage Service: The File Storage Service offers scalable, distributed, enterprise-grade network file systems that are accessible across different subnets and VCNs or through VPN or FastConnect connections. It's a fully managed shared storage. Use cases include migrating Oracle applications, general-purpose file systems, big data analytics, scaling out applications, and providing persistent storage for containers. File Storage ensures data durability through a five-way replication and offers 10,000 snapshots per file system. Data is encrypted at rest and in transit (Oracle, 2024).

In addition, bare metal instances have the option to use local NVMe storage. It consists of NVMe solid-state drives that are directly attached to the host. The common use cases for this type of storage include large databases, big data workloads, and applications requiring high local performance.

Table 2-1 (Oracle, 2020) comprehensively overviews Oracle Cloud Infrastructure (OCI) storage services. It compares various storage options available in OCI, including local NVMe, Block Volume, File Storage, Object Storage, and Archive Storage. Each storage type is detailed in terms of its type, access method, structure, durability, capacity, unit size, and use cases. This comparison aims to help users understand each storage service's key features and benefits, enabling them to choose the most appropriate solution for their specific needs.

Table 2-1. *OCI Storage Services*

OCI Storage Services	Local NVMe	Block Volume	File Storage	Object Storage	Archive Storage
Type	NVMe SSD temporary storage	NVMe SSD block storage	NFSv3 compatible file system	Highly durable object storage	Long-term archival and backup
Access	Block	Block	File	Object	Object
Structure	Block level	Block level	Hierarchical	Unstructured	Unstructured
Durability	Nonpersistent	Durable (multiple copies in an AD)	Durable (multiple copies in an AD)	Multiple copies across ADs[4]	Multiple copies across ADs
Capacity	Terabytes+	Petabytes+	Exabytes+	Petabytes+	Petabytes+
Unit Size	51.2 TB (BM), 6.4–25.6 TB (VM)	50 GB to 32 TB/vol, 32 vols/instance	Up to 8 exabytes	10 TB/object	10 TB/object
Use Cases	OLTP, NoSQL, data warehousing	Database, VMFS, NTFS, boot and data disks for instances	Oracle apps (EBS), HPC, general-purpose file systems	Unstructured data (logs, images, videos)	Backups and long-term archival (DB backups)

OCI provides robust and versatile storage solutions tailored to various needs. Block Volumes offer persistent storage for compute instances, supporting boot volumes and additional storage with comprehensive backup capabilities. Object Storage, designed for managing data as objects within buckets, allows for efficient and scalable data management with life cycle policies for automated object handling. It is ideal for storing unstructured data such as logs, images, and videos. On the other hand, File Storage facilitates the creation of shared file systems within subnets, enabling compute instances to connect seamlessly and benefit from efficient backup solutions through snapshots. These storage options ensure that users can effectively manage their data with flexibility, scalability, and reliability.

---

[4] ADs: availability domains.

Identity and Access Management (IAM)

OCI Identity and Access Management (IAM) is a service that assists in managing user authentication and authorization in OCI. In Figure 2-7, you can see the high-level flow for OCI access management, which depicts a group as a collection of users who are granted access to resources or compartments using policies.

Users in OCI authenticate using methods such as username and password, which ensures that they are who they claim to be. Users are categorized based on their access requirements, and each group is assigned specific permissions to access OCI resources. In OCI, an identity domain represents a user population, associated configuration, and security settings.

Resources such as compute instances and storage volumes belong to compartments and have unique Oracle-assigned identifiers called OCIDs. Policies are created to define which resources an authenticated group can access. These policies are applied to both the tenancy and specific compartments. A policy might look like this:

- Allow group <group-name> to <verb> <resource-type> in tenancy.

- Allow group <group-name> to <verb> <resource-type> in compartment <compartment-name> [where <condition>].

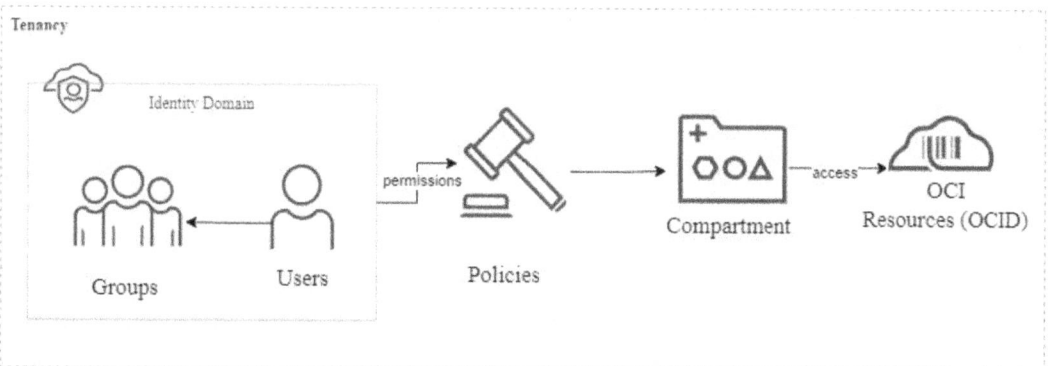

Figure 2-7. OCI's Identity and Access Management (IAM)

OCI IAM ensures secure and organized access to cloud resources by effectively utilizing compartments and their policies.

CHAPTER 2 ORACLE CLOUD FOR NLP

Oracle's AI Overview

Oracle's approach to artificial intelligence (AI) is characterized by strategically integrating AI and machine learning (ML) technologies across its cloud offerings and enterprise solutions. By embedding AI capabilities into core products and developing robust platforms for AI development, Oracle aims to enhance operational efficiency and drive innovation for its customers. This section provides an overview of Oracle's AI strategy, its comprehensive AI and ML stack, and the diverse offerings designed to meet the evolving needs of businesses across various industries.

OCI's robust infrastructure and advanced AI capabilities make it an ideal platform for developing and deploying sophisticated NLP applications, which we'll explore in detail in the upcoming sections and chapters.

AI Strategy

Oracle has strongly emphasized integrating AI technologies into its cloud offerings and enterprise solutions. The company's AI strategy involves enhancing existing products, developing new AI-powered services, and providing robust AI development and deployment platforms.

A key part of Oracle's AI strategy involves integrating AI and machine learning capabilities into its core products. For instance, Oracle Database 23ai with AI Vector Search supports retrieval-augmented generation (RAG), which combines large language models (LLMs) with private business data to provide accurate and contextually relevant responses to natural language questions.

Another important aspect of Oracle's AI strategy is its focus on "embedded AI"—integrating AI capabilities directly into business applications. This approach is evident in Oracle's suite of cloud applications, including those for customer experience, human resources, and enterprise resource planning.

Oracle is also committed to developing and providing prebuilt AI services to expedite customer adoption, such as OCI Vision for image analysis, OCI Language for Natural Language Processing, and OCI Speech for speech recognition and synthesis.

Additionally, Oracle offers solutions like OCI Data Science and Machine Learning in Database, which enable data scientists to build, train, and deploy machine learning models, streamlining the process of building and deploying ML models at scale.

In the realm of cloud infrastructure, Oracle has positioned its Oracle Cloud Infrastructure (OCI) as a high-performance platform for AI and machine learning workloads. The company has heavily invested in GPU-accelerated computing resources and has partnered with NVIDIA to offer powerful hardware options for AI model training and inference.

Furthermore, Oracle has developed industry-specific AI applications tailored for healthcare, financial services, and retail sectors. This allows the company to offer its enterprise customers more targeted and immediately applicable AI solutions.

Oracle's AI strategy aims to expand its AI services, integrate AI into core products, and explore emerging areas such as generative AI and LLMs. The company also recognizes the importance of responsible and ethical AI use, particularly in financial institutions and healthcare sectors.

AI Stack

Oracle's AI stack offers a comprehensive suite of advanced AI and machine learning solutions designed to meet the diverse needs of modern enterprises. From powerful generative AI capabilities to robust infrastructure for intensive workloads, Oracle provides tools that enable businesses to leverage artificial intelligence efficiently and effectively. Figure 2-8 illustrates the key offerings of Oracle's AI stack, including Generative AI, AI Services, Generative AI and Machine Learning for Data Platforms, and AI Infrastructure.

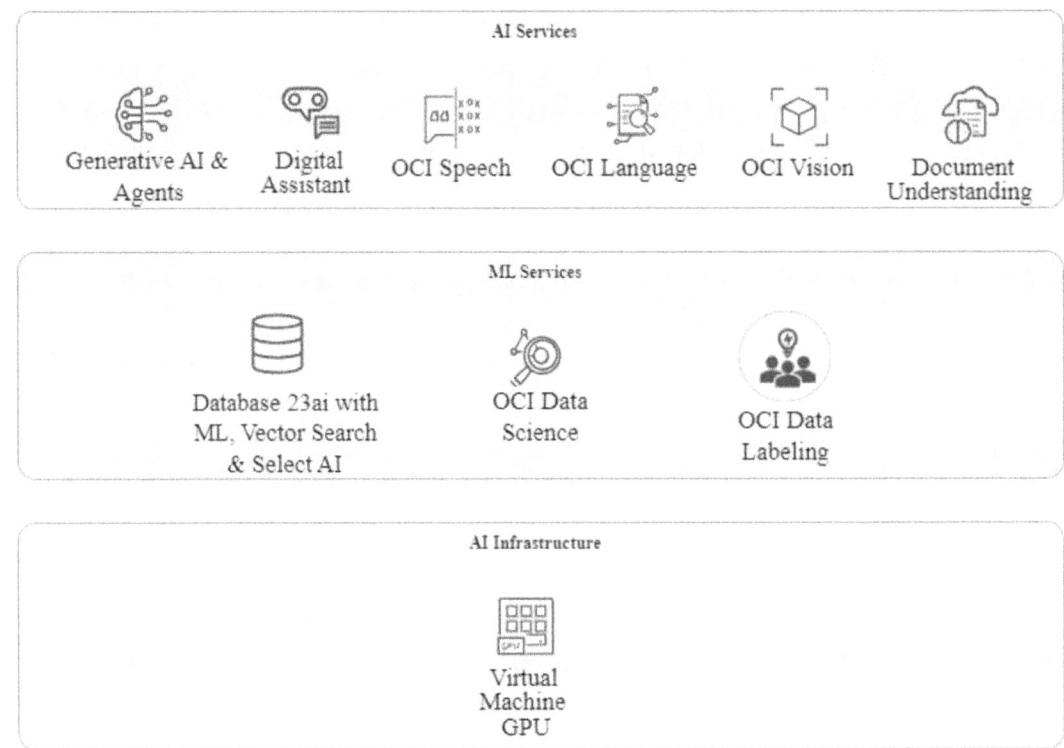

Figure 2-8. *Oracle AI stack*

Although the AI capabilities cover a wide range of areas including computer vision, various machine learning services, and diverse use cases, this section will specifically focus on products related to developing Natural Language Processing (NLP) solutions, which are relevant to our book. We will introduce these key capabilities, laying the groundwork for subsequent chapters where we will explore how to use them effectively to create advanced NLP applications that foster innovation and improve productivity.

Let's explore each category and its components and capabilities to gain a better understanding.

OCI AI Services

The AI Services include various functionalities such as Natural Language Processing, computer vision, and speech-to-text. Oracle AI Services provides pretrained models that can be easily integrated into applications to access advanced AI capabilities. These models can be customized with user data to make them more relevant. By using these prebuilt models, businesses can quickly improve their applications with advanced AI capabilities without needing extensive machine learning expertise, speeding up intelligent application deployment (Oracle).

Lately, Oracle enhanced its AI Services offering with Generative AI. This service enables users to choose from managed open source or proprietary large language models (LLMs), fine-tune them, or augment them with their enterprise data leveraging built-in vector databases retrieval-augmented generation (RAG). Generative AI is ideal for creating new content, such as text or code, based on existing data, making it a powerful tool for content creation, data augmentation, and various creative applications.

As shown in Figure 2-8, OCI AI Services provide the following services:

- Generative AI and GenAI Agents

  - OCI GenAI: Allows users to harness the power of managed open source or proprietary large language models (LLMs). It supports fine-tuning these models and augmenting them with enterprise data.

  - OCI GenAI Agents: Combines the power of large language models (LLMs) and retrieval-augmented generation (RAG) techniques with enterprise data. This allows users to converse directly with enterprise knowledge bases using natural language.

- OCI Digital Assistant: An intelligent chatbot platform that can simulate and process human conversation, whether it's written or spoken. This allows humans to interact with applications and data as if they were communicating with a real person. The platform uses deep learning algorithms to understand natural conversation, interpret intent and context, and remember user behaviors in multiple languages.

- OCI Speech: Capable of converting spoken words into written text and vice versa. It utilizes automatic speech recognition (ASR) models to transcribe audio content into text. In addition, text-to-speech (TTS) models based on neural networks synthesize natural-sounding voices from written text. These ASR and TTS pretrained models can be easily integrated into enterprise applications.

- OCI Language: Performs text analysis at scale, including sentiment analysis, entity recognition, translation, and more. It offers pretrained models that can be customized and fine-tuned on enterprise data for more relevant text analysis, enabling the development of intelligent applications for processing unstructured text.

- OCI Vision: Offers capabilities for processing and analyzing images at scale. Its pretrained models enable image recognition to be integrated into applications without requiring machine learning (ML) expertise. Enterprises can create custom vision models using their own data.

- OCI Document Understanding: Enables the extraction of text, tables, and other important data from document files through pretrained or custom trained AI models.

OCI ML Services

OCI ML Services enable users to collaboratively build, train, deploy, and manage machine learning models. This service supports the use of popular open source frameworks and in-database machine learning, providing flexibility and efficiency in model development. By leveraging these tools, data scientists and developers can create robust ML models that integrate seamlessly with their data platforms, driving insights and automation in various business processes. This offering is essential for organizations looking to harness the full potential of their data through advanced AI and ML techniques (Oracle).

Below are the main services under OCI ML Services categories:

- Oracle Database 23ai with AI Vector Search: Enhances data processing by generating and storing vectors (embeddings). One of its key features is AI Vector Search, which allows users to generate and store vectors for performing similarity searches using mathematical calculations. Another powerful feature is the ability to combine similarity searches with traditional business data searches using simple SQL. Additionally, it supports retrieval-augmented generation (RAG), combining large language models (LLMs) with private business data to deliver accurate and contextually relevant responses to natural language questions.

- ML in Oracle Database: The ML in Oracle Database supports data exploration, preparation, and machine learning (ML) modeling at scale using SQL, R, Python, REST, automated machine learning (AutoML), and no-code interfaces. It includes many in-database ML algorithms that produce models for immediate use in applications. By keeping data in the database, data scientists can automatically build models that are closely tied to their data.

- OCI Data Science: Offers a managed environment for the end-to-end for building and deploying of machine learning models.[5]

- OCI Data Labeling: A tool for labeling data, either text or images, essential for preparing datasets for machine learning, including NLP.[6]

AI Infrastructure

Oracle AI Infrastructure offers the benefits of cloud elasticity, usage-based costs, and high-performance computing, delivered through OCI's distributed cloud. This service provides access to powerful GPUs optimized for AI workloads, helping businesses use advanced computing resources for their AI projects without having to buy and manage expensive hardware themselves.

Here are the key components of OCI AI Infrastructure:

- GPU Instances: These are specialized computers with powerful graphics processing units (GPUs) from NVIDIA. GPUs are excellent at handling AI tasks because they can perform many calculations at once. OCI offers both physical (bare metal) and virtual machines with these GPUs.

- Supercluster GPU Instances: For even more demanding AI tasks, OCI provides superclusters. These are groups of very powerful GPUs working together, ideal for training large AI models or processing vast amounts of data quickly.

---

[5] A comprehensive exploration of OCI Data Science, including its features, capabilities, and integration with NLP workflows, will be provided in subsequent chapters. This service will be detailed further in the upcoming sections.

[6] Detailed information on OCI Data Labeling, its functionalities, and its role in NLP data preparation will be discussed in-depth in later sections.

- Fast Network Connections: AI often requires moving large amounts of data between computers. OCI uses a technology called RDMA to create very fast connections between machines, helping AI tasks run more efficiently.

- High-Performance Storage: AI needs quick access to lots of data. OCI offers fast storage options, including

  - NVMe storage: Very fast storage directly attached to the computing units.

  - Special file systems: These allow many computers to access the same data quickly.

The benefits of using OCI AI Infrastructure are significant. With OCI, you have the flexibility to use and pay for resources only when you need them, making it a cost-effective solution. By avoiding the expense of buying and maintaining your own AI hardware, you can redirect your resources to other critical areas of your business. Additionally, OCI offers scalability, allowing you to easily increase or decrease resources as your AI projects grow or change. This ensures that you have the necessary infrastructure to support your evolving needs. Finally, OCI provides access to cutting-edge hardware for faster AI development and deployment, thereby enhancing overall performance.

By using OCI's AI Infrastructure, companies can focus on developing their AI applications without worrying about the complex technical details of managing hardware. This can help speed up AI projects and make it easier for businesses to use advanced AI technologies.

OCI for NLP

Businesses increasingly rely on data to make informed decisions in today's landscape. From finance to social media, there is a growing need for solutions that can effectively understand and analyze language data. Cloud-based Natural Language Processing (NLP) platforms have emerged as a viable option, enabling companies to extract valuable insights from textual information, even without extensive AI expertise.

One significant player in this space is Oracle Cloud Infrastructure (OCI). OCI offers a robust infrastructure and a range of AI and machine learning solutions to support modern businesses. Its comprehensive suite of tools allows enterprises to efficiently and confidently manage the entire life cycle of AI projects.

While Oracle offers a vast array of services under its AI and ML umbrella, this book will focus on three pivotal services that empower NLP implementations:

- OCI Language: Cloud-based Natural Language Processing (NLP) platform. This service provides advanced text analysis capabilities.

- OCI Data Science: This platform offers a comprehensive environment for developing and deploying NLP models.

- OCI Data Labeling: This service is a crucial service for annotating and labeling the data required to train Natural Language Processing (NLP) models.

In the subsequent sections, we will explore how these OCI services can be leveraged to enhance your organization's NLP capabilities and drive data-driven decision-making.

OCI Language

OCI Language is a cloud-based service that enables companies to process unstructured text data for various tasks, including sentiment analysis, entity recognition, and more, using various pretrained models.[7]

Key functionalities of OCI Language include

- Language and Entity Recognition: OCI Language can identify multiple named entities, including but not limited to names of individuals, places, organizations, and product identifiers. It also allows identifying Personally Identifiable Information (PII), which is crucial for ensuring data protection and regulatory compliance.

- Sentiment Analysis: One of the primary functionalities of OCI Language is sentiment detection. The service examines the text to categorize its sentiment as positive, negative, or neutral. An associated confidence score is provided with each categorization, giving a more detailed perspective on the detected sentiment across different languages.

---

[7] Learn more about the OCI Language service pretrained models at [Oracle Docs](https://docs.oracle.com/en-us/iaas/language/using/pretrain-models.htm).

- Document Classification and Key Phrase Extraction: The service categorizes textual content into more than 600 predefined categories, which span numerous languages. It can also pinpoint key phrases within text documents through advanced Natural Language Processing techniques.

- Translation: Utilizing advanced neural machine translation mechanisms, OCI Language supports text translation across over 20 languages.

Oracle has recently announced the general availability of OCI Language 4.0. This release now supports additional languages (Arabic, French, German, and Italian) for the pretrained models. It also introduces Health Natural Language Processing (NLP) features, which provide pretrained models to extract entities from electronic health records (EHRs), progress notes, clinical trial documents, and more (Oracle, 2024). The Healthcare NLP suite includes pretrained models for

- Health Named Entity Recognition (Health NER): Identifies key entities from text, including identifying medical conditions, medications, dosages, symptoms, test results, treatments, and procedures.

- Protected Health Information (PHI) Identification and Deidentification: The PHI service extends the current PII service to detect PHI entities and provides the option to deidentify and anonymize identified entities from output.

OCI Language provides a powerful feature for customizing pretrained models to meet specific needs of different domains and industries.

Use Cases

The following use cases illustrate practical applications of the OCI Language Service, providing valuable insights into how this service can be leveraged to meet specific business needs. These examples can serve as a guide for effectively utilizing OCI Language in real-world scenarios.

- Customer Feedback Analysis
    - Understanding Customer Perception: Dive into how customers view your brand, extracting sentiments and zeroing in on specific pain points. This allows businesses to proactively address concerns and improve the customer experience.

- Keeping a Pulse on the Discourse: Monitor discussions on platforms like social media or within your customer support knowledge base. By understanding which topics dominate the conversation, you can prioritize interventions effectively.

- Gleaning Insights from Named Entities: Extracting named entities from customer feedback helps identify crucial players in the discourse, whether they're individuals, products, or organizations. This can provide businesses with actionable intelligence on what products or features might require attention.

- Elevating Customer Support
  - Prompt Issue Escalation: Real-time identification of dissatisfied customers ensures that seasoned agents can intervene timely, mitigating issues and potentially salvaging the customer relationship.
  - Enhanced Ticket Classification: By automatically extracting key phrases from incoming support tickets, similar tickets can be grouped together. This allows for patterns to be identified quickly, facilitating faster resolution.
  - Efficient Ticket Routing: The automatic language detection capability ensures that support tickets are channeled to agents proficient in the customer's language, promoting effective communication.

- Ensuring Customer Data Privacy
  - Regulatory Compliance: In the era of stringent data privacy regulations like GDPR, OCI Language assists businesses in identifying Personally Identifiable Information (PII) or Protected Health Information (PHI). This crucial feature allows data to be redacted or anonymized prior to publication, ensuring compliance and safeguarding customer privacy.

Here's an example of how a company can use OCI Language to enhance the efficiency of processing customer feedback:

A company can leverage OCI Language to streamline the analysis of customer feedback. By using sentiment analysis, the company can quickly categorize large volumes of text-based feedback into positive, negative, or neutral sentiments, providing a clear picture of customer opinions. Additionally, with features like Named Entity Recognition and key phrase extraction, the company can automatically identify specific products, services, or recurring themes mentioned in the feedback. This allows the business to transform unstructured data into actionable insights, leading to better decision-making and improved customer satisfaction.

OCI Data Science

OCI Data Science serves as a cornerstone for ML model development within the OCI environment. This fully managed platform equips both developers and data scientists with a rich set of tools to build, train, and manage ML models seamlessly (Oracle). Projects, as shown in Figure 2-9, are central to organizing and managing model development tasks.

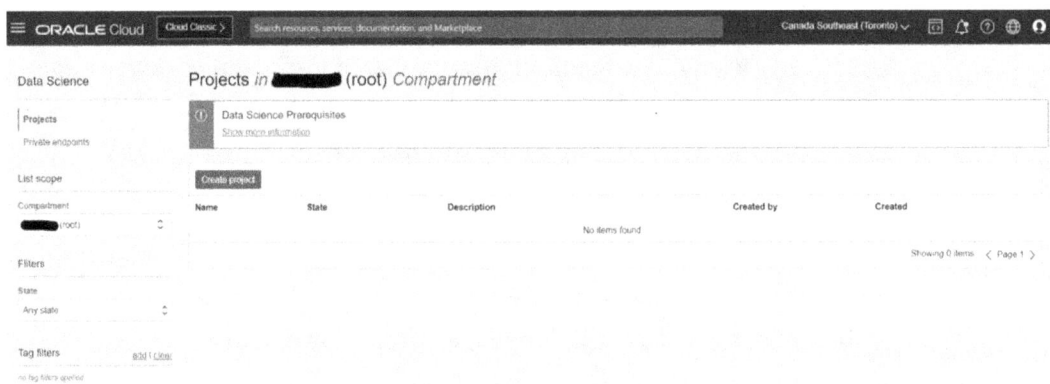

Figure 2-9. OCI Data Science Service—Projects page

Key features of OCI Data Science include

- Notebook Sessions: This collaborative environment, based on Jupyter, allows for intuitive model building and testing.

- Model Catalog: An organized system to store and version ML models.

- Accelerated Training: It taps into GPU resources, ensuring efficient and rapid model training.

- Model Deployment: Models can be seamlessly deployed, whether as REST APIs or batch processes.

OCI Data Science value revolves around several foundational elements:

- Fully Managed Service: Recognizing the complexities inherent to data science workflows, OCI Data Science offers a platform where data scientists can remain focused on deriving insights and honing algorithms, relieved from infrastructure management concerns.

- Comprehensive Toolset: The platform's integration with JupyterLab notebooks facilitates an interactive environment conducive to model development, data visualization, and real-time evaluations. This integration fosters an atmosphere of iterative development.

- Scalability: With ever-expanding data volumes, businesses require services that can scale without hitches. The auto-scaling capabilities of OCI Data Science ensure efficient handling of data, irrespective of its size.

A distinct advantage of OCI Data Science is its deep integration within the OCI ecosystem. In an era where enterprises leverage a multitude of tools, from data storage solutions to advanced BI platforms, OCI Data Science offers a unified platform. This ensures a streamlined workflow across data storage, preprocessing tools, and model deployment, enhancing the efficiency of the entire AI/ML project life cycle.

AI Quick Actions

The AI Quick Actions feature, which is part of the OCI Data Science Service, is designed for users who want to leverage AI capabilities quickly without needing extensive coding knowledge. This feature aims to make foundation models accessible to a broader audience by offering a streamlined, code-free, and efficient environment for working with these models.

AI Quick Actions allows users to quickly deploy, fine-tune, and evaluate various foundation models, as shown in Figure 2-10. The platform provides comprehensive information about each model, including fine-tuning instructions, code samples, model architecture descriptions, troubleshooting tips, and limitations.

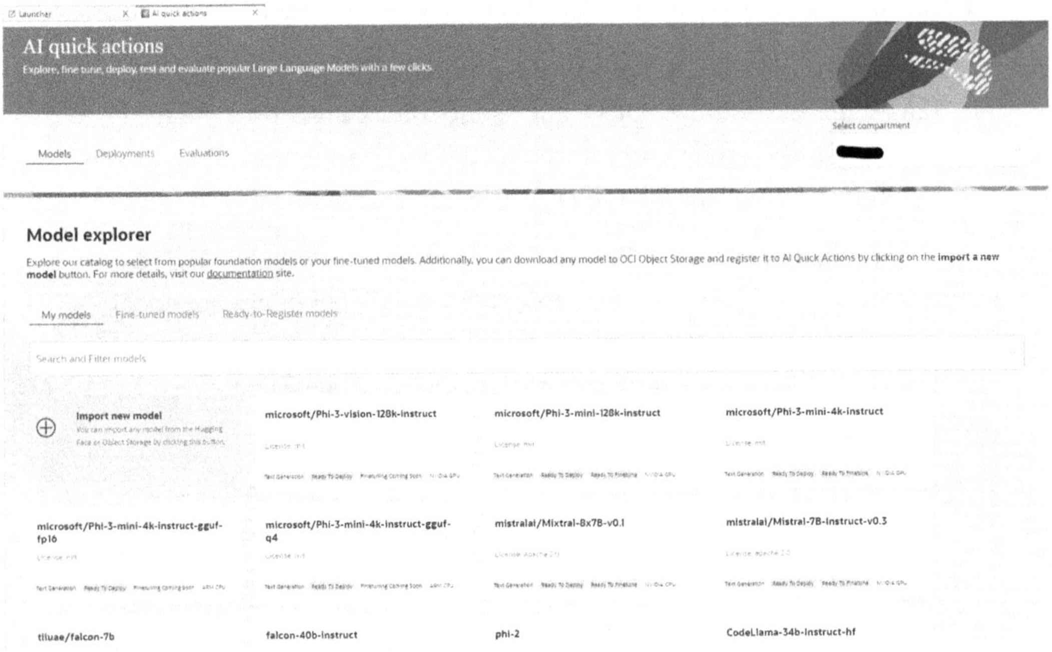

CHAPTER 2 ORACLE CLOUD FOR NLP

Figure 2-10. AI Quick Actions Model explorer

The latest release of AI Quick Actions introduces support for a "bring your own model" feature through OCI Object Storage, expanding the selection of models users can utilize. To bring your own model into AI Quick Actions, users need to download the model artifacts to an OCI Object Storage bucket and register the model within AI Quick Actions.

OCI Data Labeling

The significance of high-quality data in artificial intelligence and machine learning is well known. One key service in this area is the OCI Data Labeling platform (see Figure 2-11). It is designed to create accurately labeled datasets for training ML models including NLP models (Oracle).

CHAPTER 2   ORACLE CLOUD FOR NLP

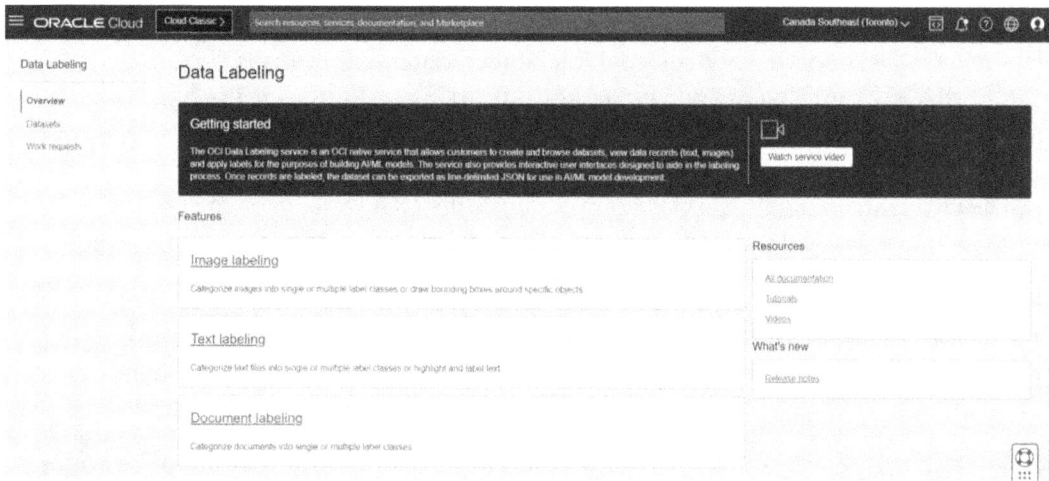

Figure 2-11. *OCI Labeling Service overview page*

The OCI Data Labeling Service makes it easy for developers and data scientists to create training datasets for machine learning models. They can quickly add labels to data using user-friendly interfaces and public APIs. Once labeled, the data can be exported and seamlessly integrated with Oracle's AI and Data Science Services, streamlining the model-building process. Key features of OCI Data Labeling Service include

- User-Friendly Annotation Interface: OCI Data Labeling focuses on making data labeling tasks easy and efficient. Whether tagging sentiments in text or highlighting objects in images, the interface is designed for accuracy and speed, enhancing the overall user experience.

- Collaborative Labeling: Enables team-based data annotation, allowing multiple users to work together efficiently on data labeling projects.

- Flexible Data Handling: OCI Data Labeling can handle various types of data, including text, images, and complex data structures. This flexibility makes it suitable for a wide range of industries and applications.

- Seamless Integration: OCI Data Labeling easily connects with other OCI AI and ML services, making labeled data readily available for various ML tasks. It works effectively with services like OCI Vision and OCI Language, enhancing their capabilities.

The OCI Data Labeling Service complements the OCI Data Science Service. Together, they form a cohesive backbone for the entire machine learning life cycle. While each service has its distinct functions, they work together seamlessly to address specific aspects and ensure robust support for NLP models' development and deployment stages.

AI Samples

Oracle provides a helpful GitHub repository[8] for those interested in using Oracle Machine Learning (ML) Services, specifically OCI Data Science and OCI Data Labeling. This repository contains demos, tutorials, and code examples that demonstrate the capabilities of these services (as shown in Figure 2-12).

Inside the repository, you'll discover a set of JupyterLab notebooks that simplify the use of the ADS SDK and various OCI Data Science features. For example, the Natural Language Processing notebook covers a variety of NLP tasks, such as part-of-speech tagging, Named Entity Recognition, and sentiment analysis. These examples offer hands-on experience, helping users take full advantage of OCI Data Science Services.

The repository also offers extensive resources for OCI Data Labeling, which focuses on creating well-labeled datasets for training machine learning models. It includes Python and Java scripts for annotating large numbers of records in OCI Data Labeling Service (DLS).

---

[8] Oracle Cloud Infrastructure Data Science and AI services Examples GitHub Repository (oracle-samples/oci-data-science-ai-samples) can be found at https://github.com/oracle-samples/oci-data-science-ai-samples

CHAPTER 2   ORACLE CLOUD FOR NLP

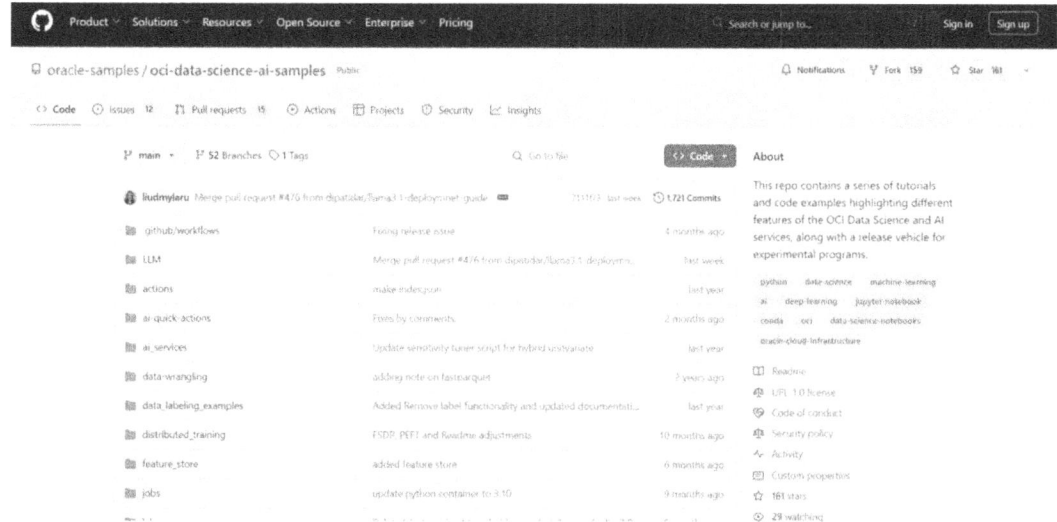

Figure 2-12. *Oracle AI sample GitHub repository*

Oracle's GitHub repository is a valuable resource for anyone using Oracle ML Services. It provides practical examples to assist developers in accelerating the development of ML solutions, including NLP solutions, using OCI Data Science and OCI Data Labeling Services.

High-Level Flow for Building NLP Models Using OCI

Figure 2-13 outlines a high-level workflow for developing Natural Language Processing (NLP) models using Oracle Cloud Infrastructure (OCI).

67

CHAPTER 2   ORACLE CLOUD FOR NLP

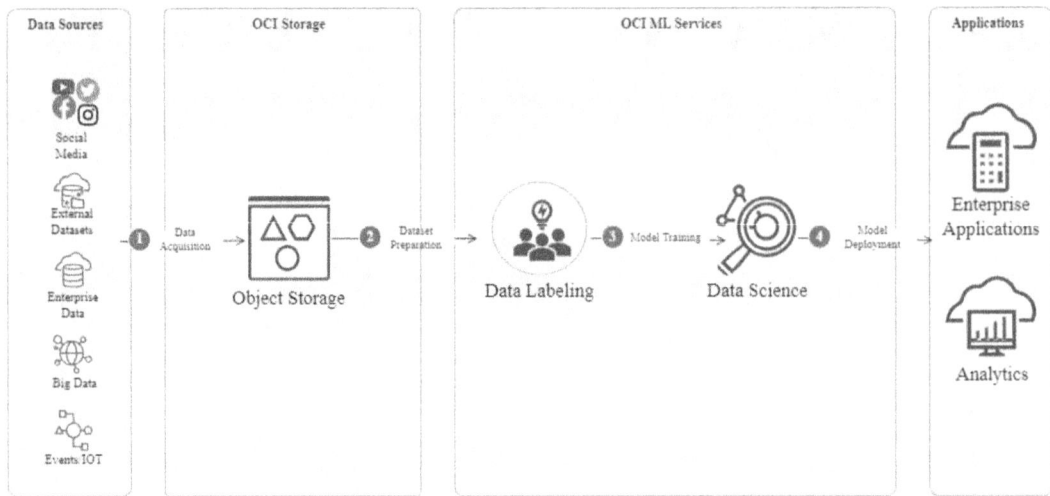

Figure 2-13. *High-level flow for building NLP models using OCI*

This process involves several key stages, supported by various OCI services:

1. Data Acquisition: The process starts by gathering data from various sources like social media, external datasets, enterprise systems, big data platforms, and IoT events. This diverse data forms the foundation for training NLP models.

2. Data Storage: The collected data is stored in OCI Object Storage, a scalable and secure solution for managing large volumes of unstructured text data. It ensures easy access and retrieval for further processing and integrates seamlessly with OCI Data Labeling.

3. Dataset Preparation: OCI Data Labeling plays a crucial role in preparing the dataset by creating and annotating data for specific NLP tasks like sentiment analysis and entity recognition.

4. Model Training: Once the dataset is ready, it's used in OCI Data Science to train NLP models. OCI Data Science provides the necessary tools and environment to build, test, and refine models efficiently.

5. Model Deployment: After training, the NLP models are deployed to enterprise applications or analytics platforms, where they can generate insights, automate processes, and enhance the user experience.

Below are some general guidelines that can help streamline NLP projects when developing solutions on Oracle Cloud Infrastructure (OCI):

- Data Management: Utilize separate OCI Object Storage buckets to efficiently manage and back up datasets as well as NLP models for different teams and stages of NLP projects.

- Model Development: Leverage OCI Data Science's integration capabilities with OCI Object Storage buckets and Git repositories to streamline model development.

- Cost Optimization: Optimize costs by strategically using resources, such as deploying GPU-based VMs primarily during the model training phase.

- Security: Implement strong security measures using OCI Identity and Access Management (IAM) to control access to resources and data, ensuring compliance and data protection.

- Monitoring and Logging: Utilize OCI's Monitoring and Logging services to track the performance and health of your NLP applications.

Summary

Oracle Cloud Infrastructure provides a comprehensive ecosystem for developing, deploying, and scaling NLP solutions. From ready-to-use services like OCI Language and OCI Speech to ML development services like OCI Data Science, OCI offers a robust platform for building and deploying NLP models for various NLP tasks. The integration of these services with Oracle's enterprise-grade infrastructure ensures high performance, security, and scalability for NLP applications.

The future of AI and NLP on OCI looks promising, with Oracle continuously investing in enhancing its AI capabilities. We can expect to see advancements in areas such as GenAI, LLMs, multilingual NLP support, more sophisticated pretrained models, and improved integration with Oracle OCI services and Applications. Oracle's focus on industry-specific solutions suggests that we may also see more tailored NLP offerings for sectors like healthcare, finance, and retail.

Below are key takeaways for organizations considering OCI for NLP solutions:

- Comprehensive Platform: OCI provides an end-to-end platform for NLP, from data storage and processing to model development and deployment.

- Performance and Scalability: With its high-performance infrastructure and global presence, OCI can support NLP solutions at scale.

- Integration Advantages: For organizations already using Oracle products, OCI offers seamless integration, potentially reducing complexity and costs.

- Enterprise Focus: OCI's emphasis on security, compliance, and enterprise-grade reliability makes it particularly suitable for large-scale, mission-critical NLP applications.

- Continuous Innovation: Oracle's ongoing investments in AI and NLP suggest that the platform will continue to evolve and improve, offering adopters access to cutting-edge technologies.

While OCI entered the cloud AI market later than some competitors, it has rapidly developed into a powerful and comprehensive platform for NLP solutions. Its combination of robust infrastructure, specialized AI services, and integration capabilities positions it as a strong contender for organizations looking to develop and deploy sophisticated NLP applications, particularly in enterprise contexts.

Having discussed the fundamentals of NLP and the features of OCI, Chapter 3 will provide an in-depth introduction to our case study. In this chapter, we will outline the motivation, challenges, and preliminary stages of creating and deploying an NLP-based solution for healthcare on OCI. This chapter lays the groundwork, setting the context for subsequent chapters where theory will be transformed into a practical NLP-based application.

References

Oracle. (2018, 10 22). *Introducing the Generation 2 Cloud at Oracle OpenWorld 2018*. Retrieved from Oracle Cloud Infrastructure Blog: https://blogs.oracle.com/cloud-infrastructure/post/introducing-the-generation-2-cloud-at-oracle-openworld-2018

Oracle. (2019, 06 05). *Microsoft and Oracle to Interconnect Microsoft Azure and Oracle Cloud.* Retrieved from Oracle Press Release: https://www.oracle.com/corporate/pressrelease/microsoft-and-oracle-to-interconnect-microsoft-azure-and-oracle-cloud-060519.html

Oracle. (2020, 02). *Oracle Cloud Infrastructure Storage Services.* Retrieved from Oracle: https://www.oracle.com/a/ocom/docs/cloud-training-storage-services.pdf

Oracle. (2022, 10 22). *Oracle and NVIDIA Partner to Speed AI Adoption for Enterprises.* Retrieved from Oracle Press Release: https://www.oracle.com/news/announcement/ocw-oracle-and-nvidia-partner-to-speed-ai-adoption-2022-10-18/

Oracle. (2023, 06 02). *Account and Access Concepts.* Retrieved from Oracle Cloud Infrastructure Documentation: https://docs.oracle.com/en-us/iaas/Content/GSG/Concepts/concepts-account.htm

Oracle. (2023, 06 13). *Oracle to Deliver Powerful and Secure Generative AI Services for Business.* Retrieved from Oracle Press Release: Oracle to Deliver Powerful and Secure Generative AI Services for Business

Oracle. (2023, June 02). *Physical Architecture Concepts.* Retrieved from Oracle Cloud Infrastructure Documentation: https://docs.oracle.com/en-us/iaas/Content/GSG/Concepts/concepts-physical.htm

Oracle. (2024, May 6). *Announcing the general availability of OCI Language 4.0.* Retrieved from Oracle AI & Data Science Blog: https://blogs.oracle.com/ai-and-datascience/post/oci-ai-language-4-0

Oracle. (2024, 5 6). *Announcing the general availability of OCI Language 4.0.* Retrieved from Oracle AI & Data Science Blog: Announcing the general availability of OCI Language 4.0

Oracle. (2024, 06 04). *Overview of Block Volume.* Retrieved from Oracle Cloud Infrastructure Documentation: https://docs.oracle.com/en-us/iaas/Content/Block/Concepts/overview.htm

Oracle. (2024, 06 14). *Overview of File Storage.* Retrieved from Oracle Cloud Infrastructure Documentation: https://docs.oracle.com/en-us/iaas/Content/File/Concepts/filestorageoverview.htm

Oracle. (2024, 04 13). *Overview of Object Storage.* Retrieved from Oracle Cloud Infrastructure Documentation: https://docs.oracle.com/en-us/iaas/Content/Object/Concepts/objectstorageoverview.htm

Oracle. (2024, May 30). *Press Release.* Retrieved from Oracle: https://www.oracle.com/my/news/announcement/oracle-plans-to-open-two-public-cloud-regions-in-morocco-2024-05-30/

Oracle. (n.d.). *AI Services.* Retrieved from OCI Artificial Intelligence: https://www.oracle.com/ca-en/artificial-intelligence/ai-services/

Oracle. (n.d.). *Data Science Service.* Retrieved from OCI: https://www.oracle.com/ca-en/artificial-intelligence/data-science

Oracle. (n.d.). *Machine Learning Services.* Retrieved from Oracle Cloud Infrastructure (OCI): https://www.oracle.com/ca-en/artificial-intelligence/machine-learning/

Oracle. (n.d.). *OCI Data Labeling.* Retrieved from OCI: https://www.oracle.com/ca-en/artificial-intelligence/data-labeling

Oracle. (n.d.). *oci-data-science-ai-samples.* Retrieved from GitHub: https://github.com/oracle-samples/oci-data-science-ai-samples

CHAPTER 3

Healthcare NLP Case Study

Starting from this chapter to the end of the book, the case study of MedTALN Inc. will guide readers through various stages of constructing an NLP project on OCI. The project example will give readers valuable insights into the practical aspects of NLP model development, training, and deployment on OCI.

This chapter lays the groundwork for implementing our case study solution. From understanding the business drivers that led to this project to grasping our implementation blueprint, we provide a comprehensive view that equips you with the understanding needed to tackle upcoming chapters.

MedTALN Inc. Case Study

In this section, we are laying the groundwork for our case study by outlining the context and the problem we are addressing: creating an NLP-based solution for healthcare on OCI. We delve into the fundamental concept of Healthcare NLP models (such as Healthcare NER models), explaining their importance in addressing the initiative of MedTALN Inc., which seeks to expand its service offerings and venture into the growing market for unstructured data analytics in healthcare.

Company Background

MedTALN Inc. is based in Montreal, Canada, and is a leading advanced healthcare analytics solutions provider. MedTALN Inc. is committed to improving its analytics services to meet the needs of its diverse client base, which includes researchers,

CHAPTER 3　HEALTHCARE NLP CASE STUDY

universities, and private companies. The company aims to empower them with analytics and insights for various applications to enhance research and drive innovation in the healthcare sector.

Note　MedTALN Inc. is a fictional company created for the purposes of this case study.

MedTALN Inc. specializes in analyzing structured healthcare data using advanced statistical algorithms. Their expertise lies in transforming this data into actionable intelligence, which nonhealthcare providers utilize for various purposes, such as medical research, academic studies, and medical and pharmaceutical market analysis.

MedTALN Inc. recognizes the increasing importance of unstructured data in healthcare and is expanding its capabilities to include Natural Language Processing (NLP) for the French language. This strategic move aims to meet the growing demand for NLP-based analytics that can extract insights from French unstructured text data, such as clinical notes, patient records, and medical research publications. By integrating NLP into its service offerings, MedTALN Inc. is positioning itself at the forefront of healthcare data analytics for French, particularly for its clients in Quebec, where French is the primary business language.

MedTALN Inc. leverages Oracle Cloud Infrastructure (OCI) for its solutions, ensuring scalability, security, and performance. The company's reliance on OCI, as depicted in Figure 3-1, allows it to offer robust, cloud-based analytics solutions that are capable of handling large volumes of data efficiently. This technological backbone is crucial for supporting the advanced analytics and NLP capabilities that MedTALN Inc. plans to develop.

CHAPTER 3   HEALTHCARE NLP CASE STUDY

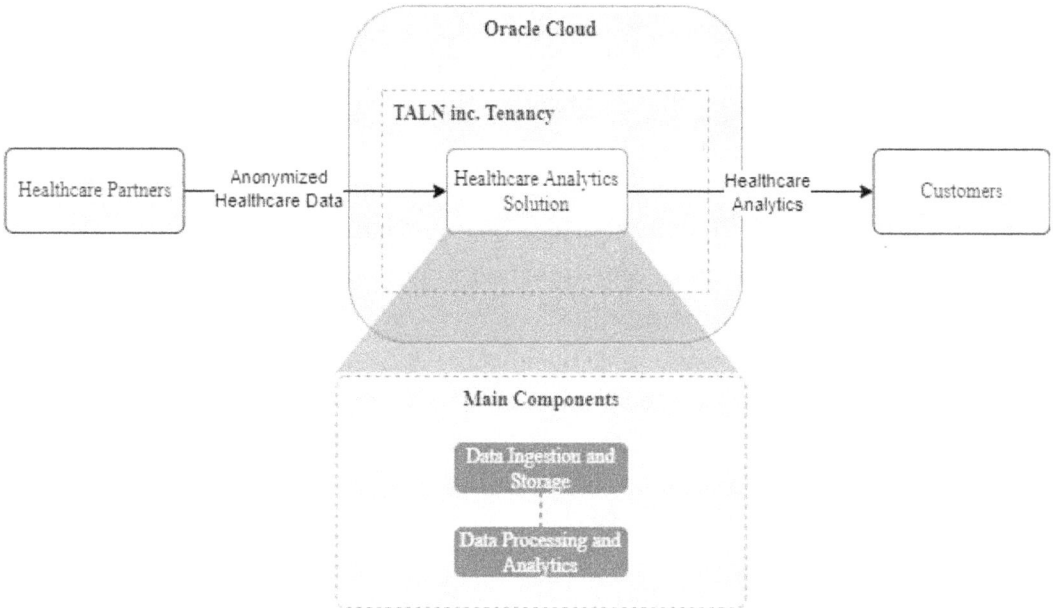

Figure 3-1. *MedTALN Inc.'s Healthcare Analytics Solution current state*

Figure 3-1 depicts the current state of MedTALN Inc.'s Healthcare Analytics Solution, which transforms large volumes of healthcare data into actionable insights. The main components of this system are as follows:

- Data Ingestion and Storage (ETL, Data Lakes, Databases): This component handles the intake and storage of deidentified health data from multiple healthcare institutions across Canada. The ETL (Extract, Transform, Load) processes clean, structure, and store the data in centralized data stores (e.g., data lakes).

- Data Processing and Analytics (Statistical Analysis, Visualization, etc.): This component involves statistical analysis and data visualization. It utilizes various data mining and statistical algorithms to transform raw data into actionable insights. Visualization tools like dashboards and reports make the data accessible and interpretable for the solution's customer.

75

Robust security and compliance measures are implicit in the design. These ensure that all data handling adheres to regulatory standards such as Quebec's Law 25[1] and Canada's federal privacy law PIPEDA,[2] incorporating data encryption, access control, and continuous compliance monitoring.

The system also provides user interfaces, including web applications, mobile applications, and reporting tools, designed to be user-friendly and accessible. These interfaces enable healthcare professionals to make informed decisions based on the results of the analytics.

Having already established a robust healthcare data analytics solution, MedTALN Inc. has a clear vision for the future: harnessing unstructured textual health data through NLP. By expanding into unstructured text analytics, the company can provide clients with even more comprehensive and valuable insights.

Healthcare NLP

Healthcare NLP is a broad field that involves the application of NLP techniques to process and analyze vast amounts of unstructured text data in the healthcare domain.

The healthcare industry generates enormous amounts of unstructured data daily. Traditional structured data analytics methods are insufficient to capture the rich, nuanced information contained in free-text medical documents (such as clinical notes, medical reports, research publications, and patient correspondence). Healthcare NLP addresses this gap by enabling the extraction of valuable insights from unstructured text, thereby transforming raw data into actionable intelligence.

---

[1] For an overview of Quebec's Law 25 and its significant changes to personal information protection laws, visit the Commission d'accès à l'information du Québec's website at https://www.cai.gouv.qc.ca/protection-renseignements-personnels/sujets-et-domaines-dinteret/principaux-changements-loi-25

[2] Canada's federal privacy law is the Personal Information Protection and Electronic Documents Act (PIPEDA). For more information, visit https://www.priv.gc.ca/en/privacy-topics/privacy-laws-in-canada/the-personal-information-protection-and-electronic-documents-act-pipeda/

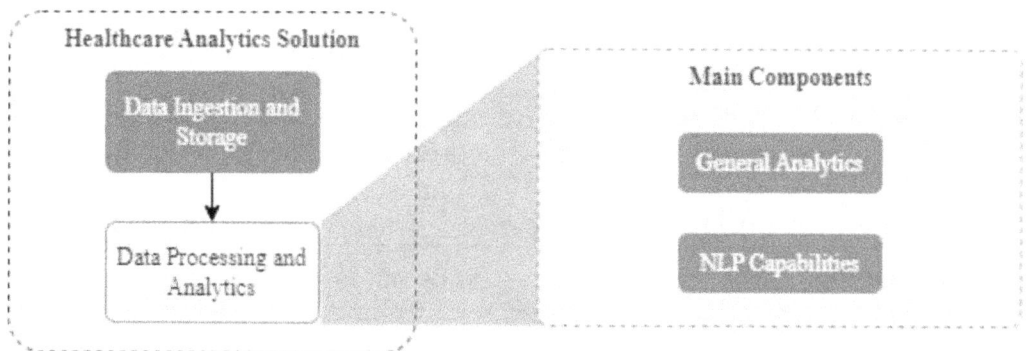

Figure 3-2. *MedTALN Inc.'s Healthcare Analytics Solution future state with NLP capabilities*

MedTALN Inc. is planning to broaden its services by adding Healthcare NLP. Figure 3-2 illustrates the projected state of MedTALN Inc.'s Healthcare Analytics Solution once Healthcare NLP capabilities like Named Entity Recognition (NER), sentiment analysis, topic modeling, and Relationship Extraction are incorporated to extract insights that are usually difficult to process using statistical analytics.

Business Drivers

An increasing number of technology companies, ranging from startups to industry leaders, aim to introduce new NLP-based services for the healthcare sector. MedTALN Inc. is one of these companies. The primary factors driving MedTALN Inc. in this direction are market expansion, customer demand, and regulatory compliance.

MedTALN Inc. seeks to diversify its service offerings and tap into the expanding market for unstructured data analytics in healthcare. By offering advanced NLP capabilities, MedTALN Inc. can attract new clients in the healthcare industry, including researchers, pharmaceutical companies, and insurers, thus increasing its market share and establishing itself as a comprehensive healthcare analytics provider.

Importantly, there is a significant and growing demand from both existing and potential clients of MedTALN Inc. for more advanced analytics and insights derived from unstructured text data. By meeting this demand, MedTALN Inc. enhances customer satisfaction and retention, providing clients with the tools to gain deeper insights from their data, such as precise entity extraction for research and market analysis.

Moreover, Quebec's Law 25, enacted in 2021, has set stringent standards for safeguarding personal information. Compliance with this law is not just a choice but an obligation for organizations operating within Quebec, with penalties for noncompliance.

This law is particularly significant for MedTALN Inc. as it processes a vast amount of health or biometric data that could include Personally Identifiable Information (PII) and sensitive personal information.

Table 3-1 provides an overview of the business drivers for MedTALN Inc.

Table 3-1. *Business drivers summary*

Business Driver	Description	Benefit
Market Expansion	Diversifying service offerings to tap into the growing market for unstructured data analytics in healthcare	Attract new clients and increase market share.
Revenue Growth	Creating new revenue streams with premium NLP services and solutions	Increase revenue and profitability through value-added services.
Customer Demand	Meeting the growing demand for sophisticated analytics and insights from unstructured text data	Enhance customer satisfaction and retention.
Enhanced Client Relationships	Strengthening relationships by providing solutions that address client pain points	Build long-term partnerships and increase client loyalty and satisfaction.
Competitive Advantage	Differentiating MedTALN Inc. by offering state-of-the-art NLP solutions	Establish a competitive edge and leadership in innovation.
Operational Efficiency	Automating data extraction and structuring to streamline processes	Increase efficiency and reduce manual effort and errors, leading to cost savings.
Regulatory Compliance	Ensuring compliance with healthcare regulations and standards	Minimize legal and compliance risks and build trust and credibility (Quebec's Law 25 and Canada's federal privacy law PIPEDA).
Innovation and R&D	Fostering a culture of innovation and continuous improvement	Stay ahead of industry trends and technological advancements, enhancing NLP capabilities.

These business drivers reflect the strategic importance of implementing Healthcare NLP capabilities for MedTALN Inc., aligning with the company's goals for growth, innovation, and client satisfaction in the healthcare analytics market.

Healthcare NER Initiative

MedTALN Inc.'s decision to initiate its Healthcare NLP vision with a Healthcare Named Entity Recognition (NER) initiative is strategically sound and aligns with its broader goals. This choice serves as a fundamental building block for more complex NLP tasks, laying the groundwork for advanced analytics such as Relationship Extraction, sentiment analysis, and predictive modeling.

Furthermore, starting with Healthcare NER, a well-defined and widely applicable NLP task, allows MedTALN Inc. to manage risks associated with expanding into new technological territories such NLP. It provides a clear, manageable scope for the initial project while paving the way for more advanced AI-driven initiatives in the future.

Before diving into Healthcare NER, let's first overview Named Entity Recognition (NER) and how it is done at a high level. NER is a critical task in Natural Language Processing (NLP). By conducting a detailed text analysis, NER can accurately identify and classify various named entities, such as persons, locations, or organizations. Its capability to understand and interpret the context and nuances of human language is essential for achieving this accuracy.

What Is Named Entity Recognition (NER)

NER specifically involves identifying entities within a sentence, as exemplified in Figure 3-3, and is a specialized application of the broader NLP task known as token classification or sequence labeling. Token classification refers to any problem where the goal is to assign a label to each token in a sentence. Besides NER, this category also includes tasks such as Part-of-Speech (POS) tagging—where each word in a sentence is labeled according to its part of speech (e.g., noun, verb, adjective)—and chunking, which involves grouping tokens that form a single entity.

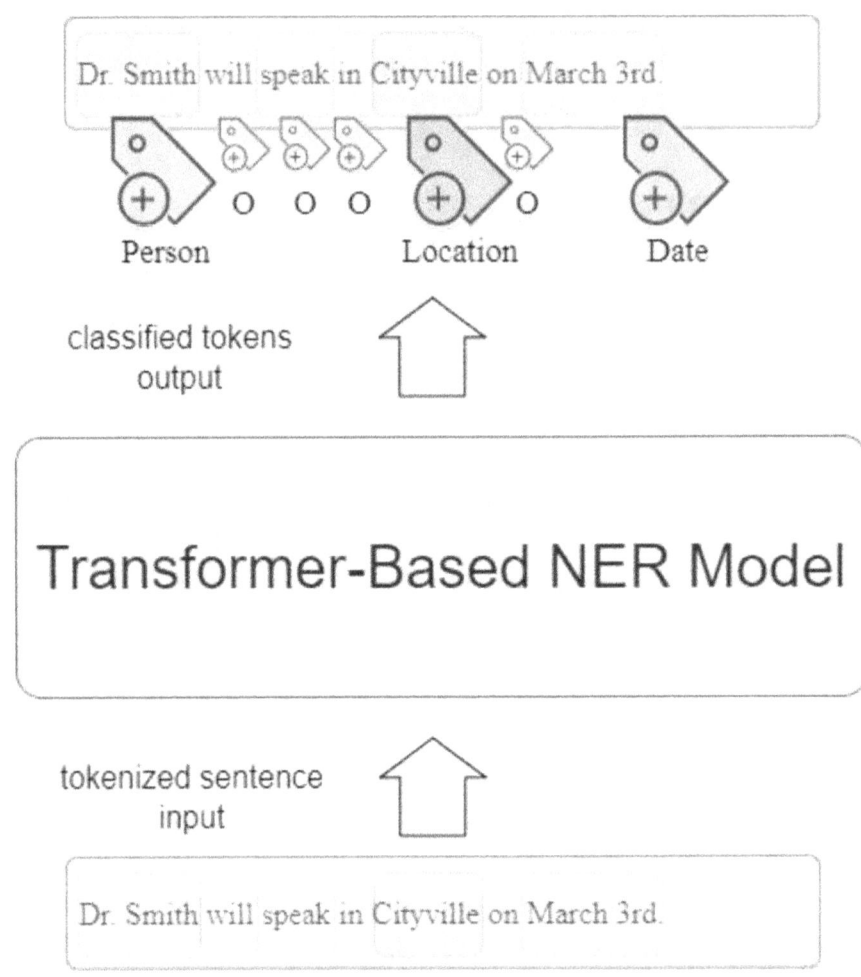

Figure 3-3. Transformer-based fine-tuned NER model prediction example

Developing NLP models, including NER, is an iterative process that begins with quality labeled data. Initial models are refined through repeated training cycles, often adding or adjusting labeled data to enhance accuracy. For example, an initial dataset with insufficient representations of certain entities might produce a model with lower precision, requiring additional labeling to improve entity detection. After deployment, continuous monitoring for data drift is crucial and may necessitate further retraining and labeling to maintain the model's performance.

Given its ability to identify and categorize key information from text, Named Entity Recognition (NER) plays a vital role in various industries, including healthcare. In this context, Healthcare NER focuses on extracting entities specific to medical and clinical text, such as diseases, medications, and procedures. The domain-specific nature of

Healthcare NER, which involves comprehending medical literature and terminology, makes it a challenging task within Natural Language Processing (NLP).

Healthcare NER Benefits

Healthcare Named Entity Recognition (NER) identifies and classifies entities within unstructured medical text, such as conditions, medications, symptoms, and procedures. Implementing a Healthcare NER solution offers several benefits for MedTALN Inc.'s NLP vision.

Firstly, it addresses immediate client needs. MedTALN Inc.'s clients, including researchers, pharmaceutical companies, and insurers, require precise and structured data extracted from unstructured text to support their activities. Implementing NER addresses this immediate need and adds significant value to their operations. By providing clients with accurate and structured data, MedTALN Inc. can help them improve research accuracy, streamline operations, and make data-driven decisions. This approach allows MedTALN Inc. to provide immediate value to clients by accurately extracting critical entities from unstructured text.

Secondly, developing a robust NER solution lays the foundation for more advanced NLP tasks. It enables MedTALN Inc. to offer a comprehensive suite of NLP services in the future, such as Relationship Extraction, which identifies relationships between different medical entities (e.g., extracting the relationship between medication and dosage). Additionally, it opens the way for Medical Entity Linking, addressing specific, high-demand needs from clients, such as improved clinical documentation, automated medical coding, and enhanced research capabilities. Starting with NER allows MedTALN Inc. to establish a strong technical foundation and gradually expand its capabilities in the Healthcare NLP domain.

Use Cases

Here is a list of ten specific use cases that MedTALN Inc.'s customers could implement using MedTALN Inc.'s Healthcare NER solution. These use cases will focus on practical applications that leverage the capabilities of a Healthcare NER model.

- Clinical Documentation Improvement: MedTALN's NER solution can analyze clinical notes to identify and extract key medical entities such as diagnoses, procedures, and medications. This assists healthcare providers in improving the accuracy and completeness of medical records.

- Pharmacovigilance and Adverse Event Detection: Pharmaceutical companies can use the NER system to scan medical literature, clinical trial reports, and social media for mentions of drugs and associated adverse events. This enables early detection of potential safety issues and supports regulatory compliance.

- Clinical Trial Patient Matching: Research organizations can employ the NER solution to analyze patient records and identify suitable candidates for clinical trials based on specific inclusion/exclusion criteria, streamlining the recruitment process and improving study efficiency.

- Automated Medical Coding: Healthcare providers and billing departments can utilize the NER system to automatically extract relevant medical information from clinical narratives, facilitating faster and more accurate medical coding for billing and administrative purposes.

- Population Health Management: Healthcare systems can apply the NER solution to large sets of patient data to identify trends in diseases, treatments, and outcomes across populations, supporting targeted interventions and public health initiatives.

- Drug–Drug Interaction Screening: Pharmacies and healthcare providers can use the NER system to analyze prescription data and medical records, flagging potential drug interactions and improving patient safety.

- Systematic Literature Review: Researchers can leverage the NER solution to quickly process large volumes of medical literature, extracting relevant entities and relationships to support systematic reviews and meta-analyses.

- Real-World Evidence Generation: Pharmaceutical companies and researchers can apply the NER system to analyze real-world data sources like electronic health records and claims data, generating insights on drug effectiveness and safety in actual clinical practice.

- Clinical Decision Support: Healthcare providers can integrate the NER solution into their electronic health record systems to automatically highlight key clinical information, supporting more informed decision-making at the point of care.

- Regulatory Compliance Monitoring: Healthcare organizations can use the NER system to scan internal documents and communications for potential compliance issues related to Protected Health Information, ensuring adherence to regulations like HIPAA.

Table 3-2 provides a broader view of Healthcare NER use cases, including use case applications, their benefits, and associated challenges.

Table 3-2. Summary table of broader Healthcare NER use cases

Use Case	Input Data	Extracted Entities	Primary Application	Key Benefits	Challenges
Medical Research Analysis	Clinical trials, publications, patient records	Medical conditions, treatments, outcomes, demographics, biomarkers	Meta-analyses, research gap identification	Enhanced research efficiency and depth	Data privacy concerns, handling of diverse medical terminologies
Pharmaceutical Market Intelligence	Clinical trials, adverse event reports, literature	Drug names, dosages, side effects, patient responses	Market trend analysis, drug repurposing, regulatory compliance	Data-driven product development and marketing, improved compliance	Keeping up with rapidly evolving medical knowledge, data integration
Health Economics and Outcomes Research	Health records, insurance claims, patient surveys	Treatment costs, hospital stay duration, quality of life measures, comparative effectiveness	Cost–benefit analyses, comparative effectiveness research	Informed healthcare policy and resource allocation	Standardizing diverse data sources, handling complex economic models

(continued)

Table 3-2. (*continued*)

Use Case	Input Data	Extracted Entities	Primary Application	Key Benefits	Challenges
Insurance Claims Processing	Claims documents	Medical conditions, treatments, patient demographics, procedure codes	Automated processing, fraud detection	Reduced costs, improved risk assessment	Ensuring accuracy in automated systems, adapting to changing healthcare policies
Patient Feedback Analysis	Surveys, social media, online reviews	Symptoms, treatment satisfaction, service quality, adverse events	Service improvement identification, pharmacovigilance	Enhanced patient experience and service quality, improved safety monitoring	Managing unstructured data, ensuring patient privacy in social media analysis

The examples of use cases presented above unequivocally underscore the critical importance and relevance of developing Healthcare NER for MedTALN Inc.

Healthcare NER Inception

In this section, we explore the inception phase of the Healthcare Named Entity Recognition (NER) project at MedTALN Inc., a fictional company created to exemplify how such a project might unfold in the real world. This phase is crucial because it sets the foundation for everything that follows, from defining the project's scope to assembling the right team to drive the initiative forward.

To start, we need to define the Healthcare NER project's scope clearly. For MedTALN Inc., this means pinpointing exactly what aspects of healthcare data the NER solution will address—such as identifying medical conditions, medications, and procedures—

and ensuring that it supports both French and English, reflecting the linguistic needs of the company's clients. By defining the scope early on, we can prevent scope creep and keep the project focused on its strategic objectives.

Next, gathering the detailed requirements is essential. This involves understanding the needs of MedTALN Inc.'s diverse client base—researchers, universities, and private companies—to ensure that the NER solution meets their specific challenges, whether ensuring high accuracy, complying with healthcare regulations, or integrating smoothly with existing systems.

Finally, though fictional, MedTALN Inc. serves as an example of how a company would recruit and organize a multidisciplinary team of NLP experts and IT specialists. This team is responsible for driving the project forward and ensuring that all technical and domain-specific expertise is available to tackle the complexities of Healthcare NER. By the end of the inception phase, MedTALN Inc. will have a well-defined scope, a clear set of requirements, and a dedicated team ready to move into the next phase of development.

Scope and Requirements

The scope of the Healthcare Named Entity Recognition (NER) project at MedTALN Inc. is defined by the following key objectives:

- Accurate Medical Entity Extraction: The project's primary focus is on developing a Health NER model capable of accurately identifying key medical entities from unstructured text. These entities include medical conditions, medications, dosages, symptoms, test results, treatments, and procedures. Ensuring consistent model performance and reliability across diverse scenarios is essential.

- French Language Support: The first version of the Health NER solution will prioritize support for the French language, reflecting the linguistic needs of many healthcare providers in Quebec province (Canada). While supporting English is important, the initial focus will be ensuring that the model performs exceptionally well in French. Subsequent versions will expand to include bilingual support.

- Customization Flexibility: The project scope includes offering customization options to tailor the Healthcare NER model to the specific needs of different use cases. The model will also be designed to remain up-to-date with advancements in the Healthcare NLP field, ensuring its continued relevance and effectiveness.

- Data Sovereignty: The Healthcare NER model should be deployed within an OCI region in Canada. This is crucial for complying with local regulations and protecting sensitive healthcare data.

- Cost-Effectiveness: The project aims to deliver a cost-effective solution by leveraging Oracle Cloud Infrastructure (OCI) capabilities. This will involve optimizing resource usage to reduce operational costs and ensure the solution is both financially sustainable and scalable.

Requirements

Here's a structured breakdown of high-level requirements for our Health Named Entity Recognition (NER) solution (see Table 3-3).

Table 3-3. Pilot project requirement summary

Requirement	Description	Benefit
Medical Entity Extraction	Utilizing state-of-the-art NLP models to extract key medical entities from healthcare unstructured text.	Increased precision and recall, providing reliable and actionable insights.
Multilingual Support	Supporting multiple languages, particularly English and French, for accurate entity extraction.	Broadened applicability across Canada's diverse linguistic landscape (the solution must fully support the French language). Quebec where French is the primary language.
High Accuracy and Precision	Developing robust validation and testing frameworks to ensure model accuracy and precision.	Consistent model performance and increased trust in system outputs.

(continued)

Table 3-3. (*continued*)

Requirement	Description	Benefit
Real-Time Processing	Implementing APIs for real-time insights processing.	Real-time integration with customers' systems.
Batch Processing	Implementing batch data processing for large datasets.	Efficiently processing large datasets in batch mode for comprehensive insights.
Security and Privacy	Implementing robust security measures such as encryption and secure access controls. Legal compliance and enhanced data privacy.	Enhanced data security and compliance with international regulations. Implementing advanced anonymization techniques to comply with Quebec's Law 25 and Canada's federal privacy law PIPEDA and ensure data privacy and security.
Customization and Flexibility for NLP Models	NLP models should be customizable and flexible to handle diverse healthcare data, specific medical terminology, and different languages and integrate with various systems.	Increased relevance through better understanding of specific medical terminology. Enhanced adaptation to specific user needs, and versatile support for various languages.
Continuous Improvement and Updates	Establishing continuous monitoring and automated model retraining.	Keeping the system up-to-date with advancements in medical knowledge and technology.
Cost Efficiency	Leveraging OCI capabilities to optimize resource usage and reduce operational costs.	Financial sustainability and high value at lower cost.

These requirements ensure that MedTALN Inc.'s Health NER solution is comprehensive, reliable, and user-friendly, meeting the diverse needs of its non-healthcare provider clients while maintaining high standards of accuracy, security, and compliance.

In the next section, we will discuss the various options that MedTALN Inc.'s IT Team evaluated to implement a Healthcare NER solution quickly and efficiently.

CHAPTER 3   HEALTHCARE NLP CASE STUDY

Assembling the Team

In this case study, MedTALN Inc., a fictional company, is preparing to launch its Healthcare Named Entity Recognition (NER) project on Oracle Cloud Infrastructure (OCI). Assembling the right team is a crucial first step to ensure the project's success. The following roles are key to achieving the project's goals:

- Project Manager: The project manager will coordinate all aspects of the project, ensuring timelines are met, resources are allocated efficiently, and all stakeholders are aligned with their roles and expectations.

- Business Users: Business users will bridge the gap between the company's strategic goals and the technical implementation. They will translate business requirements into technical specifications, ensuring the NER system aligns with MedTALN Inc.'s objectives.

- OCI Specialists: Given that the project will be deployed on Oracle Cloud Infrastructure, OCI specialists will optimize resource usage, resolve technical challenges, and guide the team through the specifics of OCI.

- Data Annotators Team: The data annotators, under the guidance of the NLP consultant, will label the data accurately. This team includes annotators, who mark the data, and reviewers, who ensure the annotations are consistent and accurate, which is crucial for training the NER models.

- NLP Consultant: The NLP consultant will lead the efforts to develop and deploy a Healthcare Named Entity Recognition (NER) model. He will also transfer knowledge to MedTALN Inc.'s in-house team on the implementation of this first Healthcare NLP model, equipping the in-house team with the necessary NLP knowledge for future Healthcare NLP initiatives.

As MedTALN Inc. prepares to launch this project, the focus is on building a strong, cohesive team with the necessary skills to ensure the NER system's success.

Engaging the NLP Consultant

MedTALN Inc. made a critical decision to bring in an external NLP consultant, John Doe, who has a deep and broad knowledge in NLP on OCI. The consultant will help navigate through the complex process of developing a highly efficient deep learning–based NER model on OCI for the first time. John will join MedTALN Inc. as both an NLP engineer and an architect who will shape not only the direction of this project but also lay the foundation for future NLP projects.

Note John Doe is a fictional character created for this case study to represent an NLP consultant who assists MedTALN Inc. with implementing the Healthcare NER project.

The primary responsibility of the NLP consultant is to ensure that the Healthcare NER model meets MedTALN Inc.'s specific needs. In addition to implementing the Healthcare NER model, the consultant is responsible for creating guidelines for building future Healthcare NLP models, focusing on dataset preparation, model training, and deployment. He also plays a key role in transferring knowledge to MedTALN Inc.'s in-house teams, helping them build the skills needed for future NLP projects.

The key responsibilities of the NLP consultant include

- Implementing the Healthcare NER model for MedTALN Inc. using OCI's AI capabilities

- Facilitating comprehensive knowledge transfer to in-house teams, including data scientists and annotators

MedTALN Inc. expects that this collaboration with the NLP consultant will achieve the project's immediate goals and prepare the company for future NLP initiatives.

Healthcare NER Elaboration

The elaboration phase will detail the planning and design of the Healthcare NER solution. This section will include selecting the right approach for building Healthcare NER models, developing the architectural baseline, and addressing any critical risks identified during inception. The focus is on ensuring the solution's feasibility and laying the groundwork for successful implementation.

CHAPTER 3 HEALTHCARE NLP CASE STUDY

Architectural Design

Given the scope of the Healthcare NER project and its specific requirements, the NLP consultant began the elaboration phase by carefully selecting the most suitable approach for building the Healthcare NER models for MedTALN Inc. The consultant employed a systematic methodology to ensure that each potential approach was thoroughly evaluated. This rigorous process was crucial in determining the best solution to meet MedTALN Inc.'s needs, ensuring that every option was thoughtfully considered before making a final decision.

Methodology

To ensure that MedTALN Inc. chose the most effective solution for their Healthcare Named Entity Recognition (NER) needs, our NLP consultant developed a methodology.

It is critical to remember that selecting the right approach should comply with our three high-level architectural decisions. Thus, we should select the potential solution options in light of those architectural decisions:

- Healthcare NER that extracts medical entities in French.
- Use SOTA NLP models, like transformer-based models and transfer learning.
- Leverage OCI and its AI capabilities for development and deployment.
- Deploy within an OCI region in Canada for data sovereignty.
- Support customization.
- Cost-effective solution.

Preselection of Candidate Solution Options

This preselection phase evaluates potential approaches for MedTALN Inc.'s Healthcare Named Entity Recognition (NER) project. This process aims to identify a list of potential solution options aligning with project requirements and architectural decisions.

CHAPTER 3 HEALTHCARE NLP CASE STUDY

Two primary approaches were identified to implement a Healthcare NER solution:

- OCI Language-based models
- LLMs and OCI Data Science AI Quick Actions
- Fully custom model

OCI Language-Based Models Option

MedTALN Inc.'s NLP consultant selected OCI Language Service's NER functionalities as candidate options for implementing Healthcare NER, particularly after Oracle announced the general availability of OCI Language 4.0. This new version includes a Health Natural Language Processing (NLP) feature. Specifically, OCI Language 4.0 offers the Health Named Entity Recognition (Health NER) model, which is capable of identifying key entities such as medical conditions, medications, dosages, symptoms, test results, treatments, and procedures (Oracle, 2024).

The pretrained Named Entity Recognition (NER) model in OCI Language supports French but does not include built-in support for medical entities or the ability to be trained with custom NER datasets for domain-specific entities. As a result of these limitations, it was not chosen for the Healthcare NER project.

The OCI Language-based models[3] that were selected and assessed are as follows:

- OCI Language Custom Named Entity Recognition (NER) Model: This is a generic NER model that allows for custom training. It can be trained to extract medical entities using a labeled dataset specifically prepared for this purpose. However, as of now, it only supports English and Spanish and does not have native support for the French language.

- OCI Language Healthcare Named Entity Recognition (Healthcare NER) Model: This specialized model is designed to handle healthcare-specific text and focuses on identifying medical entities. Even though it provides support for the French language, this model does not allow for custom entities because training data cannot be provided. As demonstrated in Figure 3-4, the identification of medical entities resulted in many false positives, and the spot-check

---

[3] Learn more about the OCI Language Service pretrained models at [Oracle Docs](https://docs.oracle.com/en-us/iaas/language/using/pretrain-models.htm).

CHAPTER 3   HEALTHCARE NLP CASE STUDY

test results were unsatisfactory for MedTALN Inc.'s NLP consultant. In terms of cost-effectiveness, this model is not efficient. It comes with a high recurring monthly cost, exceeding approximately 600 CAD per day, as shown in Figure 3-5.

Figure 3-4.   Pretrained Healthcare NLP model test

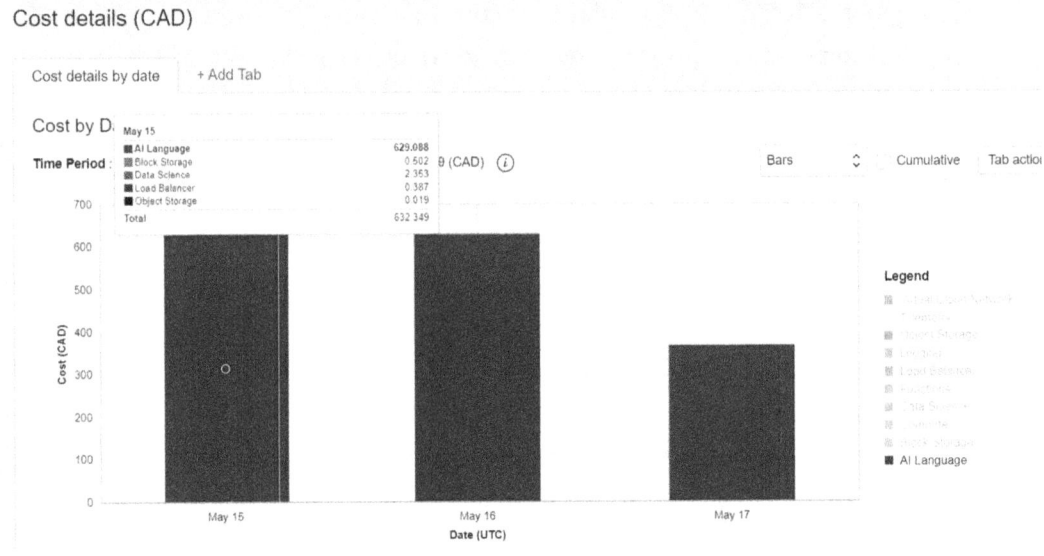

Figure 3-5.   OCI Language Healthcare NER cost per day

CHAPTER 3   HEALTHCARE NLP CASE STUDY

As shown in Figure 3-6, the OCI Language Healthcare NER option has a very high operational cost, approximately $26 CAD per inference unit hour, leading to significant monthly expenses (approximate $18,720.00 CAD[4]).

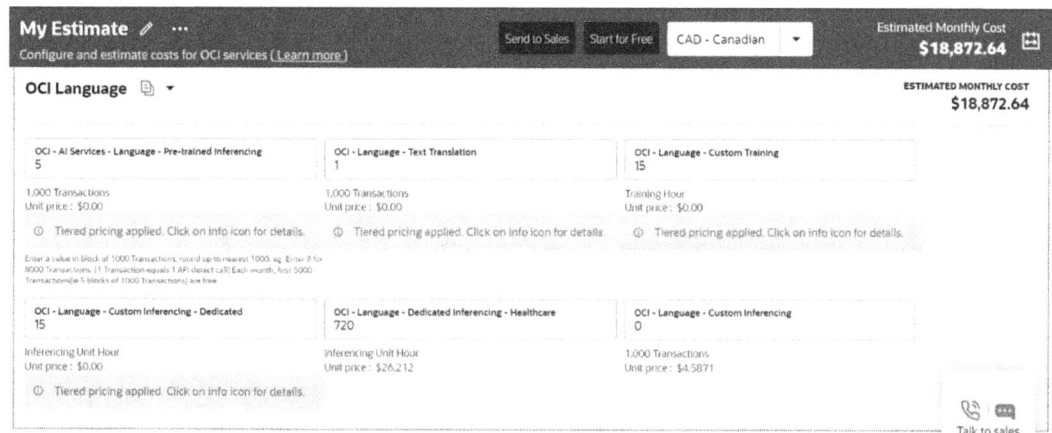

Figure 3-6. *OCI Language Healthcare model monthly cost estimate*

The preliminary assessment revealed that both models were inadequate for MedTALN Inc.'s needs.

LLMs and OCI Data Science AI Quick Actions

The option to use AI Quick Actions in OCI to fine-tune large language models (LLMs[5]) for Healthcare NER was evaluated as part of our exploration of potential solutions. AI Quick Actions enable the fine-tuning of LLMs, which could be tailored to extract medical entities effectively. However, several challenges make this approach less suitable for MedTALN Inc.'s needs.

In evaluating the feasibility of LLMs for our Healthcare Named Entity Recognition (NER) task, two major drawbacks emerged that led to their exclusion from our candidate options list: risks associated with LLMs and cost implications.

---

[4] The estimated monthly cost is $18,720.00 CAD, calculated based on 24 hours per day and 30 days per month. This is because the OCI Language Healthcare models are priced per hour and metering can only be stopped if the model is removed.

[5] LLMs are effective in performing NER task with minimal additional training, thanks to their extensive pretraining on diverse datasets. They excel in understanding context, which empowers them to perform NER in a zero-shot setting, but their performance can improve when provided with a small amount of training data, known as few-shot settings.

CHAPTER 3 HEALTHCARE NLP CASE STUDY

The first drawback for our case study is the inherent risks associated with using LLMs, even when fine-tuned with tools like OCI Data Science AI Quick Actions:

- Biases: LLMs are often trained on large, diverse datasets, which can introduce biases. These biases may manifest in the form of skewed outputs, which is particularly problematic in healthcare, where decisions need to be fair and unbiased.

- Hallucinations: LLMs can generate outputs that do not accurately reflect the input data, known as "hallucinations." In a healthcare context, this can lead to the incorrect identification of medical entities, which could have serious consequences for patient care and data integrity.

- Detection Errors: Even with fine-tuning, LLMs might misinterpret complex medical terminology or context, resulting in detection errors. This lack of precision is unacceptable in healthcare applications where accuracy is critical.

Given these risks, LLMs pose significant challenges for healthcare NER tasks, where the need for precise, reliable outputs outweighs the benefits of rapid deployment or advanced capabilities.

The second major inconvenience for our case study is the substantial cost associated with deploying and operating LLMs:

- High Resource Demands: LLMs with billions of parameters, such as a 7-billion-parameter model, require substantial GPU memory. The relationship between model parameter size and GPU memory is roughly 2× the parameter count in GB. For example, a model with 7 billion parameters will need a minimum of 14 GB of GPU memory for inference. This translates into a need for high-end GPU compute shapes, such as those offered by OCI (e.g., VM.GPU.A10.1 with 24GB GPU memory), which significantly increases operational costs.

- Increased Operational Costs: Utilizing these GPU-based resources is expensive, and maintaining such infrastructure is contrary to our cost-saving objectives, particularly since we aim to target CPU-based inference with acceptable performance and latency. This contradicts our cost-saving goals and strategies and significantly increases operational costs.

These cost implications, combined with the risks, make LLMs an impractical choice for our healthcare NER solution, particularly when more specialized, cost-effective alternatives are available.

To ensure these concerns were grounded in practical evidence, we deployed an LLM using OCI AI Quick Actions, utilizing a prompt designed for French Healthcare NER tasks. The results confirmed our concerns: the LLM produced outputs with hallucinations and inaccuracies, reaffirming that it is not a reliable option for our use case.

Considering the significant risks associated with LLMs and the high costs of their deployment, this approach is not suitable for our Healthcare NER case study. These drawbacks outweigh the potential benefits, leading us to exclude LLMs from our decision matrix in favor of more specialized, efficient, and cost-effective models that better align with our overall strategy. Consequently, the decision to discard the LLM approach in favor of fine-tuning a pretrained model is well justified based on the evaluation criteria.

Fully Custom Healthcare NER Model

The NLP consultant explored a final option: building a fully custom healthcare-specific Entity Recognition (NER) model. This means fine-tuning a model specifically for healthcare NER tasks. This approach offers the highest level of customization and control over the model architecture, potentially making it the most flexible option for evolving and taking advantage of continuous advances in the Healthcare NLP domain.

Choosing the fully custom model would allow using state-of-the-art (SOTA) NLP models that can be fine-tuned for the French medical domain. This model could be built using OCI's AI capabilities and seamlessly integrated with OCI's Compute, Storage, and AI Services, taking full advantage of the cloud infrastructure. The fully custom model can also be hosted in an OCI region within Canada, ensuring compliance with data sovereignty requirements.

However, there is a downside to creating a completely custom Healthcare NER model (fine-tuned Healthcare NER model), which is the high initial cost. Nevertheless, it's worth noting that in the long term, it can be cost-effective because there are no recurring fees, and the NLP consultant may implement cost-saving strategies during the project.

Selection of the Optimal Approach

In the preselection phase of MedTALN Inc.'s Healthcare Named Entity Recognition (NER) project, three primary approaches were evaluated: OCI Language-based models, LLMs with OCI Data Science AI Quick Actions, and a fully custom Healthcare NER model. The goal was to identify the most suitable solution that aligns with the project's requirements and architectural decisions.

Here are the detailed selection criteria we use in the evaluation process (outlined in Table 3-4):

Table 3-4. Detailed selection criteria

Criteria	OCI Language Custom Models (for Healthcare NER)	OCI Language Healthcare Models (for Healthcare NER)	OCI AI Quick Actions (LLMs for Healthcare NER)	Fine-Tuned Healthcare NER Model
Supports Training with Custom NER Dataset (Custom Domain-Specific Entities)	Yes ✔	No ✘	Fine-tuning	Yes ✔
Supports Model Architecture Selection (e.g., Transformers)	No ✘	No ✘	No ✘	Yes ✔
Supports Transfer Learning from Pretrained Language Models (BERT, RoBERTa, etc.)	No ✘	No ✘	No ✘	Yes ✔
Supports Model Training Hyper-Parameter Fine-Tuning	No ✘	No ✘	No ✘	Yes ✔
Supports Healthcare Domain	No ✘	Yes ✔	Fine-tuning	Yes ✔
Support French Language	No ✘	Blackbox	Yes ✔	Yes ✔
Accuracy for Healthcare Entities	Moderate	High ✔	Low to moderate	High ✔

> **Note** In this analysis, we have treated all criteria equally, given the importance of each in the healthcare domain. Introducing weights could refine the decision by emphasizing certain criteria over others.

The assessment revealed significant limitations in the OCI Language-based models. While these models offered some potential, such as the Health NER model, they lacked critical features like French language support and customization capabilities for medical entities. Additionally, the high recurring costs and unsatisfactory accuracy made them unsuitable for MedTALN Inc.'s needs. Similarly, the option to fine-tune LLMs using OCI AI Quick Actions was found to be impractical due to inherent risks like biases and hallucinations, coupled with substantial resource demands and high operational costs.

As a result, the fully custom Healthcare NER model emerged as the most viable option. This approach offers the highest level of customization and control, allowing the use of state-of-the-art NLP models fine-tuned for the French medical domain. Although it involves a higher initial investment, the long-term benefits, including compliance with data sovereignty requirements and potential cost savings, make it the best fit for MedTALN Inc.'s project objectives.

Solution Blueprint

In this chapter, we present the high-level architecture developed by MedTALN Inc.'s NLP consultant for the Healthcare Named Entity Recognition (NER) solution. This architecture is specifically designed to extract medical entities from French language text, addressing both the technical requirements and budgetary constraints of MedTALN Inc.

High-Level Architecture

The NLP consultant has crafted the high-level architecture for the solution, leveraging Oracle Cloud Infrastructure (OCI) AI capabilities. The focus is on building and deploying a custom Healthcare NER model tailored to MedTALN Inc.'s needs. Below are the core architectural decisions that form the foundation of this solution:

CHAPTER 3 HEALTHCARE NLP CASE STUDY

- OCI ML Services (OCI Data Science and Data Labeling Services): The NLP consultant chose OCI's Data Science and Data Labeling Services to support the entire NLP model development life cycle, from data preparation to deployment.

- Open Source NLP Resources (Hugging Face Models, Datasets, and Libraries): The consultant incorporated the Hugging Face platform to leverage preannotated datasets and pretrained models. This approach reduces the time and resources required for model development while maintaining high quality.

The diagram in Figure 3-7 illustrates the high-level architecture, emphasizing the key components involved in different stages of the Healthcare NER model development process. These components include OCI Data Science, OCI Data Labeling, and Hugging Face. They are crucial for data collection, NLP model training, and deployment preparation.

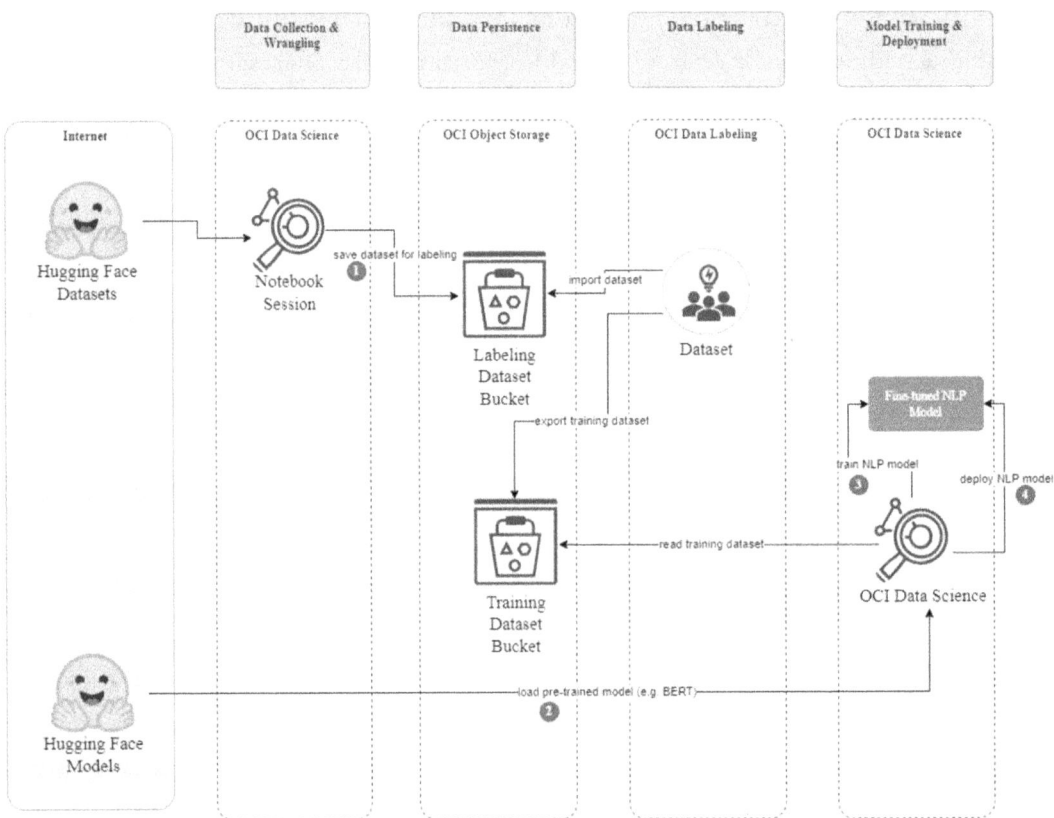

Figure 3-7. *High-level architecture for building MedTALN Inc.'s Healthcare NER model*

The following outlines how OCI Data Science, OCI Data Labeling, and Hugging Face will be used to construct a highly performant yet cost-effective Healthcare NER model:

- Dataset Preparation: The training dataset for Healthcare NER will be based on open source, ready-to-use Hugging Face datasets. This dataset will be processed and prepared using OCI Data Science and labeled using OCI Data Labeling. Throughout its life cycle, this dataset will be securely stored in OCI Object Storage.

- Model Training: GPU-based OCI Data Science Notebook Sessions are utilized for efficient model training, focusing on fine-tuning pretrained models from Hugging Face using state-of-the-art NLP models such as BERT or healthcare-specific custom models like Dr-BERT, SciBERT, or BioBERT. To further control costs, the consultant recommended deactivating notebook sessions when model training is complete. This strategy stops charges for compute resources while retaining block storage, which is particularly beneficial for GPU instances.

- Model Deployment: To streamline the deployment process, the NLP consultant recommended using OCI Data Science's model deployment capabilities. The trained model can be easily deployed from the OCI Data Science Model Catalog, with OCI handling all necessary infrastructure operations, including compute provisioning and load balancing.

This high-level architecture, developed by MedTALN Inc.'s NLP consultant, provides a robust and cost-effective solution for French language Healthcare NER. By carefully balancing OCI's AI Services with open source tools and strategic resource optimization, this solution blueprint meets both the technical and financial needs of the project.

High-Level Approach

Figure 3-8 outlines the high-level structured approach to building Healthcare NER model that the NLP consultant put in place.

CHAPTER 3   HEALTHCARE NLP CASE STUDY

The process begins with selecting a suitable open source dataset for Healthcare NER task, followed by data acquisition and preparation, which involves transforming the dataset to the format expected by the OCI Data Labeling Service (e.g., JSONL). The dataset is enriched with new annotations to meet our case study requirements during the data labeling.

Next, a French pretrained model for healthcare is selected and fine-tuned with the labeled dataset.

The final stage involves a comprehensive model evaluation, where the dataset or training hyperparameters are adjusted as necessary to ensure optimal model performance.

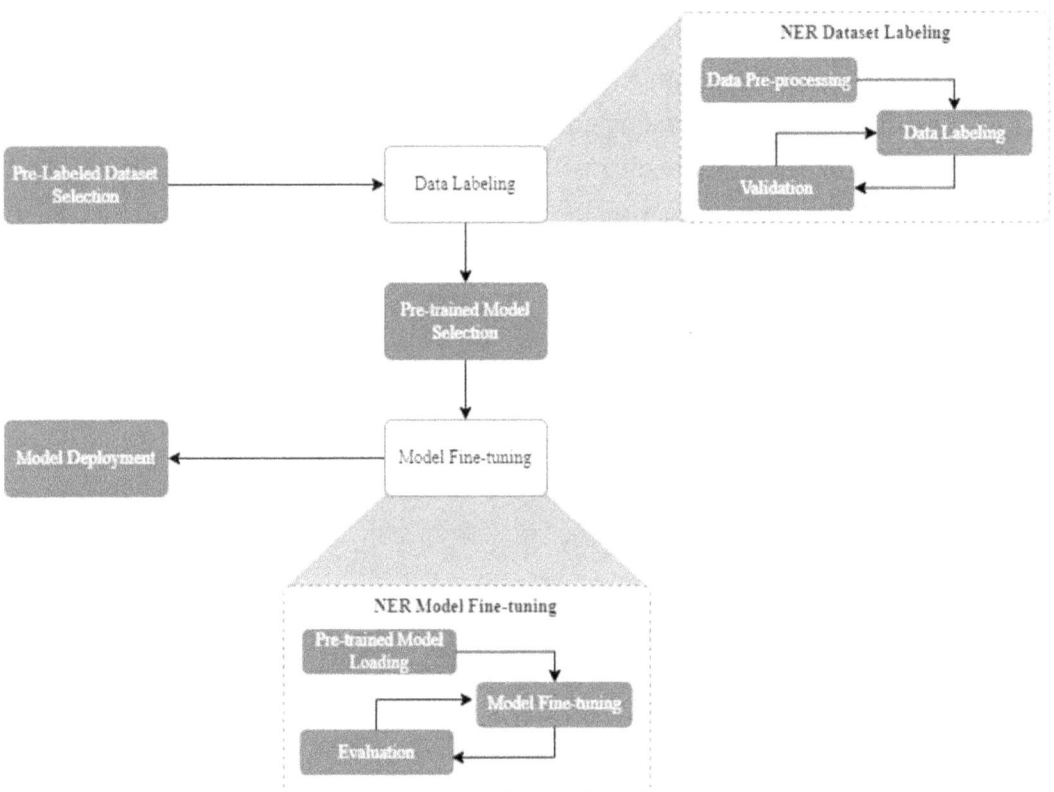

Figure 3-8. *Steps for building a Healthcare NER model for our case study*

Figure 3-9 illustrates the process of fine-tuning a BERT-based model specifically for Named Entity Recognition (NER) in the healthcare domain. It is divided into two phases: the training phase and the inference phase.

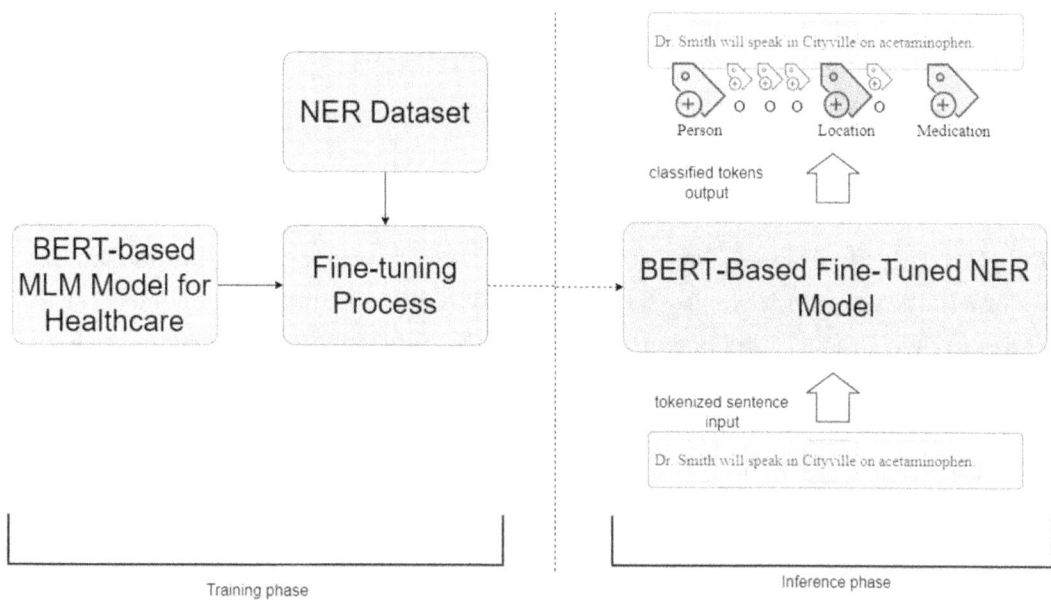

Figure 3-9. *Building and using a fine-tuned BERT-based NER model for the healthcare domain*

In the training phase, a pretrained BERT-based Masked Language Model (MLM) focused on healthcare is fine-tuned using a specially labeled NER dataset. The fine-tuning process adapts the model to recognize and classify medical entities accurately within healthcare-related texts, specializing it from a Healthcare MLM model to a Healthcare NER model.

During the inference phase, the fine-tuned Healthcare NER model is applied to new text inputs. Sentences are tokenized, and the model processes these tokens, classifying each one according to the categories learned during training. The output is a list of tokens, each tagged with an entity label, allowing the model to identify and categorize key information such as patient name, medical conditions, medications, dosages, and symptoms within the text.

Project Preparation

With the solution blueprint defined, the next steps in the elaboration phase are provisioning the Oracle Cloud Infrastructure (OCI) account and defining the roles and responsibilities within the project team. Provisioning the OCI account sets up the cloud environment where the Healthcare Named Entity Recognition (NER) solution will be developed and deployed.

CHAPTER 3   HEALTHCARE NLP CASE STUDY

Simultaneously, clearly defining the roles and responsibilities ensures that each team member understands their specific tasks, allowing the project to proceed efficiently into the construction phase, where the solution will be fully implemented.

OCI Account

While MedTALN Inc. already has an OCI account, to follow along with this book and implement the case study step by step as detailed in the subsequent chapters, you will need your own Oracle Cloud account. You can subscribe to a new trial account, which allows you to work with Oracle Cloud for 30 days at no cost. Alternatively, you can use a paid account, but be aware that some services will incur costs.

To get started, sign up for an account. OCI offers three options:

- Paid Tier: This offers access to the full range of metered services, with multiple payment options, including pay-as-you-go and leveraging existing licenses.

- Trial: At the time of writing, the trial provides a 30-day window with a $300 credit to your account.

- Always-Free Tier: This option limits the resources you can use, but these resources are available without charges for an unlimited time.

After activating your account, you can log in at cloud.oracle.com to access your cloud service dashboard (Figure 3-10).

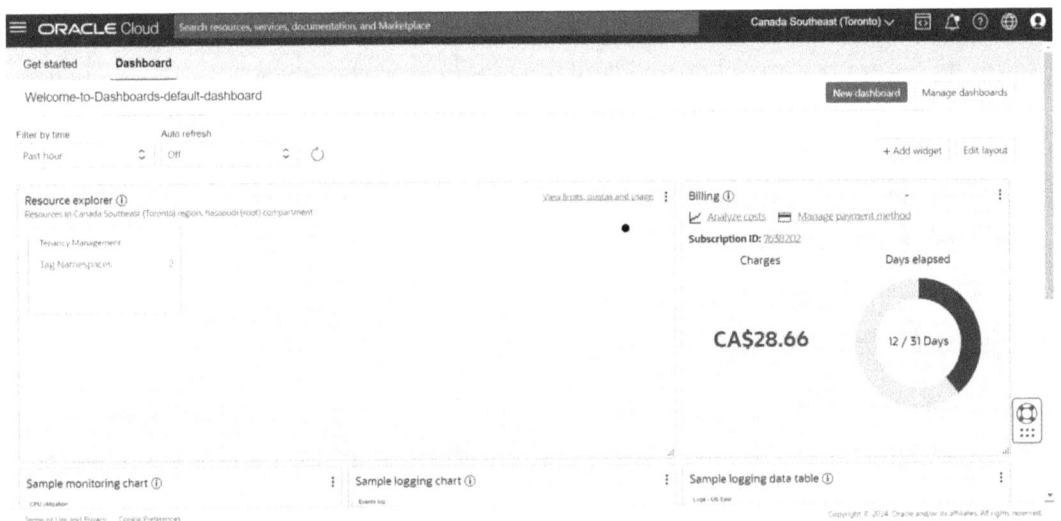

Figure 3-10. *OCI dashboard*

CHAPTER 3   HEALTHCARE NLP CASE STUDY

Before diving into the case study, readers must have an OCI account (tenancy) ready. We will not cover the initial steps of creating a tenancy, focusing instead on the specifics of the case study tasks and implementations.

Note   Although MedTALN Inc., the fictional company in this book, supposedly has an OCI account, I implemented the case study steps using my personal, paid cloud account.

Defining Roles and Responsibilities

In our case study, we have two main personas responsible for carrying out different tasks in developing the Healthcare NER solution: the NLP consultant, John Doe, overseeing the entire NLP project life cycle, and the tenancy administrator responsible for OCI tenancy administration for our project.

By outlining their roles and responsibilities, we can better understand how our case study will be carried out and how each user contributes to the overall development of our NLP solution on OCI. Let's take a closer look at each persona and their respective responsibilities.

Figure 3-11 illustrates the collaborative roles and responsibilities, as well as the high-level OCI components and services involved in the NLP solution. The OCI admin, as the manager of the overall OCI environment, works hand in hand with the NLP consultant, who focuses on leveraging the Data Science service and other resources to develop the NLP solution for MedTALN Inc.'s case study. This collaborative effort is key to the success of our project.

CHAPTER 3 HEALTHCARE NLP CASE STUDY

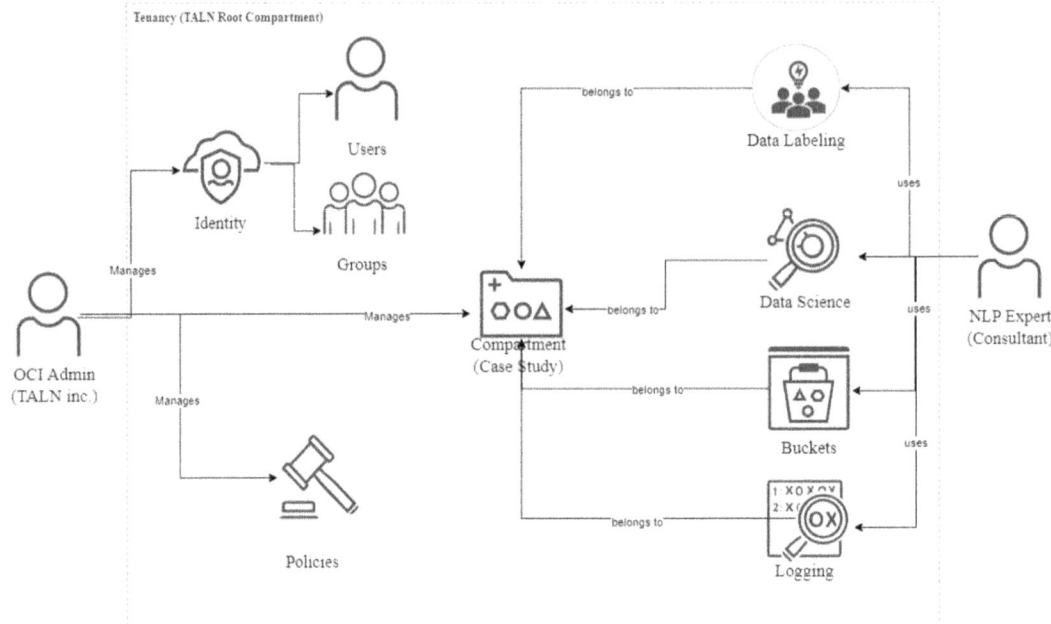

Figure 3-11. *NLP consultant vs. OCI admin roles and responsibilities for the case study*

As shown in the diagram, the OCI admin creates the project compartments. Additionally, they manage the Identity service, which handles Users and Groups, and the Policies that govern access and permissions.

On the other hand, the NLP consultant, who will use OCI ML Services, such as the OCI Data Science Service and OCI Data Labeling Service, is responsible for tasks such as building a training dataset, training the NLP model, and deploying the NLP model. These tasks are crucial in the development of the NLP solution.

Summary

This chapter introduced the case study of MedTALN Inc., a fictional healthcare analytics company. The case study is intended to be a practical example demonstrating the step-by-step implementation of a Healthcare Named Entity Recognition (NER) solution using Oracle Cloud Infrastructure (OCI). By following MedTALN Inc.'s journey, readers can gain a deeper understanding of how to design, build, and deploy custom NLP models on OCI tailored to specific needs.

This case study was imagined to provide a hands-on learning experience, illustrating real-world challenges and solutions in a controlled, fictional environment. This case study aims to provide a clear transition from theoretical concepts to building an actual NLP solution on OCI through John Doe, a fictitious persona created to represent an NLP consultant who helps MedTALN Inc.

Key Takeaways

1. This case study is intended to address the fragmented and intimidating nature of existing resources on Natural Language Processing (NLP) implementation on OCI across various sources such as books, technical documentation, and blogs.

2. The strategies, challenges, and solutions presented in the case study are based on practical experience. Although the persona of John Doe is fictional, the guidance provided is informed by genuine professional experience with NLP on OCI.

Although John Doe is not a real person, his knowledge is based on my experience in leveraging OCI to develop SOTA NLP solutions at typica.ai.[6] Through John Doe's journey, readers will discover the strategic benefits of OCI for NLP applications, showcasing how OCI enables efficiency and cost-effective NLP solution development.

In the following chapters, the second part of this book will delve into technical details, guiding you through the entire process of building an NLP model on OCI.

Reference

Oracle. (2024, May 6). *Announcing the general availability of OCI Language 4.0.* Retrieved from Oracle AI & Data Science Blog: https://blogs.oracle.com/ai-and-datascience/post/oci-ai-language-4-0

---

[6] Typica.ai is an NLP startup that I founded after my PhD. For more information, please visit the website: https://typica.ai

PART II

Case Study Implementation

Part 2 guides readers through the stages of developing MedTALN Inc.'s custom French Healthcare NER model on Oracle Cloud Infrastructure (OCI), combining theoretical insights with a detailed practical example.

Chapter 4, "Tenancy Preparation," outlines the steps to configure an OCI tenancy for our OCI Data Science Project, including the configuration of CPU- and GPU-based notebook sessions.

Chapter 5, "Dataset Preparation," focuses on dataset acquisition and preparation. This chapter highlights the importance of a systematic data annotation process for building a robust training dataset.

Chapter 6, "Model Fine-Tuning," details the fine-tuning of a pretrained model to meet the case study's specific needs.

In summary, in **Part 2**, we lay the foundation for building our Healthcare NER model. This includes preparing the OCI tenancy, creating a training dataset, and training the model for our case study.

While **Part 2** concludes with a model ready for deployment, **Part 3** will delve into its productization, covering deployment, invocation, and management, providing a comprehensive guide from setup to the operationalization of the NLP solution on OCI.

CHAPTER 4

Tenancy Preparation

Chapter 3 offered a clear direction for implementing our case study on Healthcare NER. This chapter aims to transition from the case study blueprint to tangible preparatory steps for OCI tenancy. These steps include the creation of compartments, networking, storage, and the implementation of critical security configurations.

This chapter provides valuable insights into cost-effective strategies for implementing NLP solutions using OCI ML Services and a transfer learning approach. It also discusses the responsibilities and roles of the OCI admin and data scientist teams and how they can efficiently manage and segregate the OCI tenancy preparation work for an OCI Data Science Project.

Finally, this chapter will guide readers through setting up our OCI Data Science Project, including CPU- and GPU-based notebook sessions.

Getting Started

In this chapter, we are implementing a Healthcare Named Entity Recognition (NER) solution for MedTALN Inc. for the French unstructured text. The goal is to identify medical entities, such as medical conditions, medications, dosages, symptoms, test results, treatments, and procedures. Importantly, this chapter will also tackle cost-saving strategies as a key constraint, exploring efficient approaches to develop an NLP solution within MedTALN Inc.'s budgetary constraints.

Cost-Saving Strategies

One of the biggest challenges in deploying NLP solutions is balancing cost-effectiveness with functionality and performance. However, OCI offers services for data labeling and model training that can help reduce the traditionally high costs associated with these areas. By taking advantage of OCI capabilities and adopting the transfer learning training approach, along with utilizing the open source resources available on the Hugging Face Hub, it's possible to achieve significant cost savings.

Below is an overview of the cost-saving strategies we will adopt in this case study. These strategies will be explained in more detail in the upcoming sections.

- Labeling data for NLP projects can be both time-consuming and costly due to its manual nature and the extensive human effort required. By reusing a high-quality dataset for the Healthcare NER task from the Hugging Face Hub, we can significantly reduce these costs. Additionally, OCI's Data Labeling Service, with its competitive pricing model that offers the first 1000 records for free,[1] further reduces expenses, making the task of dataset labeling more cost-effective.

- Training deep learning models, due to their need for GPU resources, can significantly raise costs, especially with intermittent training. Adopting a transfer learning approach, utilizing healthcare pretrained models for the French language from the Hugging Face Hub, and leveraging OCI Data Science Notebooks can minimize these expenses. OCI notebooks' feature to deactivate during idle times stops billing, enhancing cost efficiency. This method ensures efficient, cost-effective model fine-tuning, highlighting the potential for savings.

Through this case study, you will understand how strategic choices in using OCI's features and external resources like Hugging Face Hub can significantly reduce the financial barriers to implementing advanced NLP projects.

OCI Tenancy Preparation

The OCI administrator needs to set up the tenancy for the data labeler and data scientist teams so they can start using the Machine Learning Services, including Data Labeling and Data Science Services, within the MedTALN Inc.'s tenancy.

---

[1] As of the time of writing this book, OCI Data Labeling offers 1,000 annotated data records every month free of charge. For more information on DLS pricing, please visit the following web page: https://www.oracle.com/ca-en/artificial-intelligence/data-labeling/pricing/

First, the tenancy administrator will set up the OCI tenancy. Once this initial setup is complete, the NLP consultant, John Doe, will take over and manage all subsequent tasks and activities related to OCI ML Services.

Below are the initial setup activities for MedTALN Inc.'s tenancy:

1. Compartment Creation: Establishing a compartment to organize and isolate our case study OCI resources

2. Virtual Cloud Network (VCN) Setup: Creating and configuring a VCN with the needed elements (such as an Internet gateway, a NAT gateway, and a service gateway for the VCN) for the case study

3. Object Storage Bucket Creation: Provisioning buckets to store and manage data and models for our NLP solution

4. IAM Setup: Configuring the groups, dynamic groups, and policies for both Data Labeling and Data Science Services

By the end of this section, you'll have a comprehensive understanding of how to configure your tenancy for a successful NLP project.

Compartment Creation

For the MedTALN Inc.'s Case Study, as depicted in the diagram in Figure 4-1, we envision a hierarchy within the tenancy that could include additional compartments, such as Development, Quality Assurance, and Production, to manage resources aligned with different project life cycle stages. However, we simplify our approach for this case study by utilizing a top-level compartment—***case-study-cmpt***—instead of creating further subcompartments for streamlined focus and management simplicity.

CHAPTER 4 TENANCY PREPARATION

Figure 4-1. *Root compartment and subcompartments for the case study*

Steps to create this dedicated compartment to our case study as follows (as shown in Figure 4-2):

1. From the navigation menu, click **Identity & Security**. Under **Identity**, click **Compartments**.

2. Click the **Create Compartment** button.

3. Name the new compartment: case-study-cmpt.

4. Enter a description: *Compartment for the Case Study*.

5. Click **Create Compartment**.

6. Confirm that the compartment appears in the compartments list.

CHAPTER 4 TENANCY PREPARATION

Figure 4-2. Creating compartment

Note In OCI, tags are essential for efficiently managing project resources. They enable organization, identification, search, and filtering based on criteria such as purpose, owner, and environment, among others.

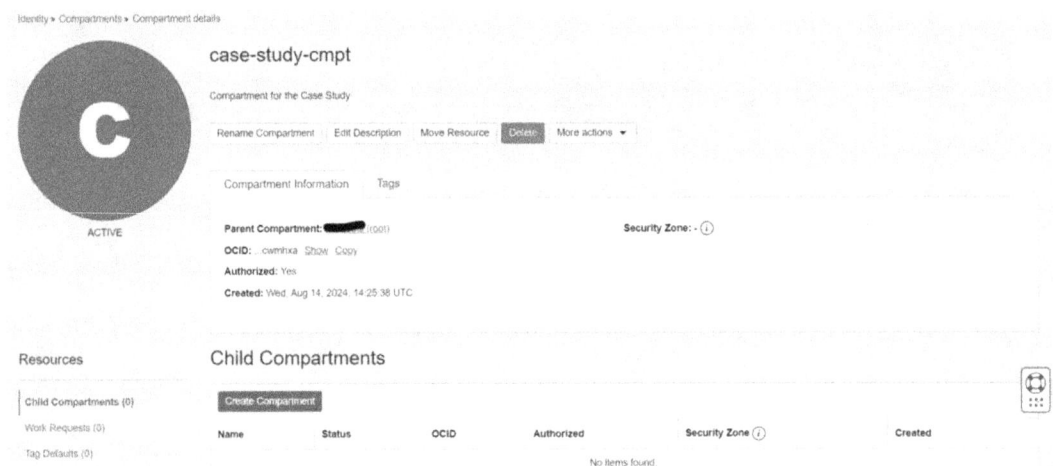

Figure 4-3. *Case study's compartment details*

For our NLP project, after completing the creation of compartments (see Figure 4-3), we will proceed with the VCN configuration.

Network Configuration

Setting up the virtual cloud network (VCN) can be an engaging task. You have two options: a detailed, manual setup where you configure each component individually or a more straightforward approach using a setup wizard. For ease and simplicity, I recommend using the wizard, and I'll guide you through that process.

Note The VCN you are currently creating can be utilized for custom networking in notebook sessions, which you will set up later in this chapter. While the default networking of notebook sessions is simpler, it results in a closed network that can only be used by the notebook session itself. On the other hand, custom networking, which makes use of the VCN you are presently setting up, provides greater flexibility. It enables you to access, modify, and utilize the network for multiple purposes beyond the notebook session, especially for your future Data Science Projects. This enhanced flexibility is a key benefit of custom networking.

CHAPTER 4 TENANCY PREPARATION

This section shows users who require access to their VCNs, how to create a VCN, and later, how to choose the recommended subnet for notebook sessions:

1. Open the navigation menu and click **Networking**, and then click **Virtual Cloud Networks**.

2. Select the compartment case-study-cmp (illustrated in Figure 4-4).

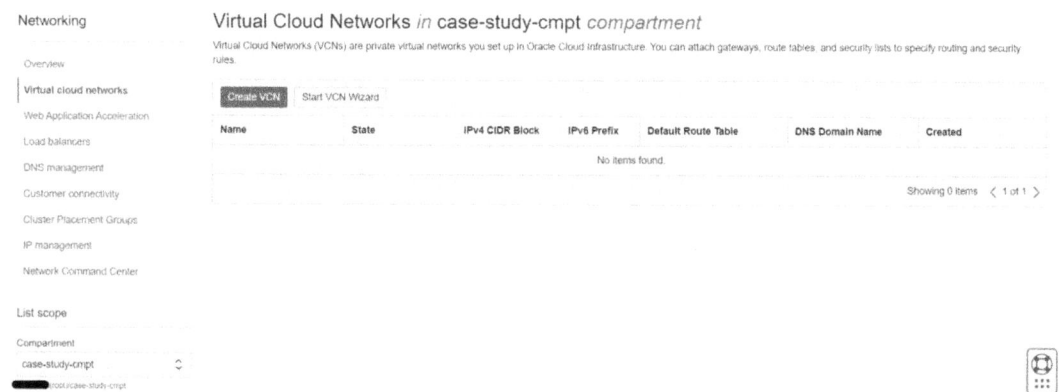

Figure 4-4. *VCNs list page*

3. Click **Start VCN Wizard** (see Figure 4-5).

115

CHAPTER 4　TENANCY PREPARATION

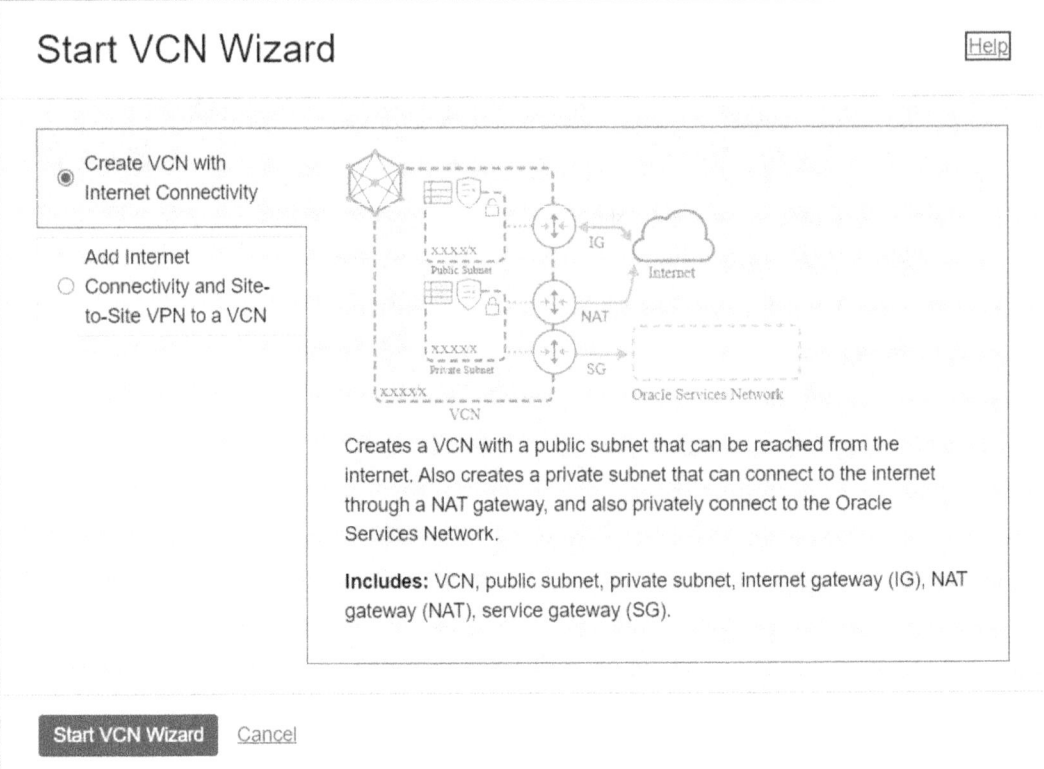

Figure 4-5. VCN wizard

4. Select **Create VCN with Internet Connectivity**, and then click **Start VCN Wizard** (Figure 4-6).

5. Enter the VCN name: cs-vcn.

6. If it is not already selected, select the case study compartment (i.e., *case-study-cmpt*).

7. For Configure VCN and Subnets, keep the default values.

CHAPTER 4 TENANCY PREPARATION

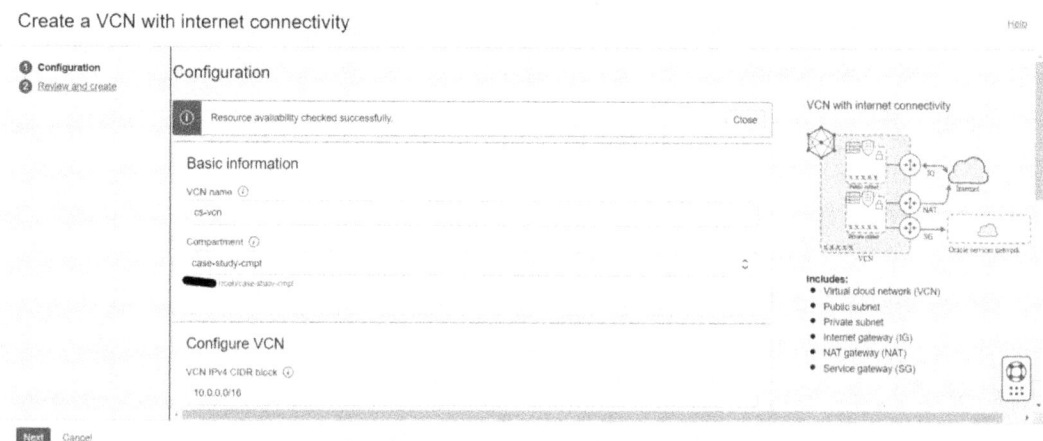

Figure 4-6. *VCN creation step 1*

8. Click **Next**.

9. Review the VCN configuration (Figure 4-7).

10. Click the **Create** button to create the VCN and the related resources such as a public and a private subnet, an Internet gateway, a NAT gateway, and a service gateway (illustrated in Figure 4-8).

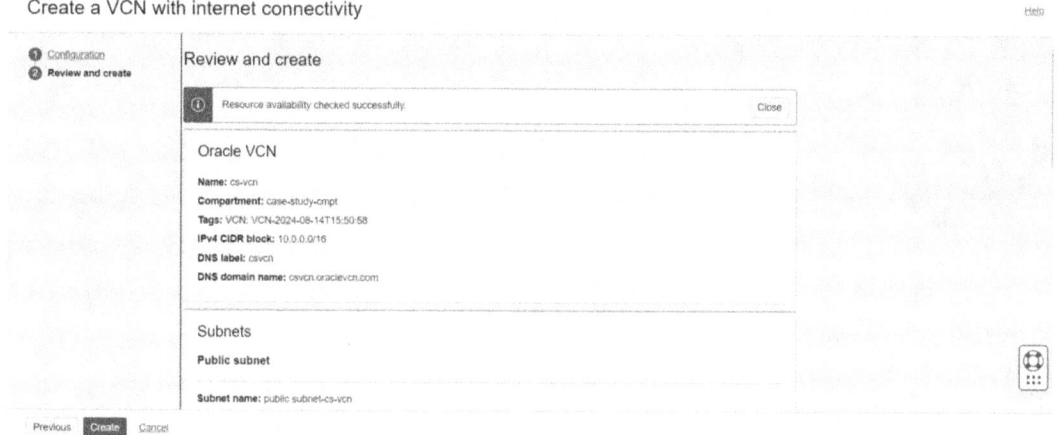

Figure 4-7. *VCN creation step 2*

117

CHAPTER 4   TENANCY PREPARATION

11. You use this VCN and its private subnet when you create your notebook session.

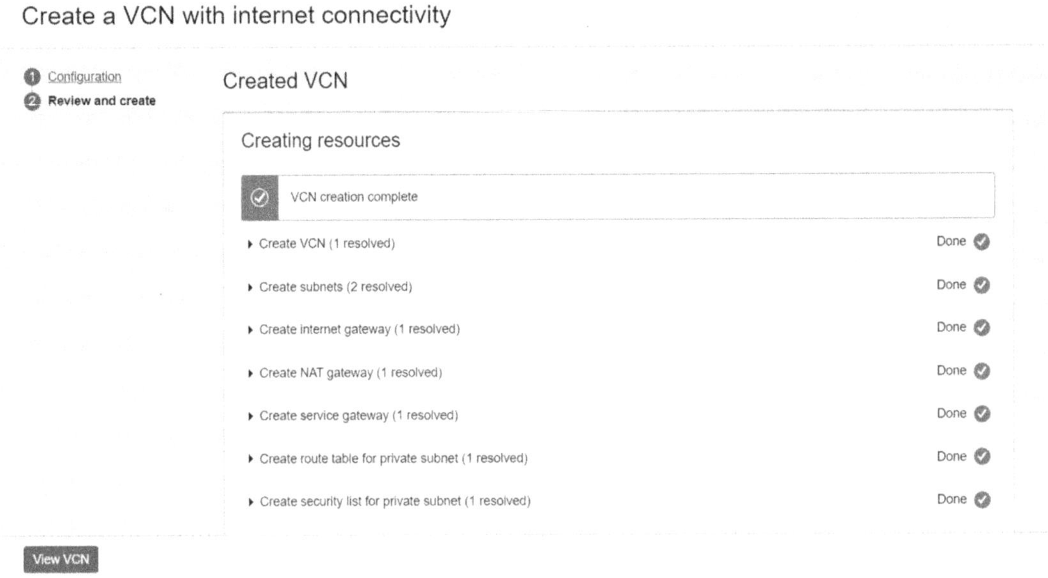

Figure 4-8. VCN creation page

12. Click the **View VCN** button to review your VCN and subnets (refer to Figure 4-9).

Figure 4-9. VCN detail page

Storage

For our case study, we've established the need for four buckets within our compartment, each crafted to fulfill a distinct role throughout the model development life cycle (as outlined in Figure 4-10). Below is a detailed description of each bucket's purpose:

- Labeling Datasets Bucket: Reserved for datasets destined for labeling via the OCI Data Labeling Service. These datasets might originate from open source datasets, such as those found on the Hugging Face Hub, and are then preprocessed and converted into the *JSONL Consolidated* format, one of the supported dataset formats for the OCI Data Labeling import process.

- Training Datasets Bucket: This bucket is specifically for holding training datasets in the Hugging Face format, derived from datasets exported from the OCI Data Labeling Service in the *CoNLL* format. It will contain all versions of these training datasets, providing clear traceability and lineage of the data utilized during the training phase.

- Model Checkpoint Bucket: The role of the third bucket is to store checkpoints created at each training epoch with Hugging Face Transformers. These checkpoints are essential for tracking progress, enabling the resumption of training from specific points, and preventing data loss in extensive training sessions, a critical consideration due to the costs associated with GPU-based training.

- Conda Environment Bucket: The fourth bucket is reserved for storing our published custom conda environments, which contain the latest versions of all the libraries needed for our deep learning–based NLP projects, such as Hugging Face Transformers, PyTorch, and Oracle Accelerated Data Science (ADS library). Our conda environments will be used by our NLP team members for model training by OCI Data Science Service for model deployment and inference, promoting consistency and enabling reproducibility for all our NLP model life cycles.

CHAPTER 4   TENANCY PREPARATION

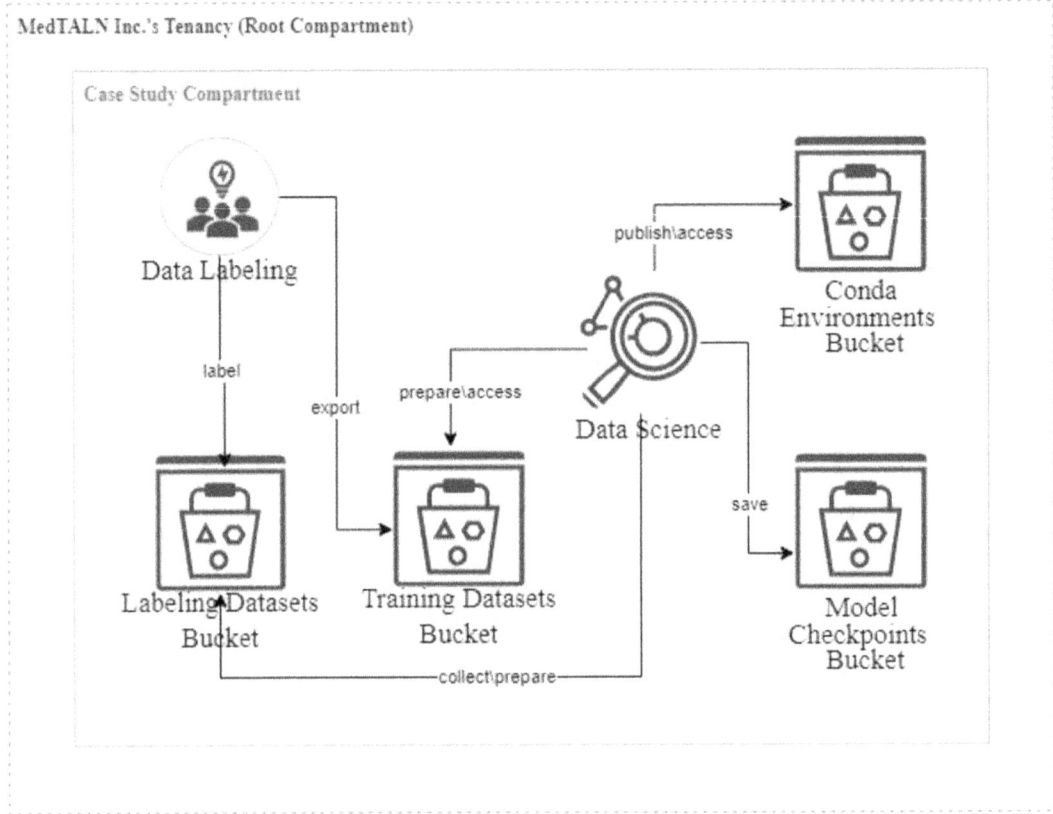

Figure 4-10. Object Storage buckets for our case study

Although creating buckets is a prerequisite for using Data Labeling, OCI Data Science doesn't require Object Storage buckets. Yet, we chose to implement distinct Object Storage buckets, recognizing their value extends well beyond mere data storage. Our proposed bucket architecture, depicted in the diagram above, is crucial for facilitating team collaboration, enhancing security and separation of concerns, and avoiding unnecessary costs during the model development process.

- Collaboration: Our buckets act as central, secure repositories, enabling direct and efficient collaboration among team members. By providing NLP engineers with access to various data versions, model checkpoints, and environments, we ensure smooth progress across different stages of model development, irrespective of the notebook sessions used (CPU- or GPU-based notebook sessions).

- Backup and Archiving: The financial and operational importance of securing model artifacts cannot be understated. Our buckets provide a robust solution for backing up and archiving valuable data, protecting against potential losses. This is especially crucial considering the high costs associated with data labeling and training sophisticated NLP models.

- Access Control and Separation of Concerns: Our architecture emphasizes strict access controls and security measures to protect sensitive data and maintain project integrity. By implementing separation of concerns between data labelers and data scientists, we assign specific access rights—such as granting exclusive access to Data Labeling buckets to the labeling teams. This approach effectively prevents unauthorized modifications, ensuring the safety and confidentiality of data throughout the project life cycle.

Furthermore, leveraging the "Mount Storage" feature simplifies our code significantly by integrating Object Storage buckets as local file systems within OCI Data Science Notebook Sessions. This integration means that when we need to read datasets or model files; it's as straightforward as dealing with local files. No longer do we have to navigate the complexities of Object Storage APIs or SDKs for basic operations. For instance, accessing our custom dataset or loading our trained model becomes a matter of using simple file paths. This dramatically reduces the amount of code needed for such tasks, allowing data scientists to concentrate on their core activities. Ultimately, this leads to cleaner, more manageable code, streamlining the development process and enhancing productivity.

The OCI administrator is responsible for creating those buckets as part of the OCI setup process.

To initiate the creation of the first bucket, perform the following:

1. From the navigation menu, go to Storage ➤ Object Storage & Archive Storage ➤ Buckets.

2. Select the compartment for our case study: *case-study-cmpt*.

    This is the compartment where the case study buckets will be located (Figure 4-11).

CHAPTER 4   TENANCY PREPARATION

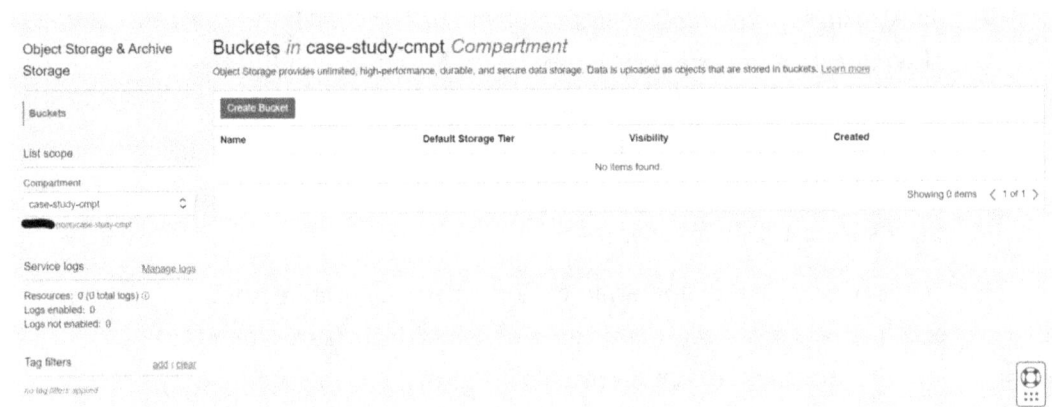

Figure 4-11. *Buckets list page*

3. Click **Create Bucket** button.

4. In the **Create Bucket** dialog box, complete the following (see Figure 4-12):

   a. Enter a value for the **Bucket Name:** *labelling-datasets-bkt*.

   b. Select the default tier in which you want to store your data. **Default Storage Tier**: *Standard*.

   We use Standard tier to store data that requires fast and immediate access. However, we use the Archive tier for storing data that requires long retention periods but not immediate access.

   c. Keep all the following check boxes unchecked:

   – Enable Auto-Tiering: Select this option if you want Object Storage to monitor and automatically move infrequently accessed objects from the Standard tier to the less expensive Infrequent Access storage tier.

   – Enable Object Versioning: Select this option if you want Object Storage to create an object version each time the content changes or the object is deleted.

   – Emit Object Events: Select this option if you want to enable the bucket to emit events for object state changes.

   – Uncommitted Multipart Uploads Cleanup: Select this option to create a life cycle rule that automatically deletes all uncommitted multipart uploads after seven days.

CHAPTER 4   TENANCY PREPARATION

d. For the Encryption option, select the option *Encrypt using Oracle managed keys*.

Buckets are encrypted with keys managed by Oracle by default, but you can optionally encrypt the data in this bucket by using your own Vault encryption key.

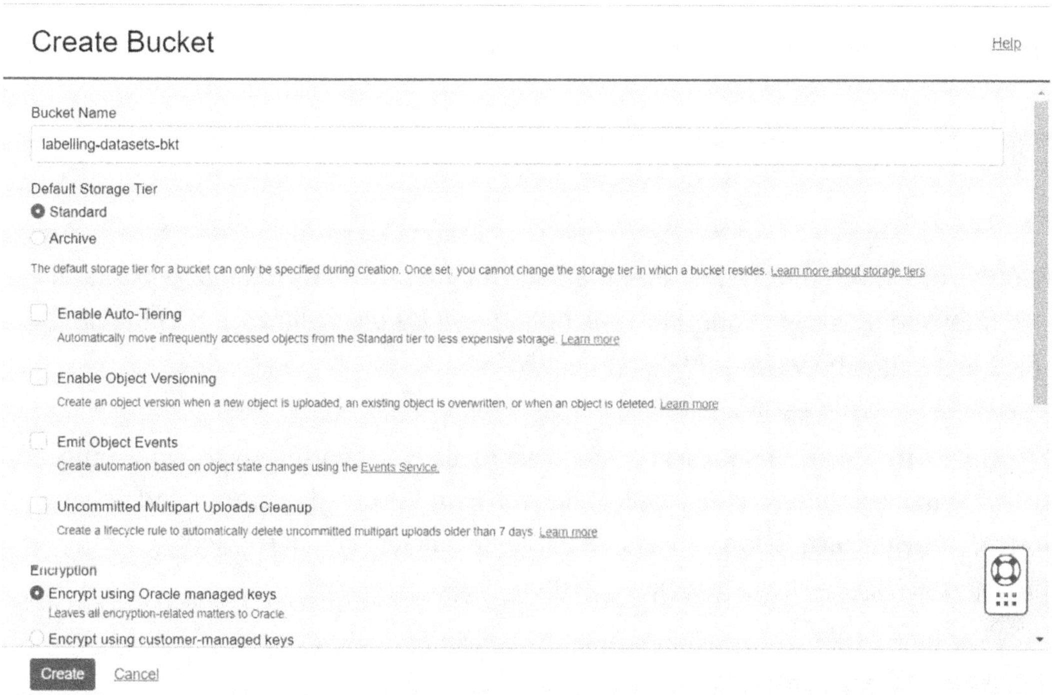

Figure 4-12. Bucket creation page

5. Click **Create**.

   Repeat the same steps to create the three remaining buckets as shown in the screenshot in Figure 4-13. Use the following bucket names:

   – The Training Datasets Bucket Name: *training-datasets-bkt*

   – The Models Checkpoints Bucket Name: *models-ckpt-bkt*

   – The Conda Environments Bucket Name: *conda-envs-bkt*

123

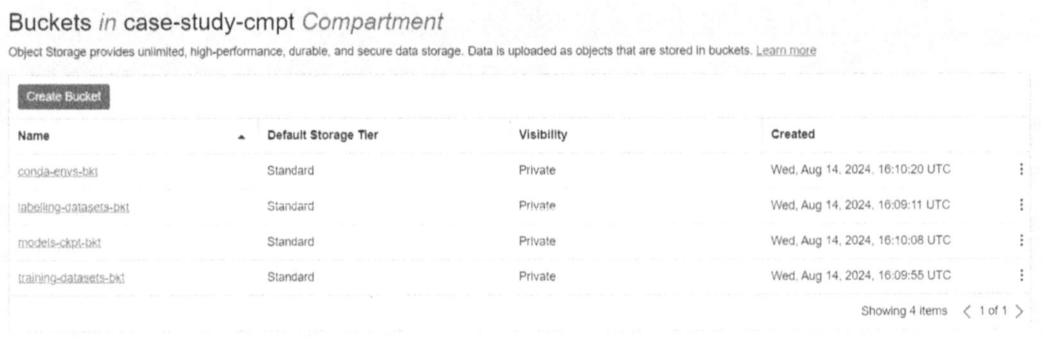

Buckets *in* case-study-cmpt *Compartment*

Object Storage provides unlimited, high-performance, durable, and secure data storage. Data is uploaded as objects that are stored in buckets. Learn more

Name	Default Storage Tier	Visibility	Created	
conda-envs-bkt	Standard	Private	Wed, Aug 14, 2024, 16:10:20 UTC	
labelling-datasets-bkt	Standard	Private	Wed, Aug 14, 2024, 16:09:11 UTC	
models-ckpt-bkt	Standard	Private	Wed, Aug 14, 2024, 16:10:08 UTC	
training-datasets-bkt	Standard	Private	Wed, Aug 14, 2024, 16:09:55 UTC	

Figure 4-13. Case study's buckets

Note It is worth mentioning that the buckets are private by default, ensuring the security and privacy of your data. The option *Edit Visibility* allows you to toggle the bucket's status between private and public.

With all the necessary buckets created, we're ready to continue with the Identity and Access Management setup (i.e., IAM). This step involves creating groups and policies that will provide permission to users such as data labelers and data scientists to access the required OCI resources and perform their assigned tasks.

Identity and Security

After successfully configuring the compartments, network, and storage for our case study, we will proceed to the security setup, specifically the IAM setup. However, before embarking on the actual setup, it would be beneficial to clarify some essential concepts related to IAM in OCI, such as user groups and dynamic groups.

User groups in OCI are groups of individual users that are granted access to data science resources within compartments. Admins can create user groups in three simple steps: creating users, creating groups, and adding users to groups. When setting up groups, admins must first decide how users will access resources in the compartments.

Dynamic groups, a unique type of group, offer a high degree of flexibility and adaptability. They contain resources that match specific rules defined by the admin. Resources such as data science notebook sessions, models, and model deployments can be included in a dynamic group. The dynamic nature of group membership, which can change as resources that match those rules are created or deleted, puts the admin in full control. These resources, considered principal actors, can make API calls to services based on the policies defined for the dynamic group, further enhancing the admin's control over resource access.

Let's take a practical example to understand the role of resources in making API calls. Consider calling the Object Storage API to read data from a bucket. This call uses the resource principle of a data science notebook session. The dynamic group associated with this session has a policy that enables Object Storage access. In other words, resources match rules, and rules are applied to dynamic groups.

OCI policies grant access to users and resources at the group and compartment levels using simple syntax with variables such as group name, verb, action, resource type, and compartment name. The syntax specifies the type of access a group has in a compartment. The group name is the user or dynamic group, the verb defines the access level, and the resource type specifies the resource or resource family. Finally, the compartment name is where access is granted.

Verbs determine the level of access to a resource or resource family. They range from least to most permissive: inspect (ability to list resources without access to user metadata), read (ability to get user metadata and the resource itself), use (ability to work with the resource, excluding creating or deleting permissions), and manage (includes all permissions, including creating and deleting).

When creating a policy, it is essential to identify the resource type for which it is intended. You can either formulate a policy for a specific resource type or for a collection of related resources. Nonetheless, it is crucial to find a balance between being too detailed and keeping things simple. This will ensure that the policies are effective in maintaining the required level of security while still being easy to manage.

IAM Setup for Data Scientists

The IAM configuration required for the data scientists' team is illustrated in this architectural diagram (Figure 4-14).

CHAPTER 4   TENANCY PREPARATION

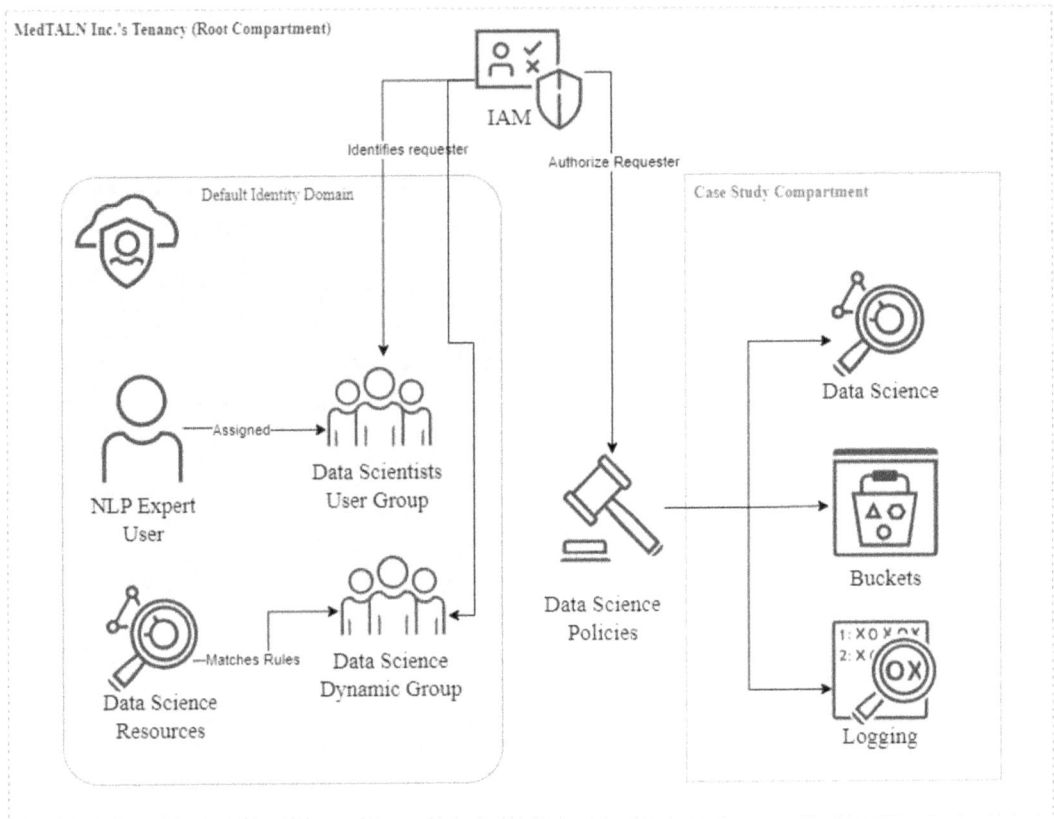

Figure 4-14. *IAM setup for data science team*

Next, we will discuss the implementation details of this data labeling-related setup. The security setup steps include creating an NLP expert user, creating a user group for data labelers, and finally, creating a dynamic group for data labeling resources (based on matching rules).

Users and Groups

We will start by creating our data science group and then create our NLP expert user.

CHAPTER 4   TENANCY PREPARATION

The following steps outline the user group creation for our data science team.

1. Go to Identity & Security ➤ *Identity* ➤ *Domains* (see Figure 4-15).

Figure 4-15. *IAM Domains list*

2. Select the identity domain named ***Default***.

3. Under ***Identity Domain***, on the left, click ***Groups***, and then click the ***Create group*** button.

4. In the ***Create group*** dialog, as shown in Figure 4-16, enter the group details:

    a. Enter a unique name for our new group ***Name***, i.e., *data-scientists-users-grp*.

    b. Enter the ***Description*** for the group (required), i.e., *Data Scientists User Group*.

    c. Click ***Create***.

127

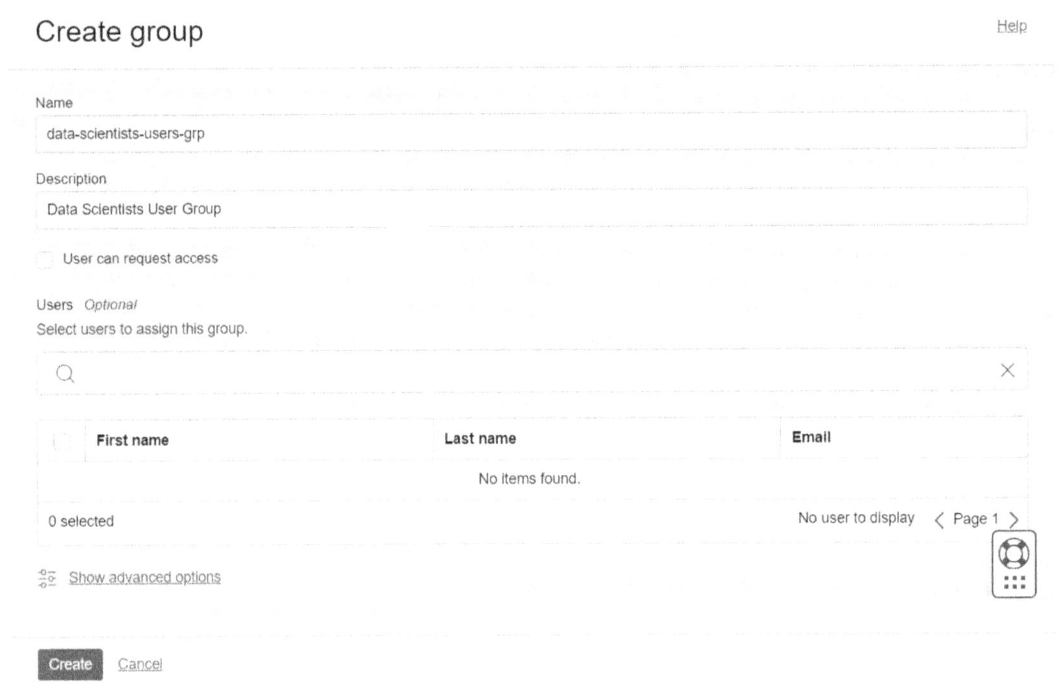

Chapter 4 Tenancy Preparation

Figure 4-16. Data scientist group creation

Now that we have our data scientist user group created, we can proceed with creating our first data science team user (i.e., John Doe the NLP consultant). The steps to create this user are as follows:

5. Go to Identity & Security ➤ *Identity* ➤ *Domains*, and select the identity domain named *Default*.

6. Under *Identity Domain*, on the left, click *Users*, and then click the *Create User* button.

7. As shown in Figure 4-17, in the *Create user* dialog:

    a. Enter the user's *First name*, *Last name*, and *Email*, e.g., John, Doe, john.doe@typica.ai.

    b. Make sure that the check box *Use the email address as the username* is selected.

    c. Assign this user to the data scientist group by selecting the check box of the group.

    d. Click *Create*.

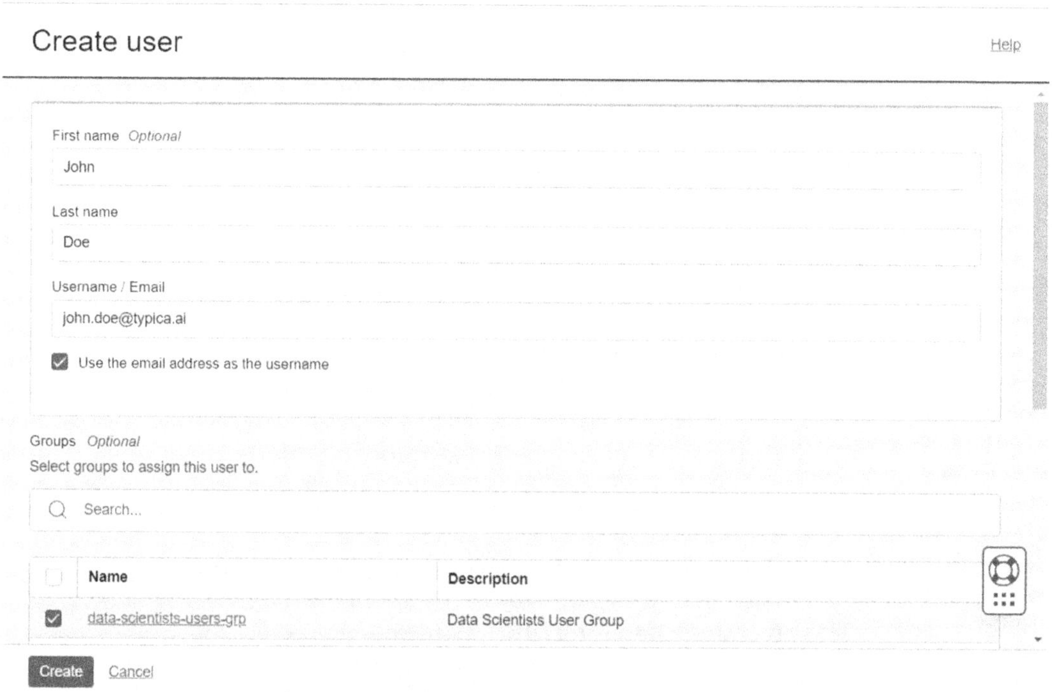

Figure 4-17. Data scientist user creation

Dynamic Groups

We will create a dynamic group for our Data Science resources. These are the steps to follow:

The following steps outline the **Dynamic groups** creation for our data science team.

1. Go to **Identity & Security** ➤ **Identity** ➤ **Domains**, and select the identity domain named **Default**.

2. Under **Identity Domain**, on the left, click **Dynamic groups**, and then click the **Create dynamic group** button.

3. In the **Create dynamic group** dialog, as shown in Figure 4-18, enter the group details:

    a. Enter a unique name for our new group **Name**, i.e., *data-science-dyn-grp*.

    b. Enter the **Description** for the group (required), i.e., *Data Scientists Dynamic Group*.

CHAPTER 4　TENANCY PREPARATION

  c. Add the following matching rules.

  d. Under the **Matching rules** section, select the option **Match any rules defined below**.

   – Enter the following two matching rules:

   Rule 1:

   ALL { resource.type = 'datasciencenotebooksession' }

   This matching rule means that all notebook sessions created are members of the dynamic group.

   – Click Additional Rule, and add the following rule:

   Rule 2:

   ALL { resource.type = 'datasciencemodeldeployment' }

   The preceding matching rule means that all model deployments are members of the dynamic group.

  e. Click **Create**.

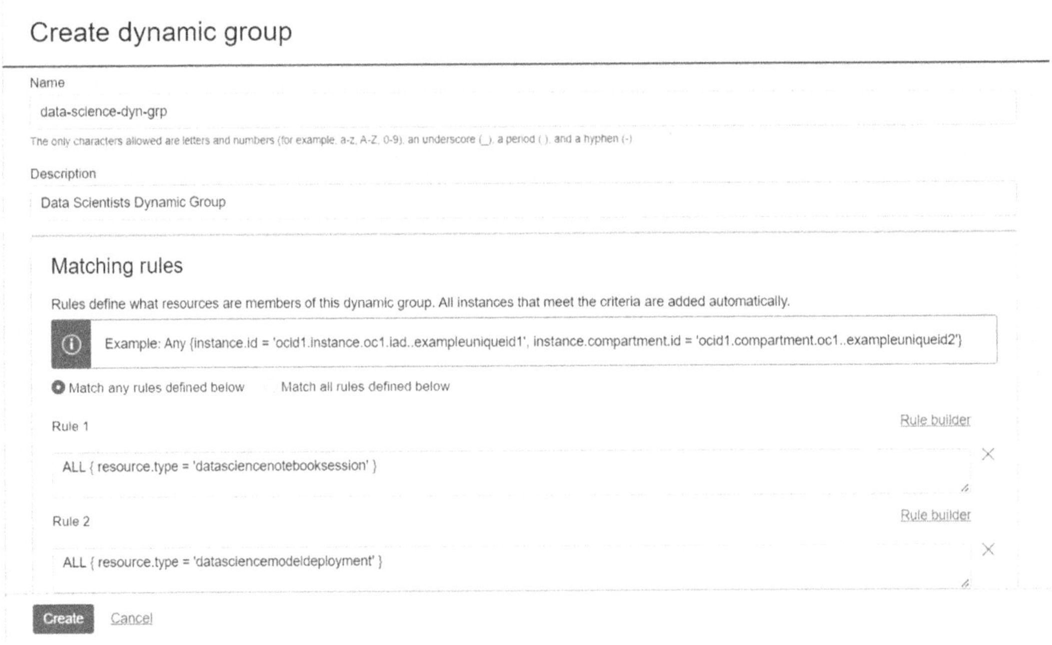

Figure 4-18. Data science dynamic group creation

CHAPTER 4 TENANCY PREPARATION

Policies

We opted for a simplified approach in creating policies for the data science team. Our policies will focus on granting access to aggregate resource types, such as *object-family*, which encompasses various individual resource types including *buckets* and *objects*, within our designated compartment.

We believe this approach provides, for our case study, an optimal balance between security, ease of maintenance, and simplicity, making it easier for readers to understand the concepts.

The following steps outline the **Policies** creation for our data science team:

1. From the navigation menu, go to **Identity & Security** ➤ **Identity** ➤ **Policies** (Figure 4-19).

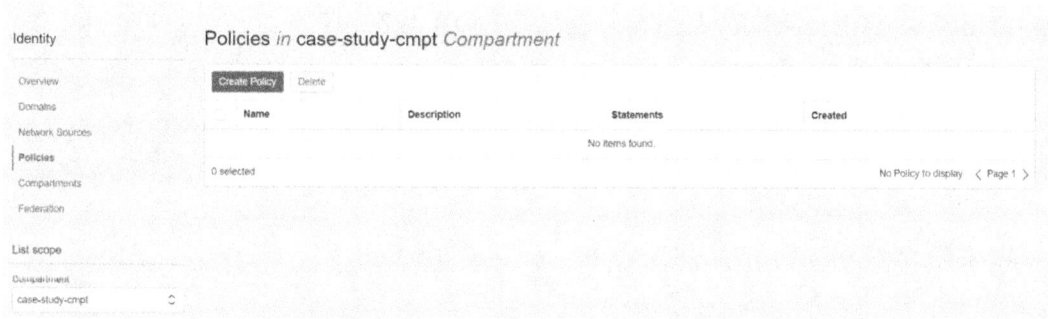

Figure 4-19. Policies list page

2. Then, click the **Create policy** button.

3. Enter the following:

   a. Name: **data-science-policies**.

   b. Description: *Policies for the Data Science team*.

   c. Select the **root compartment** (not our case study compartment).

   d. Click **Show manual editor**.

   e. Enter the policy statements in Listing 4-1 into the Policy Builder field, and then click **Create** to create the policy (Figure 4-20).

Listing 4-1. Policy statements for data science

```
allow group data-scientists-users-grp to manage data-science-family
in tenancy
allow dynamic-group data-science-dyn-grp to manage data-science-family
in tenancy
allow service datascience to use virtual-network-family in tenancy
allow group data-scientists-users-grp to use virtual-network-family
in tenancy
allow group data-scientists-users-grp to manage object-family in
compartment case-study-cmpt
allow dynamic-group data-science-dyn-grp to manage object-family in
compartment case-study-cmpt
allow group data-scientists-users-grp to use logging-family in compartment
case-study-cmpt
allow dynamic-group data-science-dyn-grp to use logging-family in
compartment case-study-cmpt
```

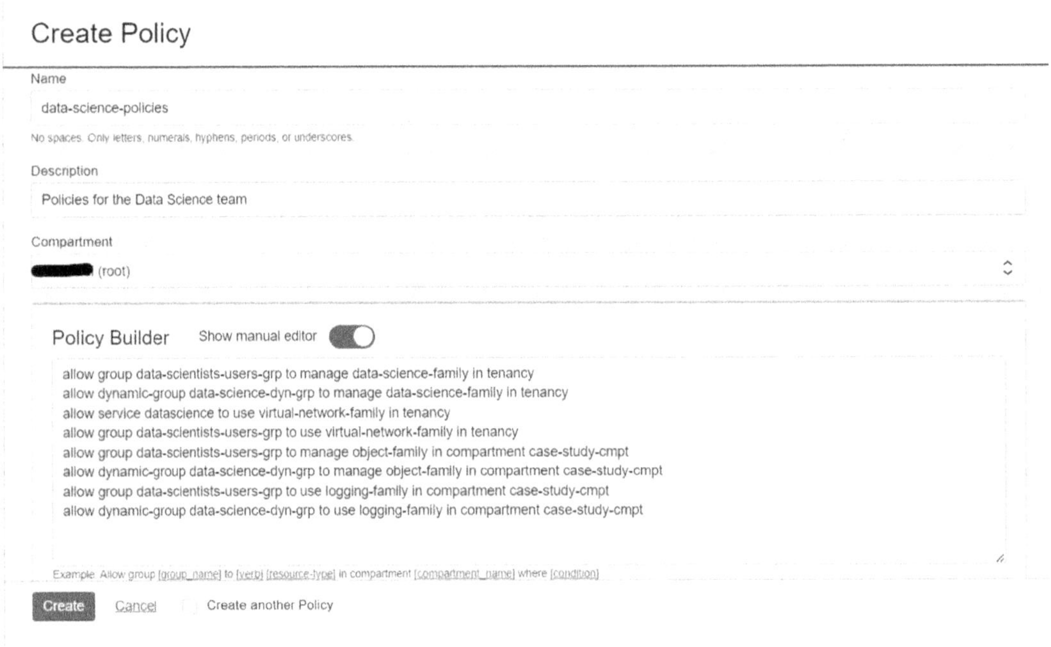

Figure 4-20. *Policies for the data science team*

Let me quickly explain the policy statements mentioned above.

With the first two policies, data scientists and data science resources (such as a notebook session) can manage all data science resources in our compartment

```
allow group data-scientists-users-grp to manage data-science-family
in tenancy
allow dynamic-group data-science-dyn-grp to manage data-science-family
in tenancy
```

We require the following policies to allow the Data Science service to utilize the virtual network family within our compartment and grant permission for our data scientist group to use it as well.

```
allow service datascience to use virtual-network-family in tenancy
allow group data-scientists-users-grp to use virtual-network-family
in tenancy
```

The following statements allow data scientists or data science resources to access Object Storage resources, such as buckets and objects, in our case study compartment. Note that the "Manage" permission is necessary to enable the creation of new objects. This access is necessary for tasks like accessing data files during model training and published conda environments during model deployment.

```
allow group data-scientists-users-grp to manage object-family in
compartment case-study-cmpt
allow dynamic-group data-science-dyn-grp to manage object-family in
compartment case-study-cmpt
```

And finally, the two following policy statements shall give model deployment access to emit logs to the Logging service.

```
allow group data-scientists-users-grp to use logging-family in compartment
case-study-cmpt
allow dynamic-group data-science-dyn-grp to use logging-family in
compartment case-study-cmpt
```

CHAPTER 4   TENANCY PREPARATION

Note While our policies restrict access to the aggregate resource types within our compartment only, their permissions could be considered permissive for real-world scenarios. If necessary, refine the scope of permissions by specifying policy statements for individual resource types, for more granular control.

IAM Setup for Data Labelers

For our case study, our NLP expert (external consultant) will act as both data labeler and data scientist. In real-world scenarios, data scientists are less likely to undertake dataset annotation tasks themselves. In fact, dataset annotation is a time-consuming task that is generally delegated to more specialized teams. This approach allows data scientists to focus on their areas of expertise, such as model development and analysis, while ensuring high-quality annotations from teams trained specifically for this purpose.

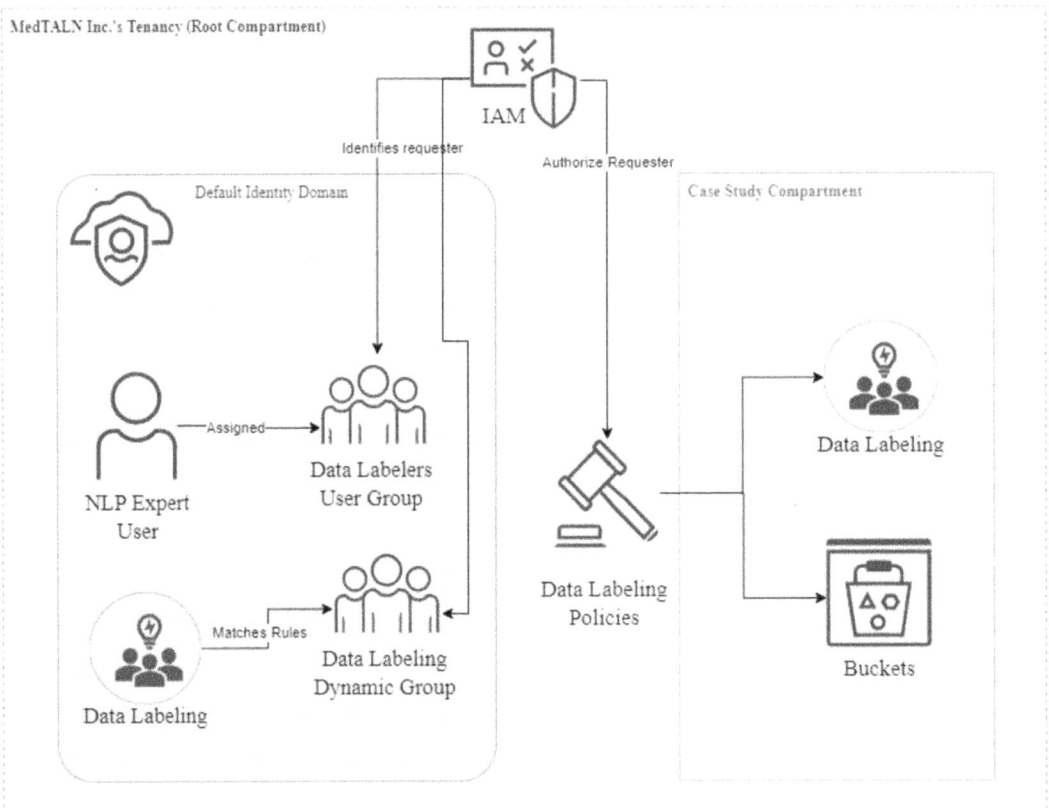

Figure 4-21. *IAM setup for data labeling team*

CHAPTER 4 TENANCY PREPARATION

The architectural diagram above (Figure 4-21) illustrates the IAM configuration necessary for the data labelers' team.

Next, we will show the implementation details of this data labeling-related setup. The security setup steps include creating the labeler group, assigning the NLP consultant user to this group, and creating a dynamic group and policies for data labeling resources.

We will start by creating our data labeler group:

1. Go to Identity & Security ➤ *Identity* ➤ *Domains* ➤ *Default Domain* ➤ *Groups*, and then click the *Create group* button.

2. In the *Create group* dialog, as shown in Figure 4-22, enter the group details:

    a. Enter a unique name for our new group *Name*, i.e., *data-labelers-users-grp*.

    b. Enter the *Description* for the group (required), i.e., *Data Labelers User Group*.

    c. Under *Users* section, select the NLP consultant, i.e., John Doe user we created earlier.

    d. Click *Create*.

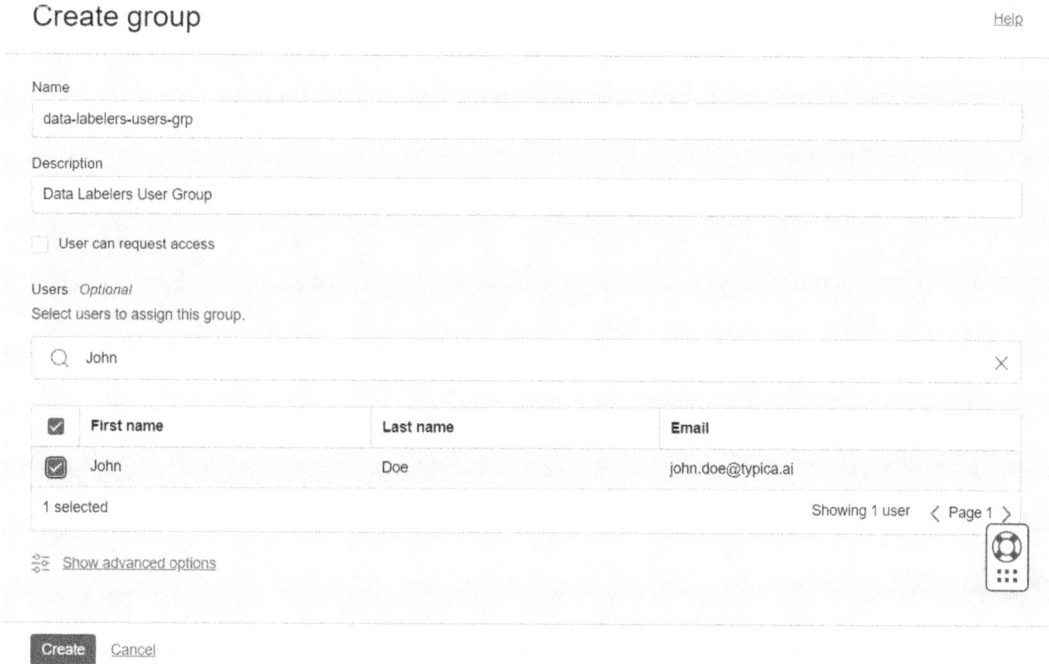

Figure 4-22. *Data labeler group creation*

CHAPTER 4   TENANCY PREPARATION

We will now continue with the creation of dynamic group for our data labeling resources:

3. Go to **Identity & Security ➤ Identity ➤ Domains ➤ Default Domain ➤ Dynamic Groups**, and then click the **Create dynamic group** button.

4. In the **Create dynamic group** dialog, as shown in Figure 4-23, enter the group details:

    a. Enter a unique name for our new group **Name**, i.e., *data-labeling-dyn-grp*.

    b. Enter the **Description** for the group (required), i.e., *Data Labeling Dynamic Group*.

    c. Add the following matching rules.

    d. Under the **Matching rules** section, select the option **Match any rules defined below**.

    Enter the following matching rules:

    Rule 1:

    ALL { resource.type = 'datalabelingdataset' }

    This matching rule means that all data labelling datasets created are members of the dynamic group.

    e. Click **Create**.

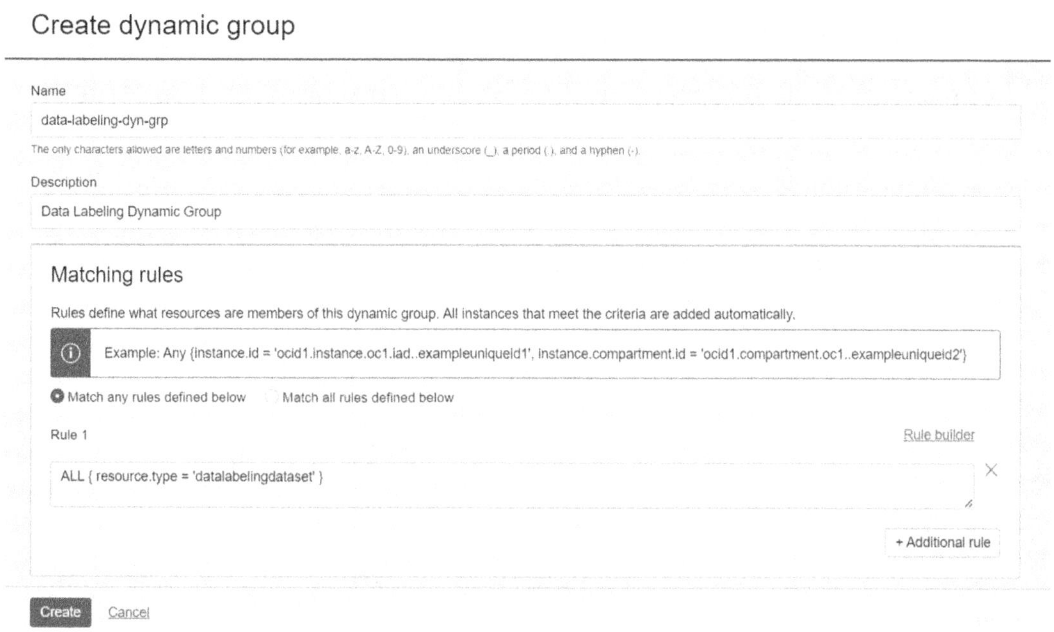

CHAPTER 4 TENANCY PREPARATION

Create dynamic group

Name

data-labeling-dyn-grp

The only characters allowed are letters and numbers (for example, a-z, A-Z, 0-9), an underscore (_), a period (.), and a hyphen (-).

Description

Data Labeling Dynamic Group

Matching rules

Rules define what resources are members of this dynamic group. All instances that meet the criteria are added automatically.

Example: Any {instance.id = 'ocid1.instance.oc1.iad..exampleuniqueid1', instance.compartment.id = 'ocid1.compartment.oc1..exampleuniqueid2'}

● Match any rules defined below   ○ Match all rules defined below

Rule 1   Rule builder

ALL { resource.type = 'datalabelingdataset' }

+ Additional rule

Create   Cancel

Figure 4-23. *Data labeling dynamic group creation*

As the final step in this IAM setup section, we will now proceed with creating policies for our data labeling principals:

5. From the navigation menu, go to **Identity & Security ➤ Identity ➤ Policies**, and then click the **Create policy** button.

6. Enter the following:

    a. Name: **data-labeling-policies**.

    b. Description: *Policies for the Data Labeling team*.

    c. Select the **root compartment** (not our case study compartment).

    d. Click **Show manual editor**.

    e. Enter the policy statements below (Listing 4-2) into the Policy Builder field, and then click **Create** to create the policy (Figure 4-24).

137

CHAPTER 4 TENANCY PREPARATION

Listing 4-2. Policy statements for data labeling

```
allow group data-labelers-users-grp to manage data-labeling-family in
compartment case-study-cmpt
allow group data-labelers-users-grp to use object-family in compartment
case-study-cmpt
allow dynamic-group data-labeling-dyn-grp to use object-family in
compartment case-study-cmpt
```

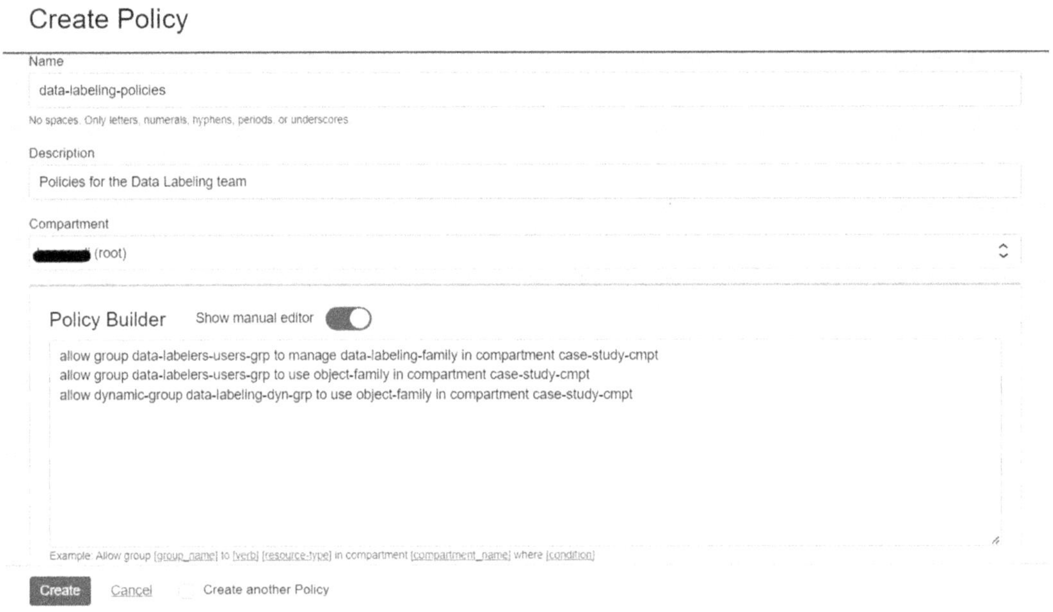

Figure 4-24. *Policies for the data labeling team*

Let's review and briefly explain the policy statements mentioned. With the first policy, data labelers can manage all data labeling resources in our compartment.

```
allow group data-labelers-users-grp to manage data-labeling-family in
compartment case-study-cmpt
```

The following statements are needed to allow data labelers or data labeling resources to access Object Storage buckets in our case study compartment.

```
allow group data-labelers-users-grp to use object-family in compartment
case-study-cmpt
allow dynamic-group data-labeling-dyn-grp to use object-family in
compartment case-study-cmpt
```

CHAPTER 4 TENANCY PREPARATION

In this section, we discussed IAM configurations such as user groups, dynamic groups, and policies needed as prerequisites to start creating using Data Science and Data Labeling Services.

Data Science Environment Setup

This section focuses on preparing the Data Science Project and notebook sessions for our case study. After you successfully sign in to MedTALN Inc.'s OCI tenancy, the NLP consultant, i.e., John Doe, shall create a Data Science Project intended for the data scientist to collaborate, organize, and document our case study NLP project's work.

To create a project, you can create notebook sessions and models and associate them with the project. You and collaborators can then organize and document data science work within the projects.

Project

we delve into the core element of any data science workspace: the project. A project in data science serves as a collaborative workspace where teams can organize and align their efforts around specific business questions or use cases. It's the central hub where all resources, including notebook sessions and models, are managed and documented.

Creating a project can be done through the console user interface (UI). Open the Navigation menu. Click **Analytics and AI**. Under **Machine Learning**, click **Data Science**. Then, click **Create project**, and select the compartment you want to add the project to. Here, you'll provide a unique name and description for the project and optionally add tags for easy identification and tracking.

Steps that our NLP consultant John Doe should follow after signing in to our tenancy to create a Data Science Project intended to organize our case study NLP project resources are as follows:

1. From the navigation menu, go to Analytics & AI ➤ Machine Learning ➤ Data Science.

2. If it is not already selected, select the case study's compartment (i.e., *case-study-cmpt*).

3. Then, click the **Create project** button (Figure 4-25).

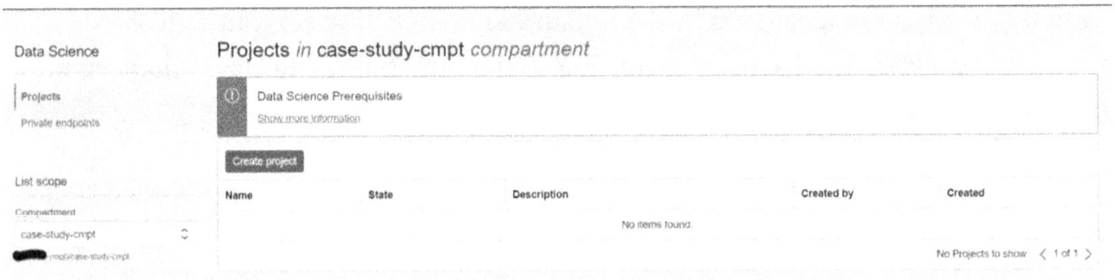

CHAPTER 4 TENANCY PREPARATION

Figure 4-25. Data Science Projects list page

4. In the Create project panel (Figure 4-26):

   a. Select a compartment for the project (i.e., *case-study-cmpt*).

   b. Enter a unique name for the project: *cs-nlp-prj*.

   c. Enter a description for the project: *Data Science Project for the Case Study (NLP)*.

   d. To view the details for the project immediately after creation, select the check box **View detail page on clicking create**.

5. Click **Create**.

Figure 4-26. Data Science Project creation

CHAPTER 4 TENANCY PREPARATION

The ***Project details*** page indicates that John Doe, our NLP consultant, has successfully created our case study's Data Science Project (as shown in the ***Created By*** field in the ***Project information*** section; see Figure 4-27).

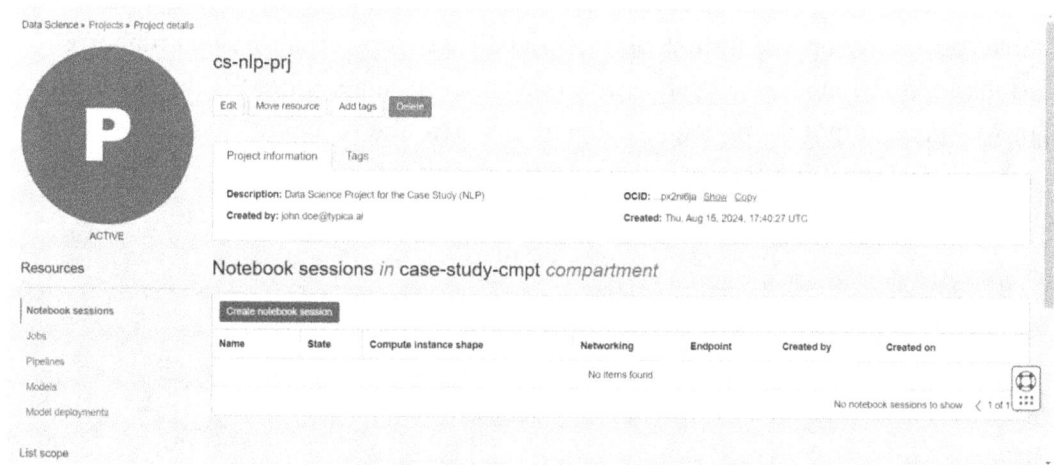

Figure 4-27. *Data Science Project detail page*

Note Alternatively, Data Science Projects can also be created programmatically using the ADS SDK, using the (ProjectCatalog object).

Managing projects is straightforward and includes viewing, editing, and deleting them. To delete a project, ensure it's empty of any data science resources.

The following sections will explore setting up Data Science notebook sessions. These sessions provide interactive coding environments that our NLP team, specifically our NLP consultant, will use for the entire life cycle of our Healthcare NER model, from the initial dataset preparation to the final steps of training and deploying the NLP model.

Notebook Sessions

In configuring our OCI Data Science environment, the NLP consultant intends to create two notebook sessions, each adapted for the distinct computing demands of our NLP project's different stages. The first, a CPU-based session, will be dedicated to every aspect of model development—spanning dataset acquisition, preprocessing, to model deployment—except for the model training activity. The second, a GPU-based session, is

CHAPTER 4 TENANCY PREPARATION

specifically set aside for training our deep learning model. This approach is strategically designed to minimize the costs related to GPU utilization, a key consideration we'll explore in more detail.

OCI Data Science Notebook Sessions are interactive coding environments that enable you to develop and train your AI models. These JupyterLab-based notebook sessions are fully managed, which means that the OCI Data Science Service takes care of provisioning compute, managing software updates, and patching. Additionally, the notebook sessions support both CPU and GPU shapes and offer persistent storage for saving your code, data, and files.

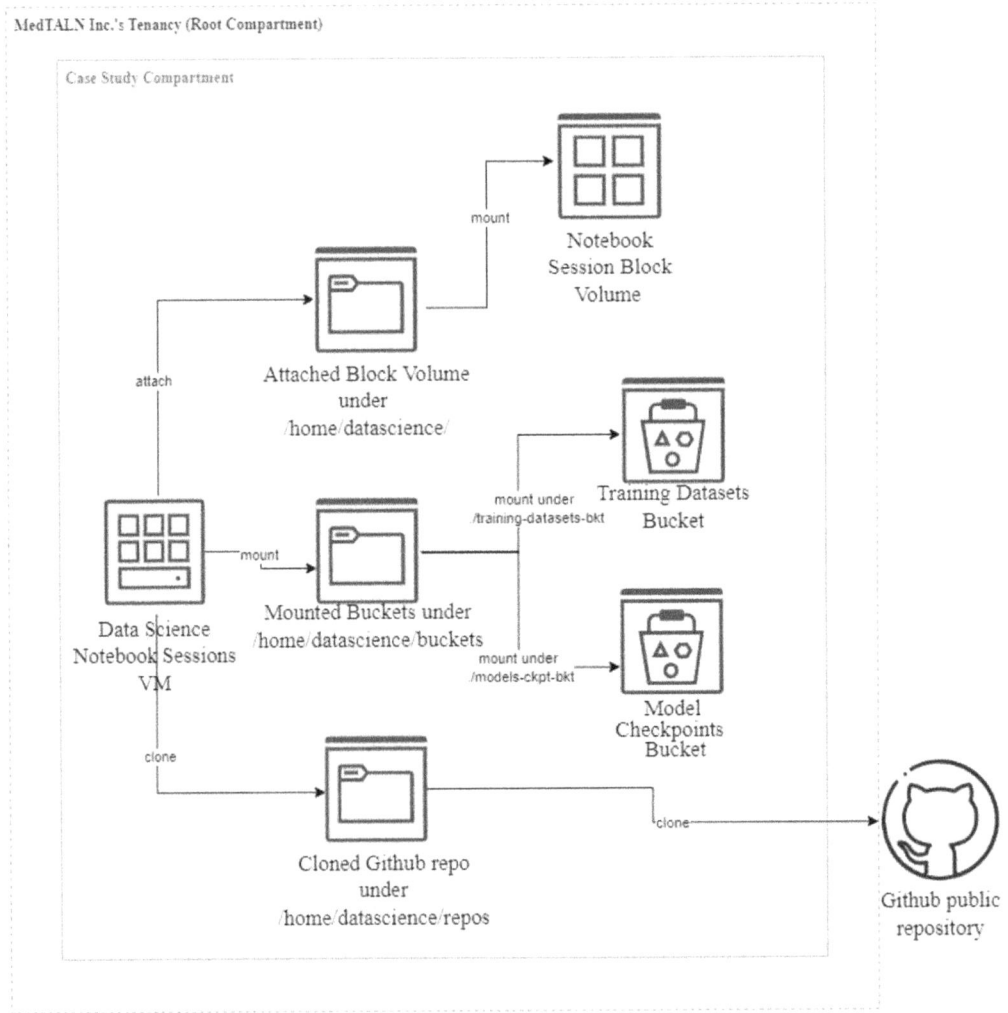

Figure 4-28. Notebook session's local directories from attached block volume, mounted buckets, and cloned Git repository

CHAPTER 4   TENANCY PREPARATION

Figure 4-28 shows how a data science notebook session storage-related out-of-the-box features simplify data and code access. The automatic attachment of block volumes ensures that any code, data, or files saved to the block volume will be preserved, even if the compute instance is shut down. The direct mounting of Object Storage buckets as local file system directories streamlines data access. At the same time, the cloning capabilities for public Git repositories enable easy access and version control of code repositories. These features, when combined, provide a seamless and powerful platform for data science teams to work together effectively.

- Automatic attachment of block volumes for persistent storage, accessible via JupyterLab, maintaining data between active sessions

  When you create a notebook session, a block volume is automatically attached and mounted in the directory /home/datascience. This directory is displayed as the root directory in the JupyterLab file explorer. If you deactivate the notebook session, the compute instance will be shut down, but the block volume will be retained. This ensures that any data, notebooks, and conda environments you save to the block volume will be saved, thus preserving your progress. However, please note that any data on the boot volume of the notebook session will be deleted upon deactivation.

- Direct mounting of Object Storage buckets as local file system directories within the notebook environment, streamlining data access

  The service allows for the mounting of up to two buckets that are configured during the "Mount Storage" step when activating a notebook session. This feature greatly simplifies our code by integrating Object Storage buckets as local file systems within OCI Data Science Notebook Sessions. We have chosen to mount the two buckets under the directory /home/datascience/buckets to ensure that they are easily accessible and visible from the JupyterLab file explorer.

- Cloning capabilities for public Git repositories, integrating code repositories into the notebook for easy access and version control

143

When a notebook session is activated, a public Git repository is cloned (if configured in the runtime configuration section). This process clones the remote repository under the local directory /home/datascience/repos. It eases access, collaboration, and version control among the data science team, thanks to the Git extension and JupyterLab interface integration.

Overall, these three features significantly enhance the functionality of OCI Data Science Notebook Sessions, making it easier for data scientists to work efficiently and collaboratively.

Caution Please be aware that any files saved on the notebook session block volume that have not been backed up will be permanently lost when the notebook session is deleted. This is because the compute instance will be terminated and the block volume will be destroyed. So, it's important to make sure that you have backed up all the necessary files before deleting the notebook session.

CPU-Based Notebook Session Setup

Regarding the setup of our Data Science environment, our NLP consultant, who was responsible for creating the Data Science Project, will also handle the creation of the project's notebook sessions. Here are the steps to create our first notebook session with a CPU-based compute shape:

1. Open the navigation menu, and go to Analytics & AI ➤ Machine Learning ➤ Data Science.

2. Select our case study compartment (i.e., *case-study-cmpt*). All projects in the compartment are listed.

3. Click the project we created earlier (i.e., *cs-nlp-prj*).

4. On the ***Project details*** page, click the ***Create notebook session*** button.

CHAPTER 4 TENANCY PREPARATION

5. In the ***Create notebook session*** dialog box, complete the following (Figure 4-29):

    a. Enter a unique name for the notebook session: *cs-nlp-nbs-cpu*.

    b. Keep the default Compute shape.

    c. Enter 50 for the block storage size or leave it empty to use the default value which is 100 GB.

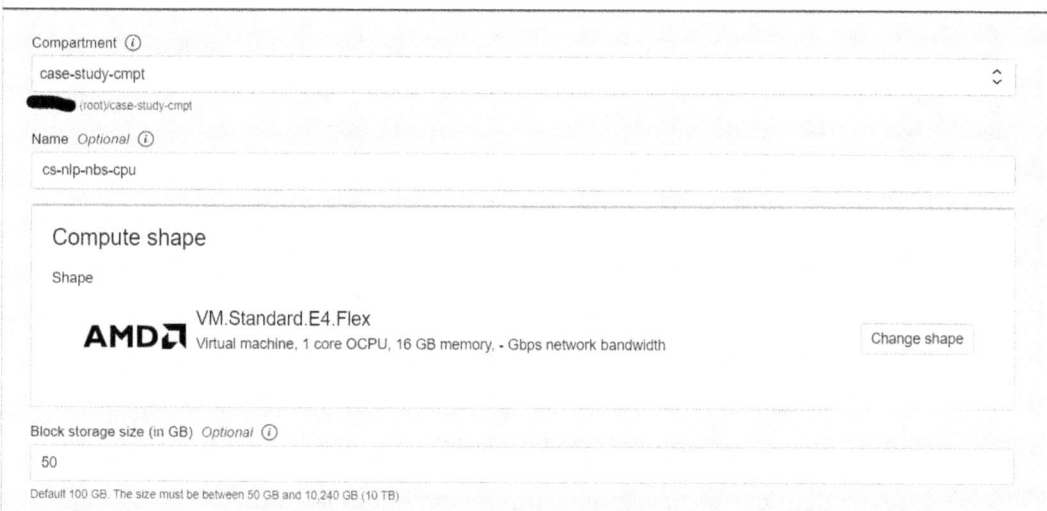

Figure 4-29. *Creating a CPU notebook session: part 1*

    d. To configure the network type, select Custom networking, and then select our case study VCN and its private subnet.

    VCN: cs-vcn

    Subnet: private subnet-cs-vcn

    e. Configure the endpoint type by selecting the option Public endpoint (Figure 4-30).

145

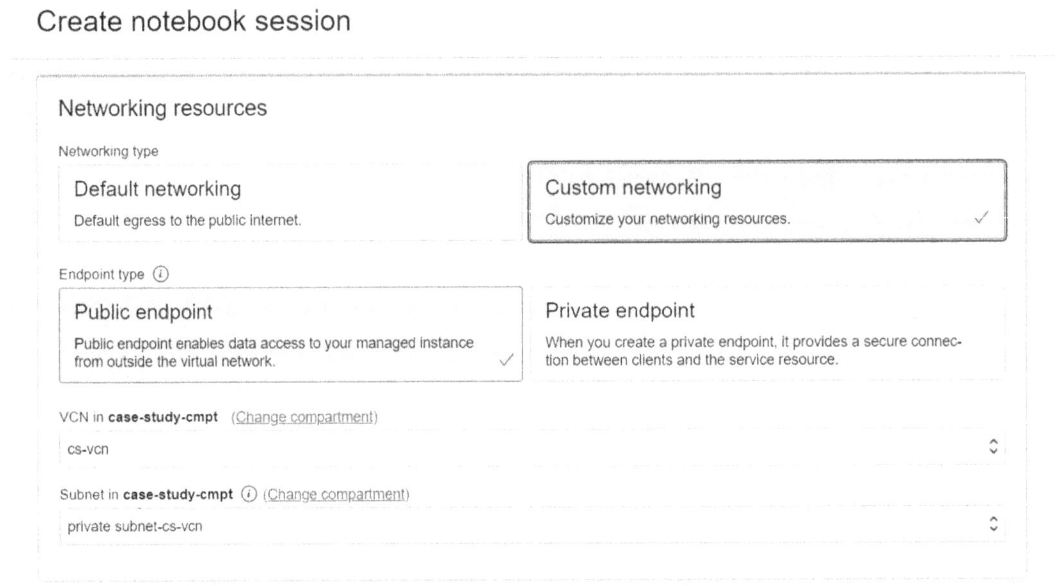

CHAPTER 4 TENANCY PREPARATION

Create notebook session

Figure 4-30. Creating a CPU notebook session: networking part

    f. To use storage mounts, click the button **+Add storage mount** (Figure 4-31).

    g. Select a storage mount type: OCI Object Storage.

    h. Select our case study compartment: case-study-cmpt.

       Select the first bucket we need to mount: training-datasets-bkt.

    i. Leave **Object Name Prefix** blank.

    j. For **Destination path and directory**, enter:

       /home/datascience/buckets/training-datasets-bkt.

    k. Click **Submit**.

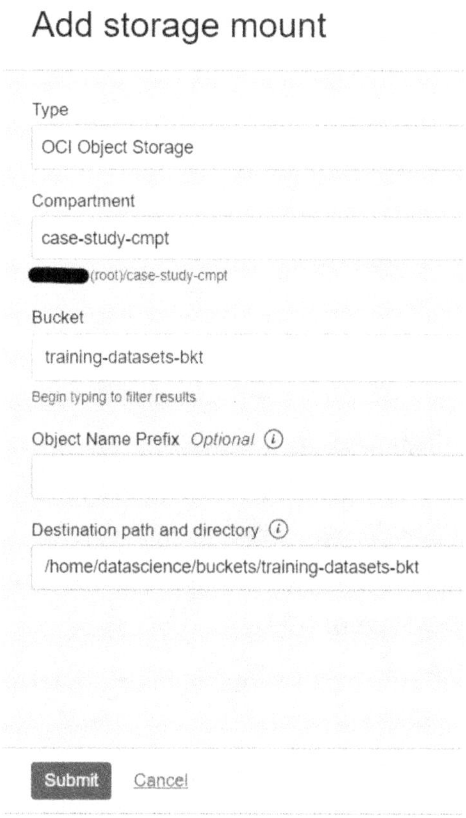

CHAPTER 4 TENANCY PREPARATION

Add storage mount

Type

OCI Object Storage

Compartment

case-study-cmpt

▇▇▇(root)/case-study-cmpt

Bucket

training-datasets-bkt

Begin typing to filter results

Object Name Prefix *Optional* ⓘ

Destination path and directory ⓘ

/home/datascience/buckets/training-datasets-bkt

Submit Cancel

*Figure 4-31. **Add storage mount** dialog box*

1. Repeat these steps to add our second storage mount for our notebook session (Figure 4-32):

   Bucket: models-ckpt-bkt

   Destination path and directory:

   /home/datascience/buckets/models-ckpt-bkt

Create notebook session

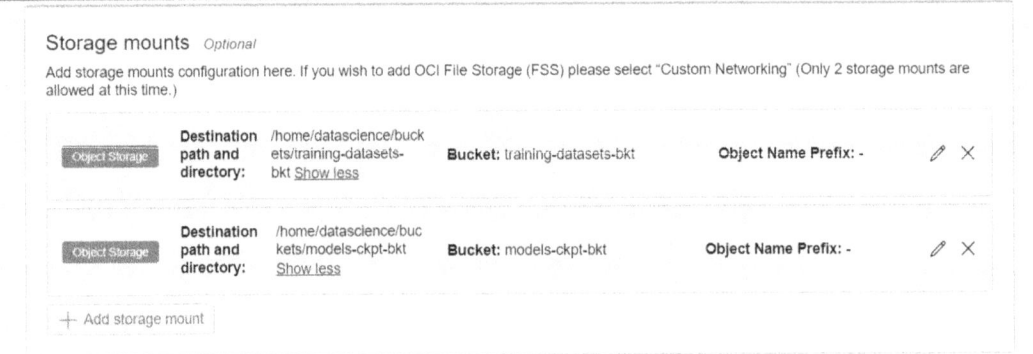

Figure 4-32. Our two Object Storage mounts

    m. Expand the Runtime configuration section to define our Git settings.

    n. Select the tab **Git settings**, and enter our case study public Git repository URL (Figure 4-33).

    Git repository URL: `https://github.com/john-doe-typica-ai/nlp-on-oci.git`

Note The source code for this book is available on GitHub via the book's product page, located at `www.apress.com/979-8-8688-1073-2`. Go to the repository GitHub and copy the repository URL.

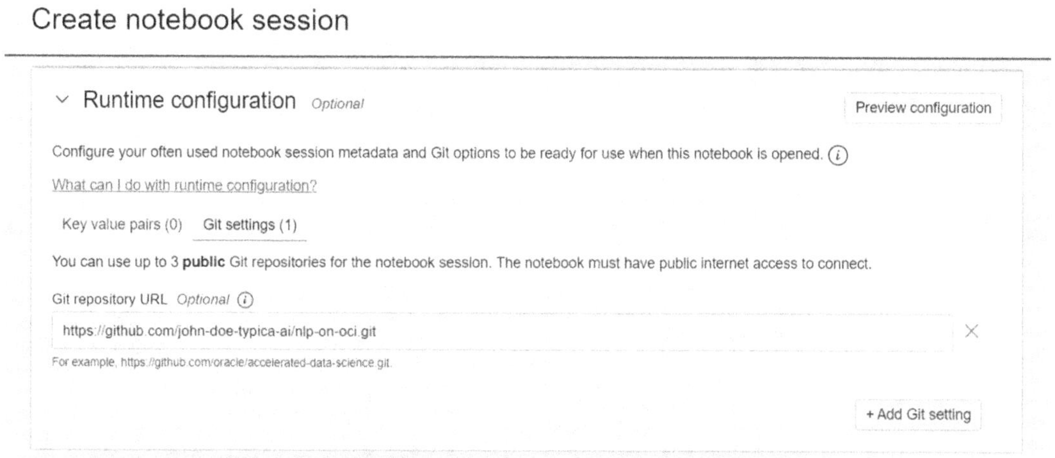

CHAPTER 4   TENANCY PREPARATION

Create notebook session

Figure 4-33. Notebook session Git settings

o. To view the details for the notebook session immediately after creation, select ***View detail page on clicking create***.

p. Click ***Create***.

6. The ***Notebook sessions*** page opens. When the notebook session is successfully created, as shown in Figure 4-34, the status turns to Active, and you can open the notebook session.

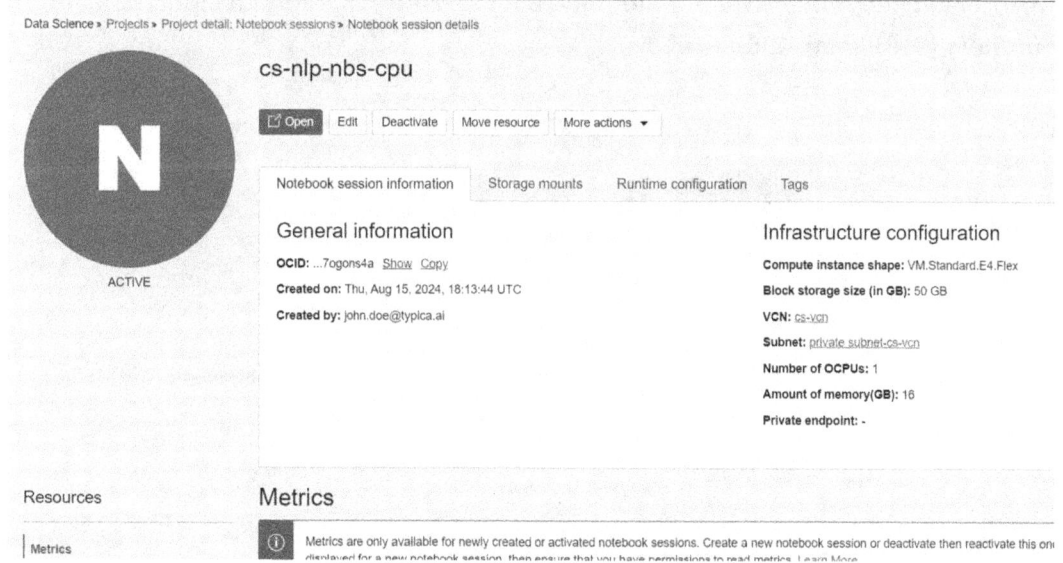

*Figure 4-34. CPU **Notebook session details** page*

149

CHAPTER 4 TENANCY PREPARATION

So far, we've discussed the steps required to create our first notebook session in OCI. Once the notebook session is created and active, you can access it by clicking **Open** and signing in with our NLP consultant credentials. This will bring you to the JupyterLab UI, as illustrated in Figure 4-35.

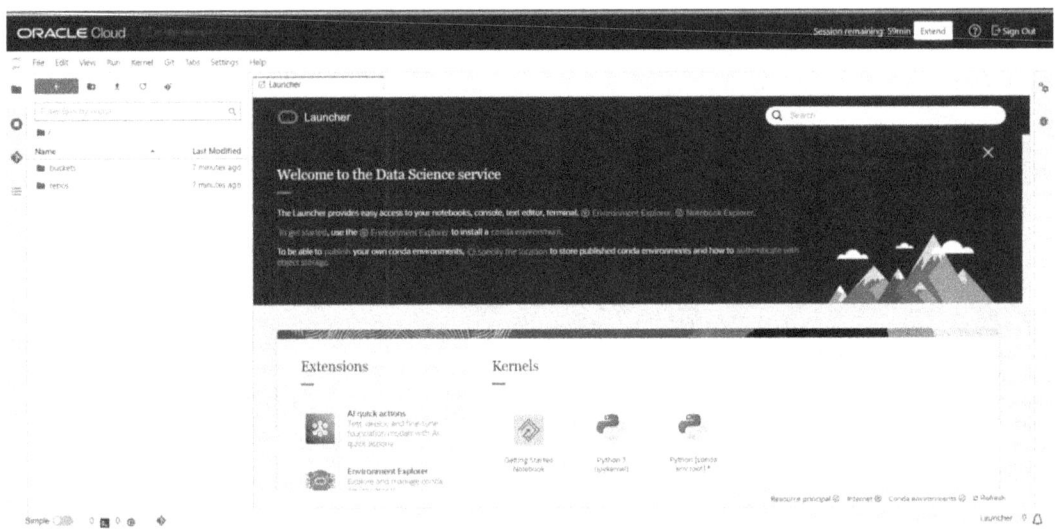

Figure 4-35. Notebook session JupyterLab UI

After launching the JupyterLab interface, we can see that the file browser tab displays two local folders: */buckets* for our mounted buckets and */repos* for our cloned GitHub repository (see Figure 4-36).

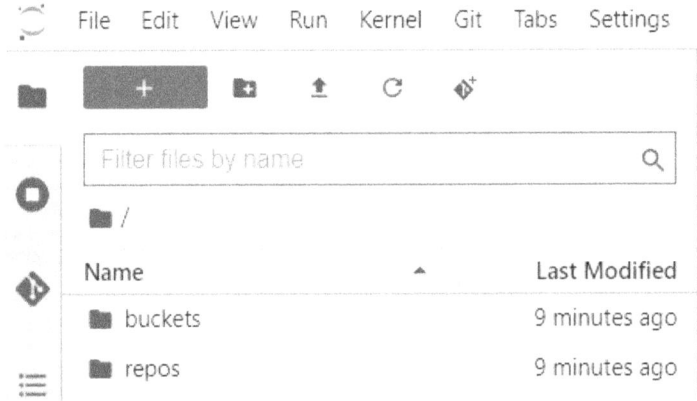

Figure 4-36. JupyterLab file browser

Conda Installation

Once the notebook session is created, the first step is to install the appropriate conda environment for our NLP project (explained in Figure 4-37). Conda environments are used to bundle Python dependencies in the notebook sessions. To develop our NLP project, we need to install a prebuilt PyTorch[2] for Python conda environment.

However, before we dive into the process of installing the conda environment on our CPU-based notebook session, let's take a moment to understand what PyTorch is. PyTorch is an open source deep learning library for Python, primarily developed by Facebook's AI research team. It is widely used for deep learning projects, including Transformer-based NLP projects. Moreover, PyTorch provides excellent support for GPUs, which is crucial for training our Healthcare NER model.

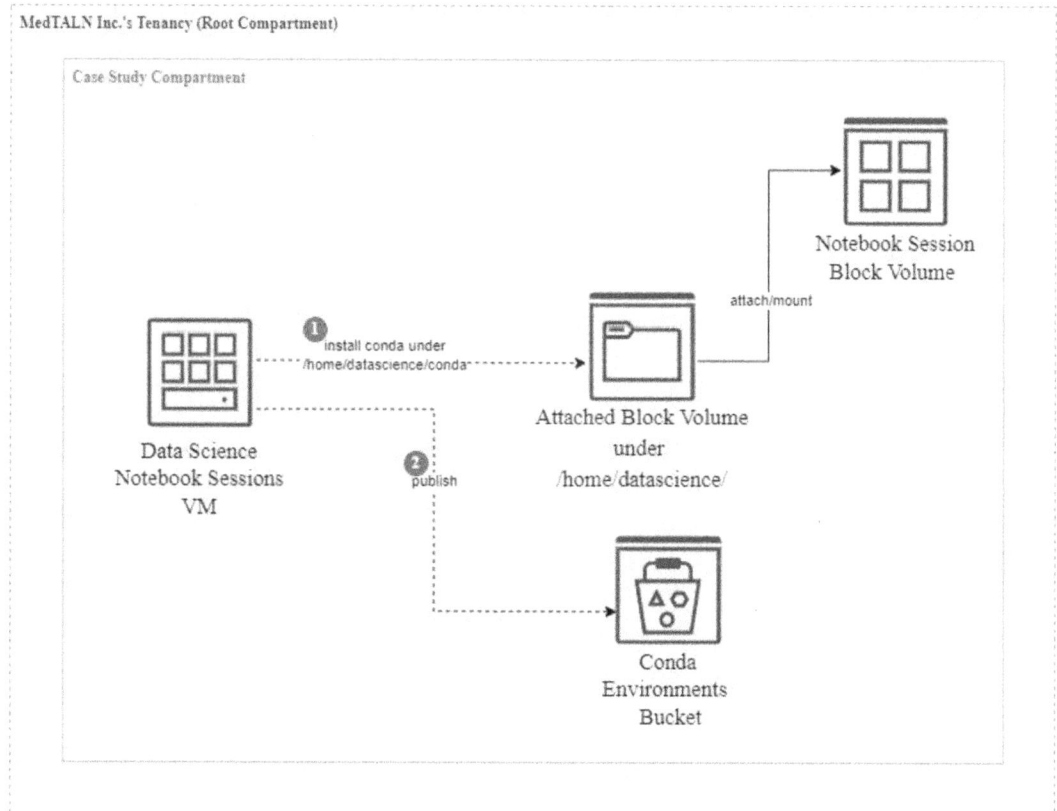

Figure 4-37. *Install and publish conda env*

---

[2] For more information on PyTorch, visit https://pytorch.org

CHAPTER 4 TENANCY PREPARATION

Steps to install and publish the appropriate conda env. for our case study are as follows:

1. In the **Launcher** tab, click ***Environment Explorer***.

2. In the ***Environment Explorer*** search box, enter PyTorch. Multiple PyTorch conda envs. are found (Figure 4-38).

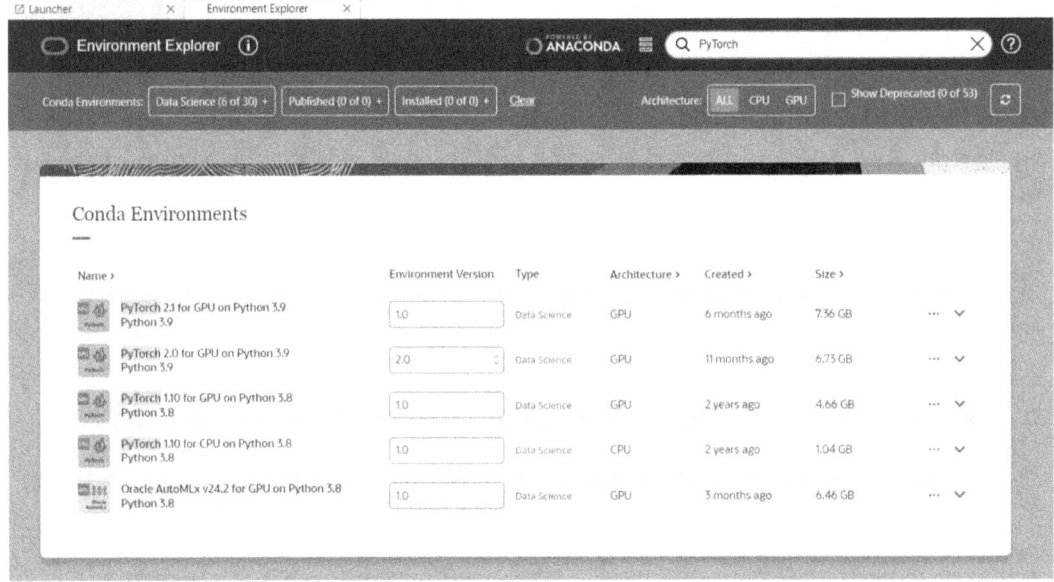

Figure 4-38. *PyTorch available conda envs.*

3. Select the most recent one with the latest library version, i.e., PyTorch 2.1 for GPU on Python 3.9. This conda env. contains the latest Transformer libraries (v4.37.2) and oracle-ads version v2.10.0 (see Figure 4-39).

CHAPTER 4 TENANCY PREPARATION

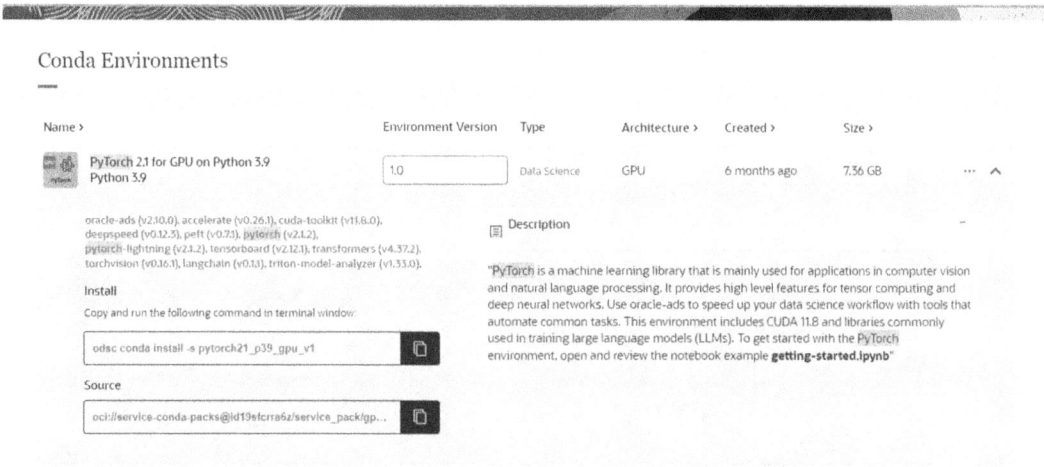

Figure 4-39. Selected PyTorch conda env. for our case study

4. To install the conda env.

   a. Copy the command line:

   ```
   odsc conda install -s pytorch21_p39_gpu_v1
   ```

   b. Open a new terminal (from the menu, select File ➤ New ➤ Terminal). Then, paste the copied install command in the terminal and hit enter (Figure 4-40):

   odsc conda install -s pytorch21_p39_gpu_v1

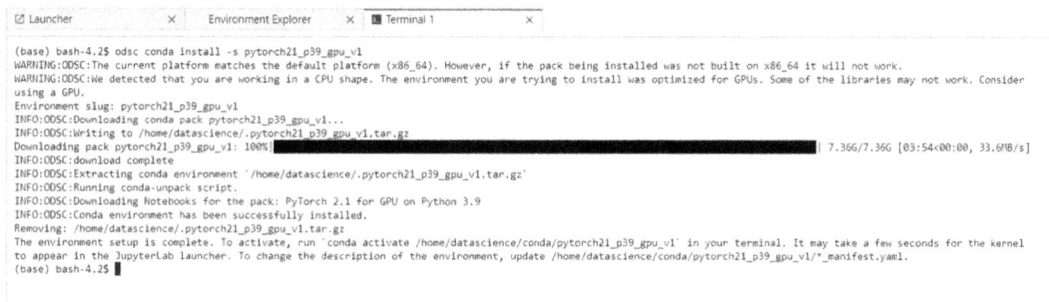

Figure 4-40. Conda env. installation from terminal

5. The new conda env. is installed in a block volume under the folder */home/datascience/conda* (see Figure 4-41).

153

Note Since all the installed conda envs. are stored on the block volume under /home/datascience, we don't need to reinstall them after deactivating the notebook session.

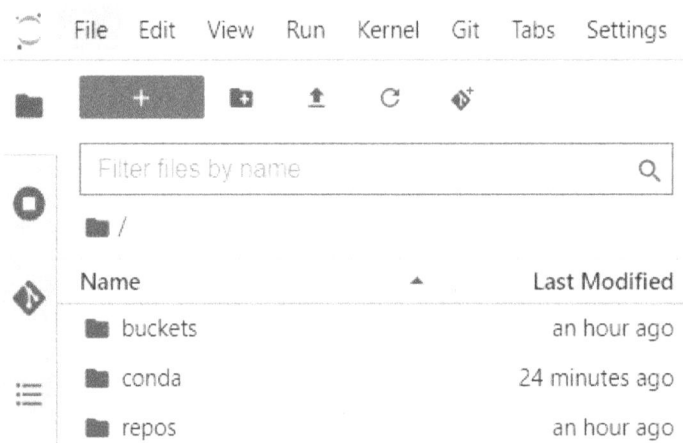

Figure 4-41. *Conda env. installation dir*

After installing our PyTorch conda env., we can notice as shown in Figure 4-42 that a new kernel for this particular conda env. is available in the JupyterLab **Launcher** tab in the Notebook category. You can start working in that conda environment by clicking the ***Environment*** kernel icon to open a new tab to open a new notebook file.

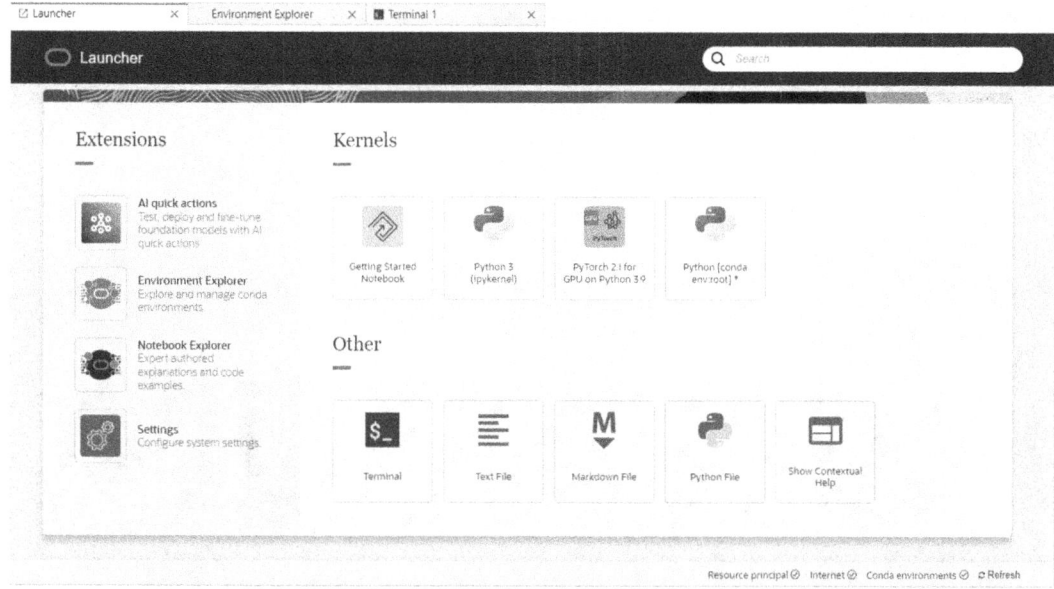

Figure 4-42. *New kernel for PyTorch conda*

We have successfully set up our PyTorch conda environment. This means we can now use it to run our Python scripts and notebooks. However, before we start using it, we need to publish it to a dedicated bucket for storing conda environments. Publishing a conda environment is useful when we need to include third-party dependencies that are not included in the prebuilt conda environment by default. It also helps when we want to share the conda environment with colleagues or across different notebook sessions or assign it as a runtime environment for model deployment.

To publish our newly installed conda env., the steps are as follows (refer to Figure 4-43):

1. Configure *odsc conda* to use an Object Storage bucket *conda-envs-bkt* using this command (replace yz2wwgkgt8eh with your Object storage namespace):

   ```
   odsc conda init -b conda-envs-bkt -n yz2wwgkgt8eh
   ```

2. Publish the conda env. to our bucket by running the following command line:

   ```
   odsc conda publish -s pytorch21_p39_gpu_v1
   ```

CHAPTER 4 TENANCY PREPARATION

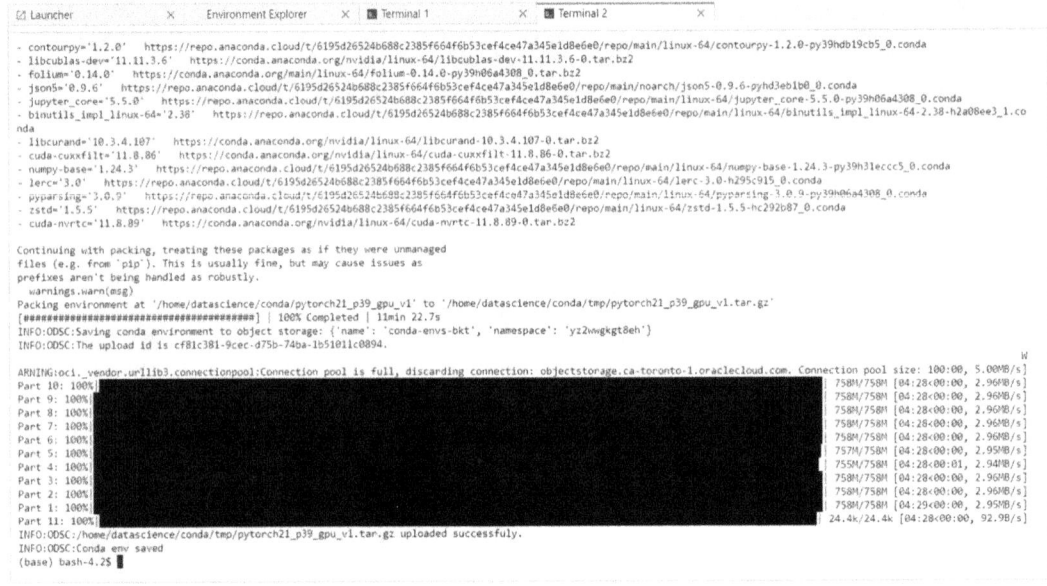

Figure 4-43. Publishing conda env.

Once the publishing process is done, we can go to our Object Storage bucket in the OCI console and confirm that our published conda pack is stored in the bucket as illustrated in Figure 4-44.

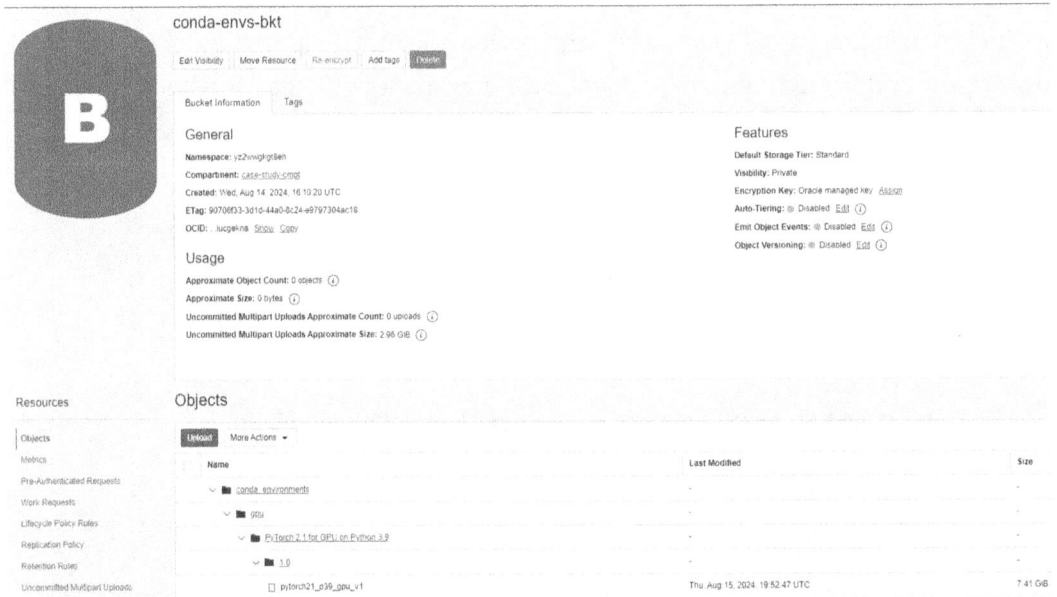

Figure 4-44. Published conda stored in the bucket

CHAPTER 4 TENANCY PREPARATION

In the *Environment Explorer* extension of the notebook session, we can also list and inspect all the condas that we have installed and published to a shared Object Storage bucket.

As illustrated in Figure 4-45, clicking the *Published Conda Environments* tab in the *Environment Explorer* lists all the conda environments that have been published. This feature is particularly helpful when team members are using the same bucket to publish their conda environments, as it allows us to view and access the conda environments that our colleagues have installed, created, and shared. It is also a great way to archive and share environments across multiple notebook sessions.

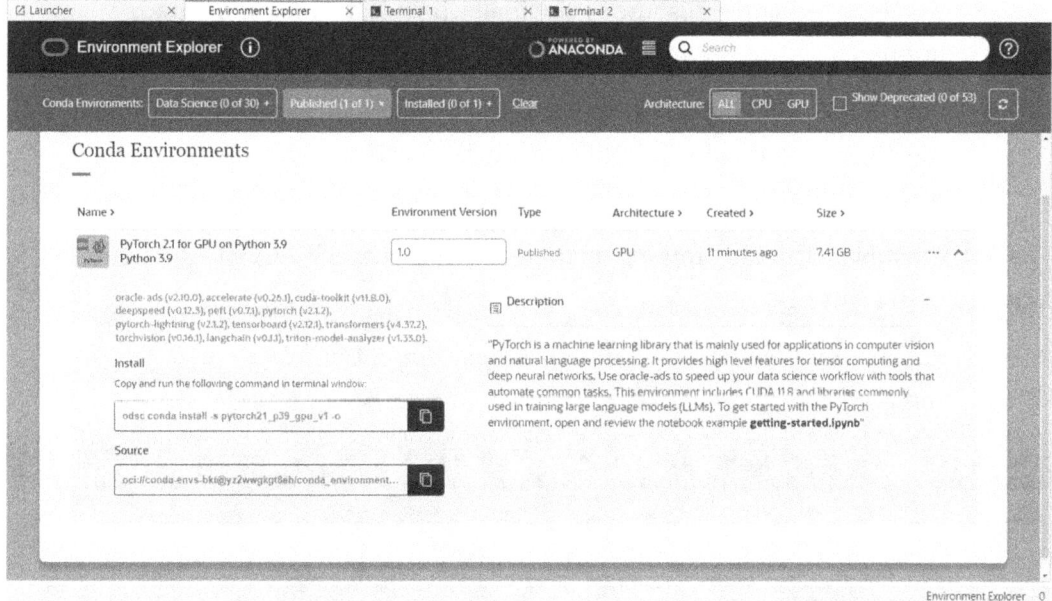

Figure 4-45. *Published conda envs.*

You can notice that the conda env. source (i.e., Object Storage path) changed from the default location to our bucket (replace yz2wwgkgt8eh with your Object storage namespace):

`oci://conda-envs-bkt@yz2wwgkgt8eh/conda_environments/gpu/PyTorch 2.1 for GPU on Python 3.9/1.0/pytorch21_p39_gpu_v1`

We are now ready to perform the setup check to ensure successful completion of the notebook session setup.

CHAPTER 4   TENANCY PREPARATION

Setup Check

We are now prepared to perform a setup check to ensure that the OCI tenancy preparation for our project has been completed successfully. We will use Python code to determine the readiness of our notebook session by performing a series of essential checks, including

- GPU Verification: Confirms whether or not this notebook session has a GPU attached

- Object Storage Authentication: Validates authentication using the notebook session's Resource Principal, which is the recommended approach

- Object Creation in Bucket: Tests the ability to create a dummy text file object in our Labeling Datasets Bucket, verifying both access and functionality.

From the JupyterLab file browser, open the notebook *check_setup.ipynb* under the folder (refer to Figure 4-46):

/repos/john-doe-typica-ai/nlp-on-oci.git/chapt-4/check_setup.ipynb

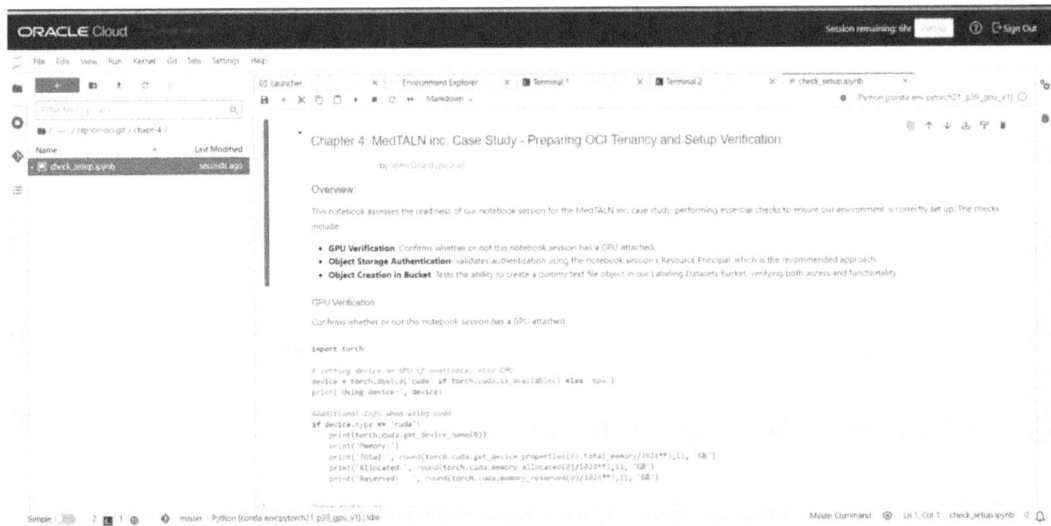

Figure 4-46. *Notebook for setup check*

158

CHAPTER 4 TENANCY PREPARATION

First, we check whether the current notebook session has a GPU attached by running the code in Listing 4-3. The current conda environment supports GPU (CUDA) but is only using the CPU. This is useful for checking the setup for the upcoming GPU-based notebook session.

This code checks if a GPU is available using *torch.cuda.is_available()*. If available, it sets the "device" variable to "cuda," otherwise to "cpu." A message indicates whether PyTorch will be using "cuda" or "cpu."

Listing 4-3. GPU verification

```
import torch

# setting device on GPU if available, else CPU
device = torch.device('cuda' if torch.cuda.is_available() else 'cpu')
print('Using device:', device)

#Additional Info when using cuda
if device.type == 'cuda':
    print(torch.cuda.get_device_name(0))
    print('Memory:')
    print('Total:', round(torch.cuda.get_device_properties(0).total_
    memory/1024**3,1), 'GB')
    print('Allocated:', round(torch.cuda.memory_
    allocated(0)/1024**3,1), 'GB')
    print('Reserved:   ', round(torch.cuda.memory_
    reserved(0)/1024**3,1), 'GB')
```

Output

```
Using device: cpu
```

As expected, the output confirms that this notebook session uses a CPU-based shape only.

Next, execute the code in Listing 4-4, to initialize Object Storage service client and validate authentication using the notebook session's Resource Principal.

Listing 4-4. Object Storage authentication

```
import oci
# Initialize OCI Object Storage Client with notebook session's resource principal
signer = oci.auth.signers.get_resource_principals_signer()
object_storage_client = oci.object_storage.ObjectStorageClient(config={}, signer=signer)
object_storage_client
```

Output

```
<oci.object_storage.object_storage_client.ObjectStorageClient at 0x7fb8554f9b20>
```

Authentication, distinct from authorization, verifies the identity of a user recognized by OCI, enabling them to perform specific actions. The IAM Resource Principals feature allows OCI resources, such as notebook sessions or jobs, to act as authorized entities, interacting with other OCI services securely. Resource principals authenticate using internally managed certificates, eliminating the need for external credential storage and manual rotation.

The Data Science Service leverages resource principals, offering a secure and convenient method for your data science environments to authenticate and interact with other OCI resources. This method is particularly advantageous over the traditional OCI configuration and API key approach, especially in noninteractive environments like job runs where managing configuration files is impractical.

Note It's important to note that resource principal tokens are cached for a duration of 15 minutes. Any changes to policies or dynamic groups will only take effect after this cache period expires.

Code, in Listing 4-5, tests the authorization policies by creating a dummy text file object in the Labeling Datasets Bucket.

Listing 4-5. Dummy file creation in bucket

```
# Initialize Object Storage bucket infos
namespace = object_storage_client.get_namespace().data
bucket_name = "labelling-datasets-bkt"

# Initialize Object Storage object's name and body
object_name = "dummy.txt"
object_data = b"Dummy text file for setup check - to be deleted."

#create the dummy file object in the bucket
obj = object_storage_client.put_object(
    namespace,
    bucket_name,
    object_name,
    object_data)

# Check the creation of the dummy file in the bucket
list_objects_response = object_storage_client.list_objects(
    namespace_name=namespace,
    bucket_name=bucket_name,
    fields="timeCreated"
)

# Get the data from response
print(f"File named {list_objects_response.data.objects[0].name} created on {list_objects_response.data.objects[0].time_created}")
```

Output

```
File named dummy.txt created on 2024-08-15 20:07:17.629000+00:00
```

Successful execution of the cell confirms the creation of our test file in the Object Storage bucket. To double-check, we can navigate to the bucket to confirm its existence (Figure 4-47).

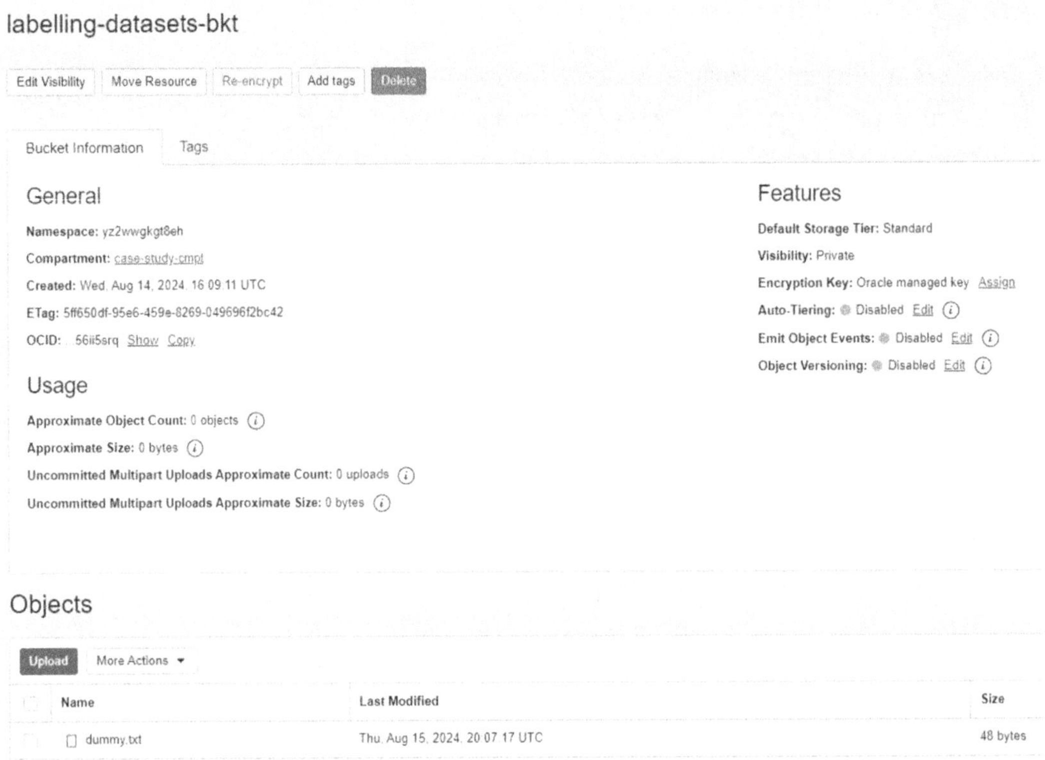

Figure 4-47. *Dummy file in the bucket*

We have finished setting up a notebook session that runs on CPU. The next section will explain the steps to create a notebook session with a GPU attached. Although the process is quite similar to setting up a CPU-based configuration, we will highlight some key differences to ensure a successful setup process for a GPU-based notebook session.

GPU-Based Notebook Session

As we transition into the model training phase of our NLP case study, the computational demands escalate significantly. To accommodate the intensive processing requirements of deep learning algorithms, we embark on creating a GPU-based notebook session. This section is dedicated to guiding you through the process of setting up a notebook session optimized for training our bespoke Healthcare NER model. Utilizing a GPU configuration is not merely a choice but a necessity for handling the complex computations and large datasets characteristic of NLP tasks efficiently. This introduction sets the stage for the detailed steps that follow, ensuring you have a robust environment ready to tackle the challenges of model training within our project's framework.

For our GPU-based notebook session, we've chosen the VM.GPU.A10 shape (also known as VM.GPU.GU1). This decision reflects a strategic consideration, offering an excellent balance between performance and cost. The VM.GPU.A10 shape is particularly well suited for our needs—training models as required without incurring unnecessary expenses. This is because once model training completes, we can deactivate the notebook session, effectively stopping the VM instance behind it. A key advantage of the VM.GPU.A10 series is that billing pauses when the instance is stopped, though the instance still counts toward our service limits. This functionality is critical for managing costs efficiently, especially for workloads like ours, where model training is intermittent.

> *Dense I/O, GPU, and HPC Shapes: For most shapes, billing continues for stopped instances because the attached GPU and NVMe local storage resources are preserved. To halt billing, you must terminate the instance. For shapes in the VM.GPU.GU1 series (also named the VM.GPU.A10 series), billing is paused for stopped instances.*
>
> —Section "How am I billed for instances?"[3]

When preparing to use a Data Science notebook session with GPU shapes, such as VM.GPU.A10.1, it's crucial to understand and manage your resource limits within your OCI tenancy. If the default limit for GPUs is set to 0, attempting to create a notebook session with a GPU shape would lead to an error message indicating a temporary reduction in resource creation capacity. The error message advises filing a service limit increase request with OCI support to restore your resource creation capabilities.

> *Your resource creation has been temporarily reduced. To unblock resource creation, please file a service limit increase request so that our support team can assist with restoring your resource creation capability.*
>
> —**Create notebook session** dialog box

To successfully create a Data Science notebook session with a GPU shape, you must open a service request (SR) to request an increase for the specific resources required. In the case of the VM.GPU.A10.1 shape, you should specify an increase for the Resource and Limit Name: VM.GPU.GU1 and ds-gpu-a10-count (Figure 4-48).

---

[3] Oracle official FAQ: https://www.oracle.com/ca-en/cloud/compute/faq/

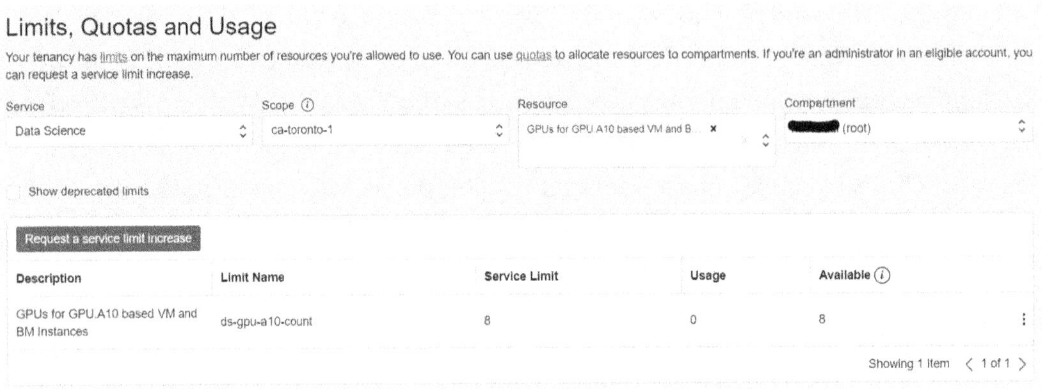

CHAPTER 4   TENANCY PREPARATION

Figure 4-48. *Limits for GPU.A10 resources in the case study tenancy*

You can check your current resource limits through the OCI console. For example, in our case study tenancy, the console reveals that the service limit for GPUs, particularly for GPU.A10-based VM and BM instances, is set to 8. This demonstrates that, unlike the default limit of 0, our tenancy already has a specified capacity of GPU resources allocated.

Launching a GPU-enabled notebook session (or reactivating a deactivated notebook session) within our case study tenancy will dynamically update the limits for the GPU. A10 resource in real time. As the screenshot below shows, the **Usage** has increased by one, which, in turn, has reduced the **Available** count from the total service limit, demonstrating the current allocation and remaining GPU resources (Figure 4-49).

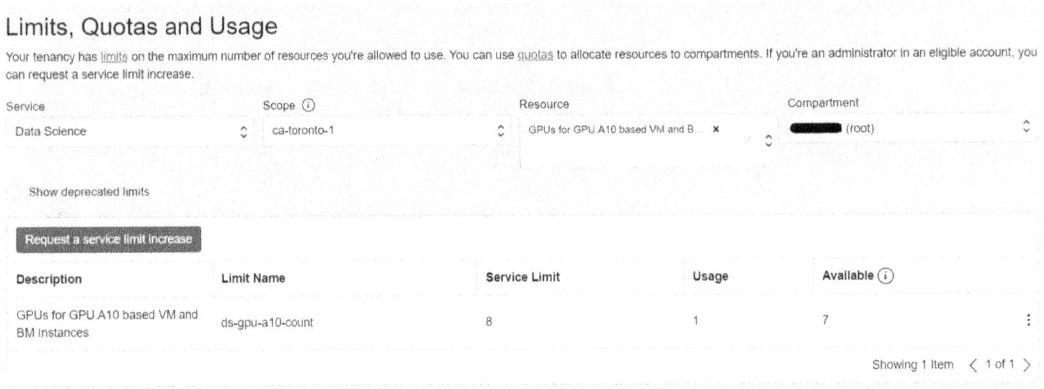

Figure 4-49. *Real-time update for our case study GPU.A10 limits*

CHAPTER 4 TENANCY PREPARATION

Now, let's create our second notebook session with GPU support. Here are the steps:

1. From the project *cs-nlp-prj* details page, click the **Create notebook session** button.

2. Create a new notebook session the same as the CPU-based one, with a few changes (Figure 4-50):

    a. Use the following name for this new GPU notebook session: *cs-nlp-nbs-gpu*.

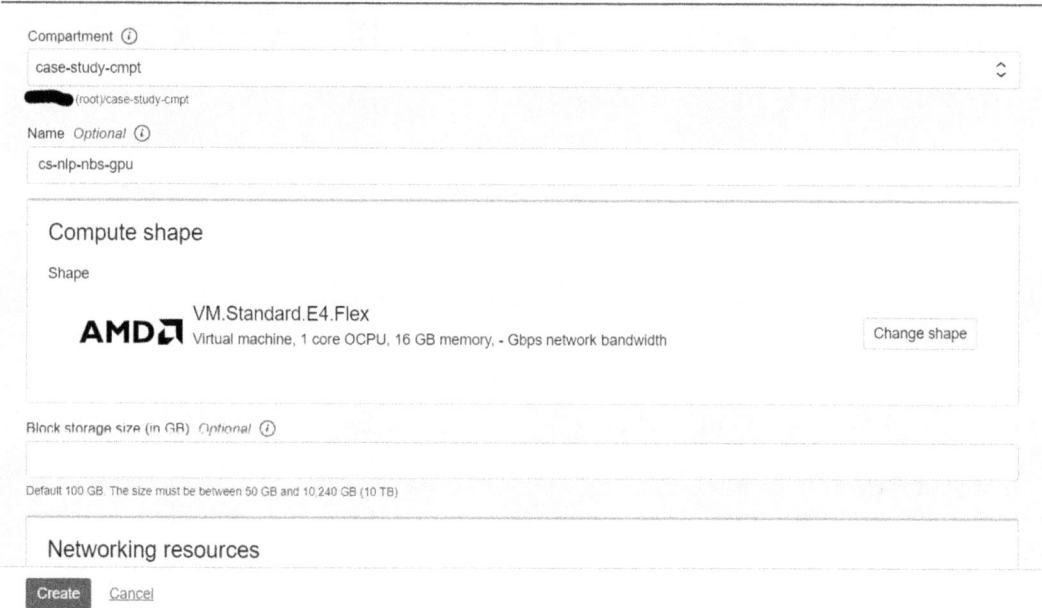

Figure 4-50. *GPU notebook session creation dialog box*

    b. In the **Compute shape** section, click the button **Change Shape** to select a GPU shape: VM.GPU.A10.1 (Figure 4-51).

165

CHAPTER 4 TENANCY PREPARATION

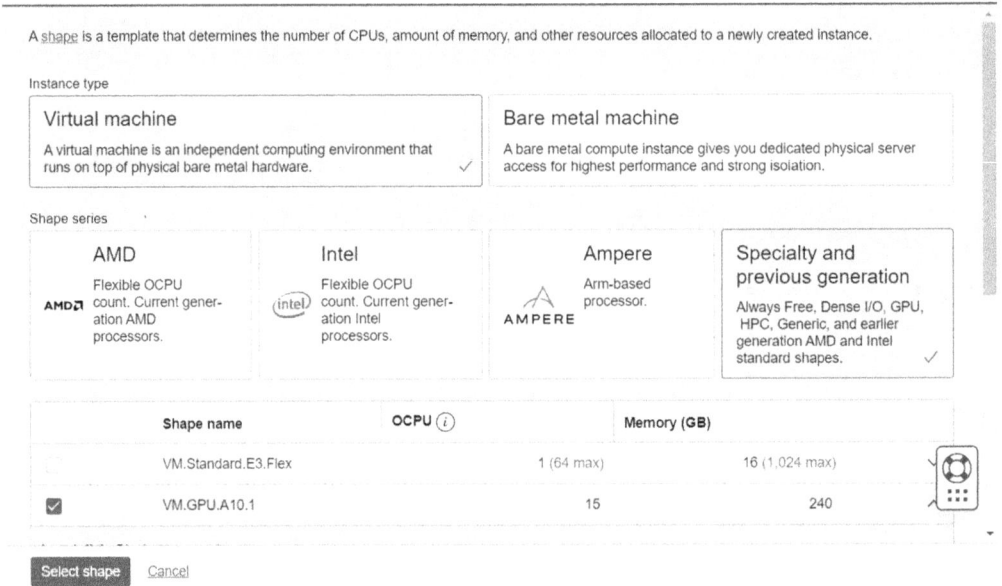

Figure 4-51. GPU compute shape selection

After selecting the Compute shape, select a GPU shape: VM.GPU.A10.1 (Figure 4-52).

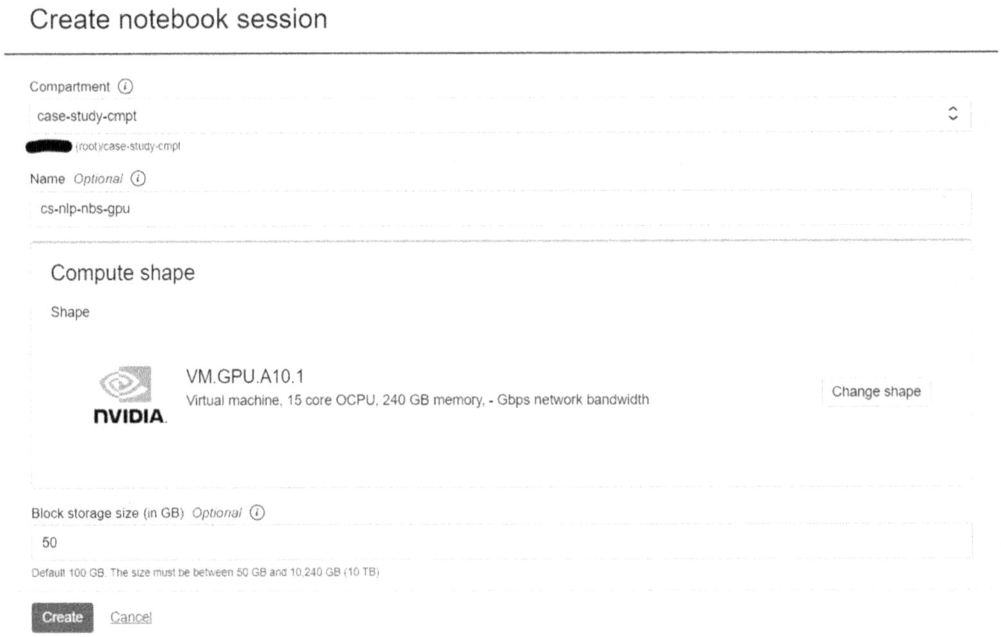

Figure 4-52. GPU notebook session creation dialog box with GPU shape

CHAPTER 4 TENANCY PREPARATION

c. Continue with Networking, Storage Mounts, and Git settings the same way as we did for our previous notebook session (Figure 4-53).

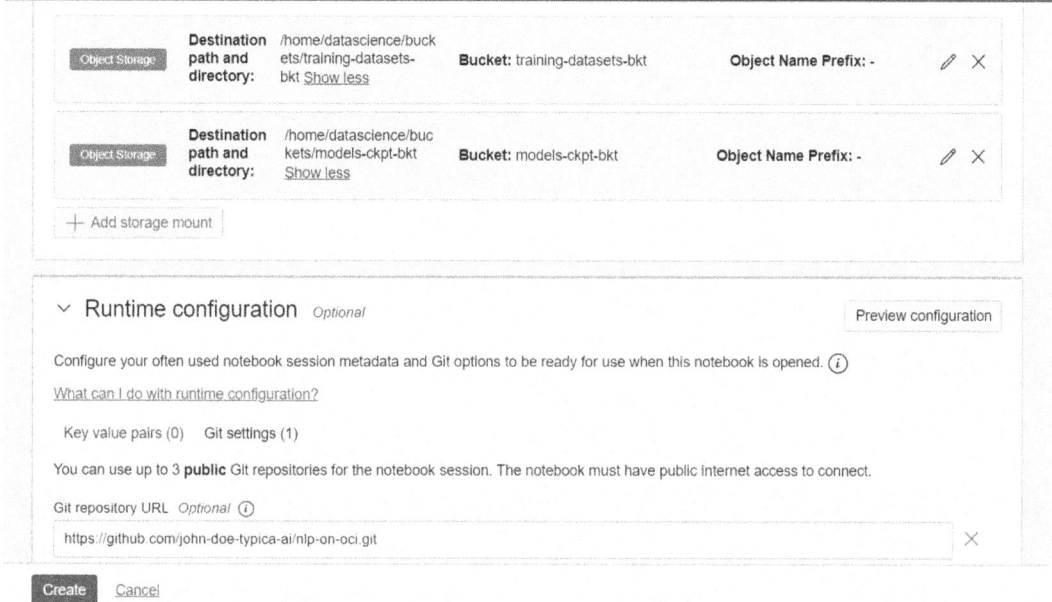

Figure 4-53. Storage mount and runtime configuration for the GPU-based notebook session

d. Click **Create**.

3. On the **Notebook session details** page, shown in Figure 4-54, you can confirm that the notebook session is active and the compute instance shape is VM.GPU.A10.1.

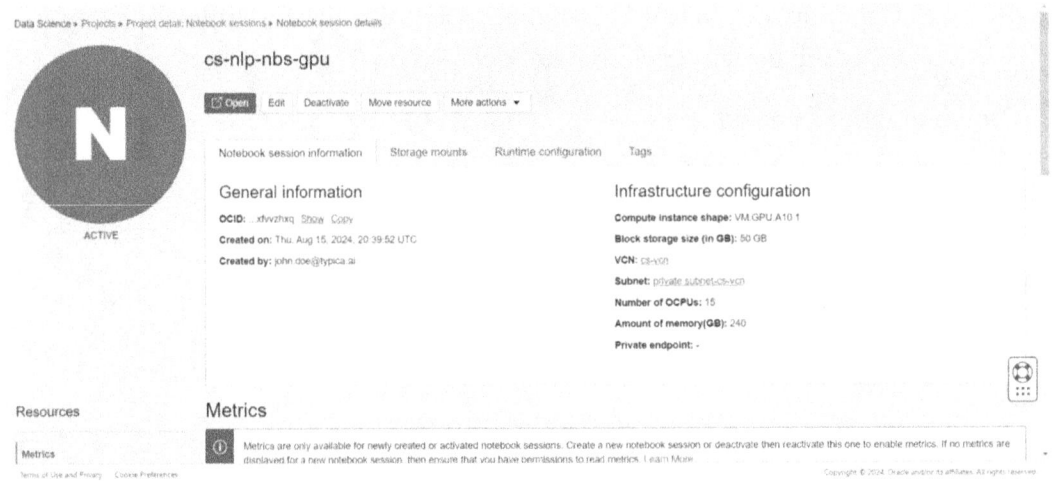

Figure 4-54. GPU Notebook session details page

Note It is important to note that OCI Data Science only supports certain compute shapes for notebook sessions. Additionally, it's worth mentioning that the compute shapes supported are not available in all regions. For instance, as of the time of writing this book, in Canada, the GPU-based shape "VM.GPU.A10.1" is only available in the OCI region "Canada Southeast (Toronto)" and not in the OCI region "Canada Southeast (Montreal)."

We have completed the necessary steps to create our GPU notebook session. Next, we will log in as John Doe (NLP consultant) and proceed with the following postcreation setup steps for the notebook session:

1. Install a Conda Environment: Repeat the same steps done for the CPU-based notebook session to install the conda env. *PyTorch 2.1 for GPU on Python 3.9*.

2. Perform the Setup Check: Repeat the same steps done for the CPU-based notebook session.

Based on Figure 4-55 we confirmed our current GPU notebook session has a GPU attached, and the PyTorch library installed via the conda environment can use the GPU (CUDA). As anticipated, the output confirms the presence of a GPU in this notebook session. Using `torch.cuda.get_device_name(0)`, we were able to retrieve the

CHAPTER 4 TENANCY PREPARATION

name of the GPU, which is *NVIDIA A10*. Additionally, we found out that the total GPU memory available for use is 22 GB by using the command `torch.cuda.get_device_properties(0).total_memory`.

Figure 4-55. *GPU verification*

There is an easy way to check if the notebook session has an attached GPU. You can use the following command from the notebook session terminal, as illustrated in Figure 4-56:

`nvidia-smi`

This command line utility is designed to monitor and manage NVIDIA GPU devices, making it one of the simplest methods for identifying the presence of a GPU.

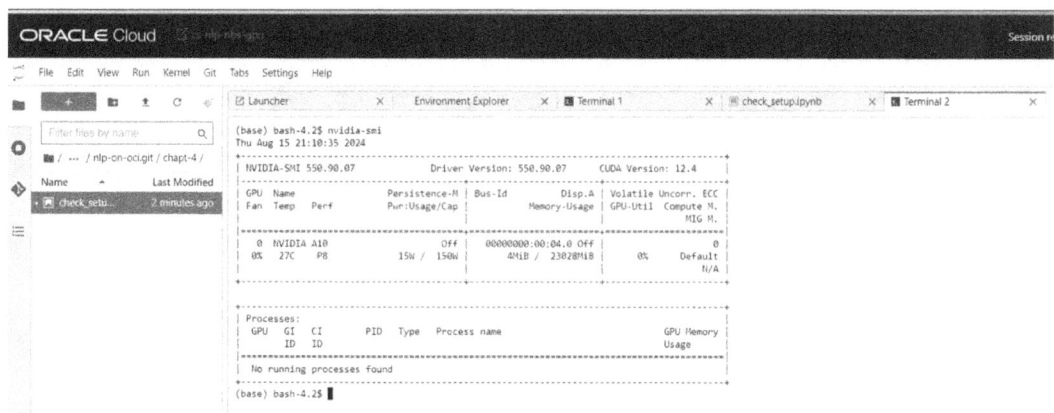

Figure 4-56. *nvidia-smi on GPU notebook session*

Summary

In this chapter, we progressed from initial configuration to programmatically assessing notebook readiness. We examined OCI tenancy preparation, including identity, security, network, and storage configurations. We also meticulously configured the data science environment, covering CPU- and GPU-based notebook sessions.

Key Takeaway   OCI Data Science Notebooks provide significant cost savings through intelligent resource management. By allowing deactivation during idle periods, these notebooks pause billing when not in use. This feature is especially beneficial for GPU-intensive tasks like model training, where workloads are often intermittent. You can activate resources for training and then deactivate them upon completion, effectively stopping all associated costs. This approach ensures you only pay for actual compute time used, maximizing cost efficiency without compromising on performance for your Data Science Projects.

In Chapter 5, we will learn how to build a training dataset. The process will be explained in a step-by-step manner, starting with data collection, followed by data preparation and creating a data labeling dataset. Next, we will label the dataset using data labeling tools and export it as a CoNLL dataset, which is ready for model training.

CHAPTER 5

Dataset Preparation

This chapter highlights the importance of creating high-quality yet cost-effective training datasets.[1] It also elucidates the curating of a dataset for NLP downstream tasks and explains the influence of high-quality annotations on the model's training effectiveness.

This chapter provides a comprehensive guide on preparing a training dataset for NLP tasks, such as Healthcare Named Entity Recognition (NER), utilizing publicly available and open-for-use community-curated datasets from Hugging Face. These preexisting datasets can be tailored using OCI ML Services, i.e., OCI Data Science and OCI Data Labeling. The chapter covers all aspects of the process, from defining the problem to selecting prelabeled datasets, managing and annotating the data, and finally creating the training-ready dataset.

The dataset preparation techniques outlined in this chapter are not limited to the practical example; they can be confidently applied to other NLP downstream tasks. Utilizing and enhancing a community-curated dataset from Hugging Face is an effective strategy that aligns with modern data science practices for assembling cost-efficient NLP datasets.

Preliminaries

In the "Preliminaries" section, we'll give an overview emphasizing the importance of labeled datasets in NLP-supervised learning tasks, e.g., Healthcare Named Entity Recognition (NER). We'll then discuss practical and effective cost-saving strategies, providing insights and methods that you can readily implement to reduce the expenses associated with dataset creation by utilizing community-curated datasets.

---

[1] Please note that the terms "training dataset" and "training data" are used interchangeably throughout the book and both refer to training, validation, and evaluation data.

CHAPTER 5 DATASET PREPARATION

In the "Dataset Life Cycle" section, we'll describe activities, from defining the problem to selecting suitable prelabeled datasets from trusted sources such as Hugging Face. This involves evaluating the quality and relevance of the dataset to the task at hand and ensuring it aligns with the project's requirements. Once a suitable dataset is identified, we'll discuss the process of preparing it for use, which may involve cleaning, balancing, and enriching the data. This will enable us to clarify all the important concepts before starting the actual work of dataset preparation using OCI ML Services.

Labeled Datasets

Hand-labeled data is crucial for supervised learning, much like teaching a child. The accuracy and quality of what they learn directly affect their subsequent actions and knowledge. Similarly, NLP models rely on the data they're trained on.

When building an NLP model for a downstream task such as Named Entity Recognition (NER), it is important to understand that the training phase significantly determines the model's effectiveness. However, it is equally important to note that the dataset's quality is essential for the model's success. Large, diverse, and well-labeled datasets are necessary for transfer learning to achieve optimal model performance.

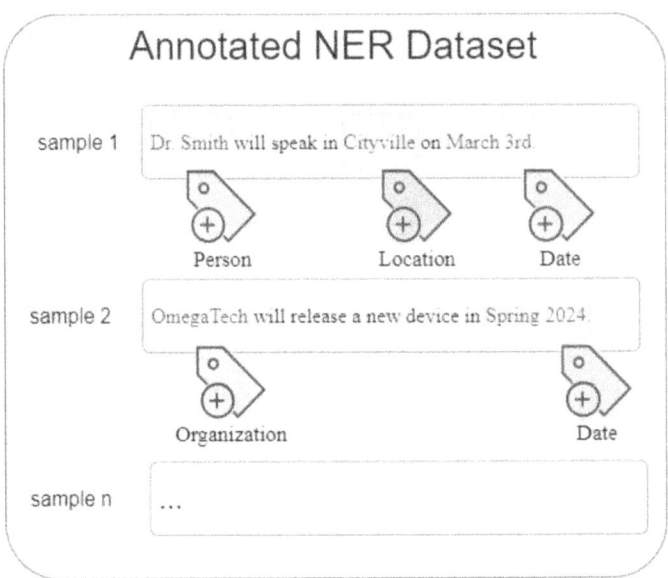

Figure 5-1. *Toy example of an annotated NER dataset*

To train models for NLP tasks, we need a labeled dataset, as illustrated in Figure 5-1 for the NER task. The size of the dataset is a key factor that determines the model's accuracy. Generally, larger datasets result in better performance, as they allow the model to learn from diverse examples and make more precise predictions. With more labeled data, a supervised learning model can learn better and generalize more effectively. A larger dataset provides a more comprehensive representation of real-world scenarios, which helps reduce overfitting and makes the model more robust.

Dataset quality for an NLP task can be improved by ensuring diversity. A diverse dataset could include examples from different contexts, such as various domains, sources, and writing styles. This makes it possible for the model to make accurate predictions on unseen data and generalize better. A diverse dataset includes variance in domains such as medical, legal, and journalistic sources such as social media, formal publications, spoken transcripts, and writing styles such as informal, technical, and literary. This diversity challenges the model to adapt to its learned patterns across different contexts, which is essential for its ability to perform well in practical applications.

One crucial aspect of an NLP downstream task dataset is the accuracy of its labelling/annotations.[2] Low-quality annotations can lead to incorrect predictions by models, resulting in decreased performance. Ensuring that the annotations are consistent and accurate throughout the dataset is vital. Reliable annotations in the training data are critical. Incorrect or inconsistent labeling can mislead the model during the training phase, leading to poor performance and unreliable outputs in real-world scenarios. It is also essential to ensure that the different annotators follow consistent guidelines to avoid introducing noise into the data, which can confuse the model and harm its performance.

To sum up, for an NLP model to be effective, it needs to have a high-quality dataset that is large, diverse, and accurately annotated. Building such a dataset requires adhering to best practices in dataset preparation. By doing so, organizations can significantly enhance their NLP models' predictive accuracy and generalizability. While a robust training phase is essential, the foundation of any successful model lies in its dataset. Therefore, it is crucial to prioritize the quality of the dataset.

---

[2] In this book, "labeling" and "annotation" are used interchangeably. "Labeling" is typically applied to tasks such as sentiment analysis, where individual examples, instances, or records are categorized, while "annotation" is used for more detailed tasks such as Named Entity Recognition (NER), which involves fine-grained tagging within these instances.

CHAPTER 5 DATASET PREPARATION

Cost Saving

Data preparation is widely known to be one of the most time-consuming and resource-intensive parts of machine learning projects, including those involving Natural Language Processing (NLP). An article published by O'Reilly Radar raises an interesting point: the commonly held belief that data scientists spend 80% of their time on data preparation is closer to the truth (Bowne-Anderson, 2020).

It's important to keep in mind that the actual costs can vary depending on factors such as the availability of prelabeled datasets, the complexity of the data, the quality requirements for labeling, and the use of automated or crowdsourced labeling techniques.

Before embarking on the dataset creation process for our Healthcare NER model, we must address a pressing question: How do we create a high-quality labeled dataset while being cost-effective?

Our NLP consultant will guide us through different strategies, as shown in Figure 5-2, to solve this challenge, ensuring we can obtain the labeled data without sacrificing quality or incurring excessive costs.

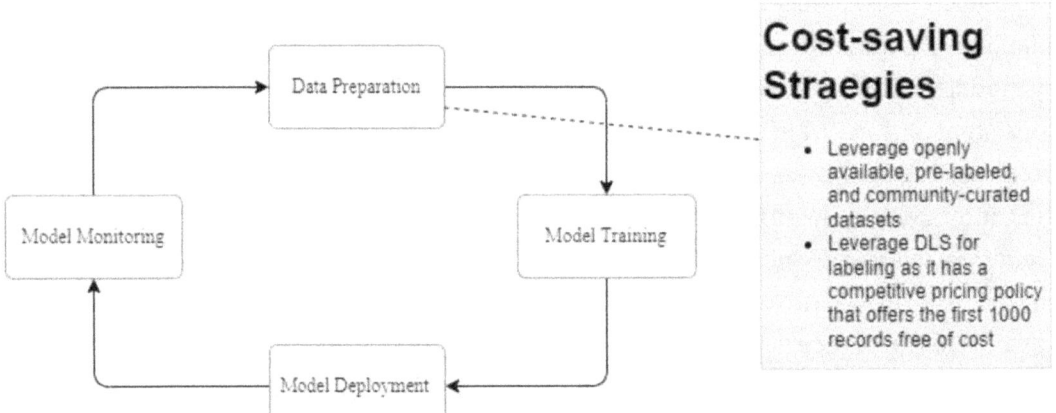

Figure 5-2. Cost saving strategies for the dataset preparation

Off-the-Shelf Datasets

This section discusses how leveraging ready-to-use datasets can help balance quality and cost-effectiveness while building our training dataset.

Leveraging openly available, prelabeled, and community-curated datasets can provide substantial cost advantages, particularly for languages like French, where manual annotation can be prohibitively expensive.

Open public datasets can significantly reduce data acquisition costs when developing NLP solutions. Those community-curated datasets, freely accessible through platforms such as Hugging Face, can eliminate the financial barrier associated with purchasing proprietary datasets or collecting and annotating new data from scratch. This makes developing and deploying NLP solutions more economically feasible for small teams and organizations with limited budgets.

Using prelabeled datasets can significantly lower the costs of data annotation. These datasets already have tagged entities, so the main expense is adapting and enhancing the labels to meet specific project needs. This customization process usually involves making minor adjustments or creating new custom labels and annotations to match the requirements of the target model.

For example, while a general NER model can recognize broad entity categories, customizing it for medical or legal texts may require more detailed entity types. Prelabeled datasets can save much money, as most labeling work is already done. This allows teams to focus on refining and customizing the dataset rather than building one from scratch.

Community-curated datasets, maintained by a collective of academic and professional contributors, do not need ongoing investments in maintenance and enhancement to remain up-to-date and reliable. As these updates are community-driven, they also benefit from a diverse range of insights, which helps to improve the quality and diversity of the annotations.

Using preexisting datasets is an efficient and cost-effective way for preparing a training dataset. Here's why:

- Openly Available: Many datasets are available at no cost, especially those hosted on platforms like the Hugging Face and Kaggle. These platforms offer a variety of datasets under open source licenses, including general-purpose datasets, domain-specific datasets, and datasets for specific languages like French.

- Prelabeled: These datasets come preannotated, relieving you from the substantial costs and time commitments associated with the manual annotation process. For specific languages like French, hiring native speakers for annotation tasks can be particularly costly, especially when offshoring is not an option due to quality concerns or legal restrictions.

- Large and Diverse: Open datasets often encompass a wide range of text types and styles, which is crucial for training robust NLP models. This diversity instills confidence in the effectiveness of the models you develop, as they are not only accurate but also effective across different contexts and domains.

- Maintenance: Open datasets are frequently curated and maintained by academic or professional communities. These communities follow rigorous annotation guidelines and often employ a peer-review process to ensure the quality and reliability of the annotations.

We've decided to use and customize an open, prelabeled French dataset for the Healthcare NER task in our case study to reduce the high manual data gathering and annotation costs. This approach will help us acquire the necessary data within our proof-of-concept budget constraints.

Cost Comparative Analysis

Before deciding whether to use a prelabeled dataset to create a training dataset for our Healthcare NER project or to create a new one from scratch, the NLP consultant compared the costs of both options and determined which one was more cost-effective. Now, let's discuss this analysis and choose the best way to build a training dataset for our case study.

Tip The purpose of this comparative analysis is not only to draw a comparison between two scenarios but also to provide suggestions on how we can approximate the expenses required for creating and annotating a dataset from scratch.

Let's assume that we need to create a training dataset for the generic NER task (not a domain-specific like Healthcare NER) that can effectively identify and classify entities in textual documents. The entities of interest include persons, organizations, locations, money, etc., with a total of ten classes of generic entities. We will need a dataset of approximately 10,000 examples to train this model, each of which may contain multiple entities. Furthermore, annotations for the dataset examples shall be manually reviewed to ensure dataset annotation quality.

We will be making some assumptions for estimating the preparation of an annotated dataset. These assumptions are as follows:

- Annotation Speed: On average, a skilled annotator can label about 100 examples per hour, depending on the complexity of the text. This includes reading the text and selecting the appropriate labels.

  - Text Simplicity: The text in each record is extremely simple, i.e., single sentence or a fragment with clear, straightforward entities.

  - Number of Annotations Per Record: Each record requires annotations for approximately three entities.

  - Number of Labels: The annotator has to choose from ten possible labels, which adds a layer of decision-making but is manageable with familiarity.

- Training and Familiarization: Initial training and periodic quality checks are necessary, which will reduce the effective annotation rate.

- Tool Efficiency: Using an efficient annotation tool like OCI Data Labeling Service (DLS) could potentially enhance labeling collaboration, thus speeding up the annotation process.

Here is a comparison (presented in Table 5-1) that presents the costs and efforts involved in two different approaches to dataset creation: building and annotating a dataset from scratch, versus enhancing a prelabeled dataset. This should provide a clear and concise view of the two scenarios.

CHAPTER 5   DATASET PREPARATION

Table 5-1. *Costs and effort comparison for building a dataset from scratch vs. enhancing a prelabeled dataset*

Activity	Description	Scenario 1: From Scratch	Scenario 2: Prelabeled
Data Collection	Sourcing and acquiring data	$1,000	Included in dataset
Data Wrangling and Preprocessing	Cleaning, normalizing, and preparing data	$1,500 (50 hrs @ $30/hr)	$300 (10 hrs @ $30/hr for adjustments)
Annotation	Manual annotation of entities	$1,500 (100 hrs @ $15/hr)	$450 (30 hrs @ $15/hr)
Total Costs	Sum of all costs	$4,000	$750
Time Investment	Estimated time to complete all tasks	Extensive	Reduced
Flexibility and Customization	Ability to tailor data to specific needs	High	Moderate
Ease of Deployment	Time and effort to get the NLP model to deployment stage	Longer	Shorter
Overall Efficiency	Cost and time efficiency	Lower	Higher

This table clearly shows that using a prelabeled dataset (Scenario 2) offers significant cost and time savings compared to building and annotating a dataset from scratch (Scenario 1).

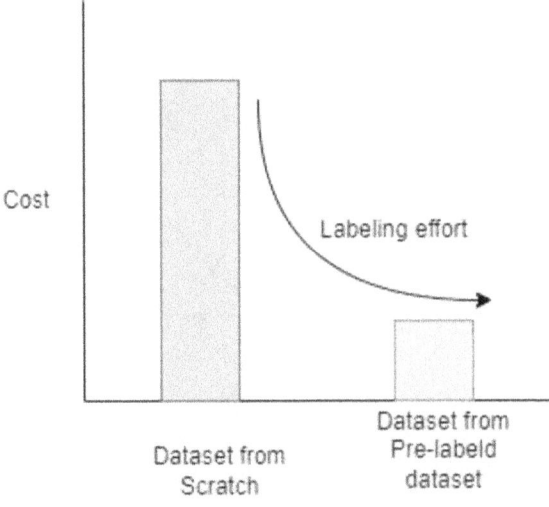

Figure 5-3. *Dataset preparation cost comparison*

Building a dataset can be costly, especially for domain-specific and language-specific datasets (as outlined in Figure 5-3). For instance, building a labeled dataset for our Healthcare NER from scratch would involve hiring annotators with medical knowledge in French, which is not easily accessible and, therefore, costly.[3] Although building a dataset from scratch can be expensive, it does provide greater control over the labeled data.

After conducting a comparative analysis, it became clear that adapting an existing dataset could save up to 75% in costs. Based on this insight, our NLP consultant, John Doe, has decided to utilize existing datasets to save time and achieve high-quality labeled data.

In the upcoming sections, the NLP consultant will go through the process of searching and selecting a suitable dataset for French Healthcare NER from the community-curated datasets publicly available on the Hugging Face platform.

Dataset Life Cycle

The data engineering life cycle comprises stages that turn raw data ingredients into a useful end product, ready for consumption by analysts, data scientists, ML engineers, and others (Joe Reis, 2022).

Constructing a robust training dataset for Natural Language Processing (NLP) projects involves a series of methodical steps to ensure that the data is comprehensive, relevant, and high quality, which is necessary to train effective models. In our case study, we will focus on Healthcare Named Entity Recognition (NER) as the targeted NLP task and cover how to build a labeled dataset for it.

As illustrated in Figure 5-4, our flow is designed to be iterative, meaning we can use insights from model performance to refine data wrangling and labeling steps, leading to continuous improvement.

---

[3] More than the estimated cost for manual annotation of $15/hr in the comparative analysis scenarios.

CHAPTER 5 DATASET PREPARATION

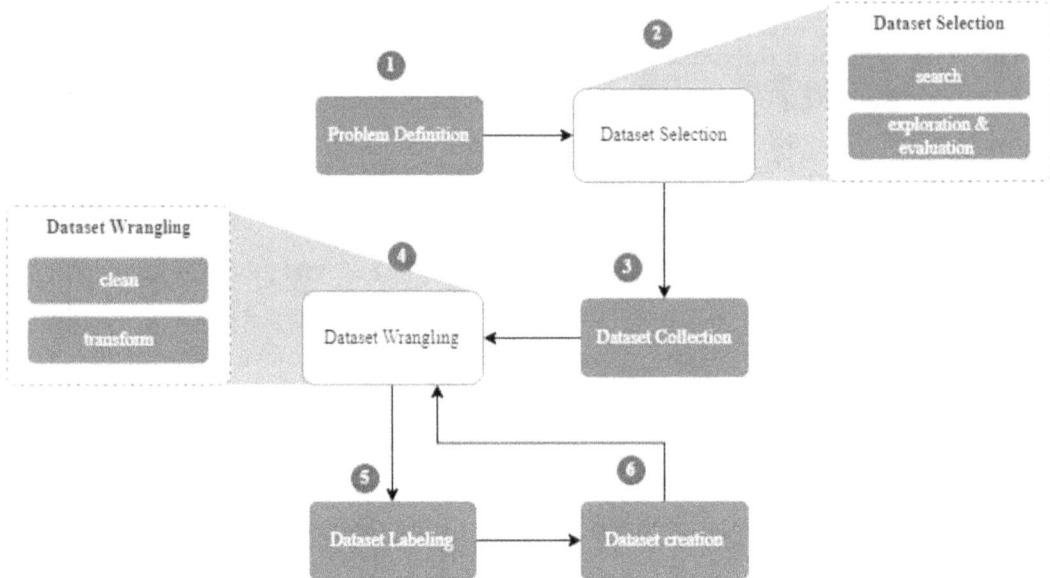

Figure 5-4. *Training dataset preparation life cycle activity flow*

Here is a list of steps for constructing a robust our training dataset:

1. Problem Definition: We need to train a Healthcare NER model using a transfer learning approach, which involves leveraging knowledge from a pretrained model on a different but related task, to identify medical entities.

2. Dataset Selection: We will leverage a ready-to-use prelabeled publicly available dataset that aligns with our objectives.

3. Dataset Collection: Storing our selected dataset to Labeling Datasets Buckets from where OCI Data Labeling will access it after the data processing step.

4. Dataset Wrangling: This step encompasses the following two main substeps:

    a. Clean: We will remove all examples from the dataset that are irrelevant to our case study, such as those that do not contain any entity of interest. We will also ensure that the labeled entities are balanced within the dataset.

    b. Transform: we will transform it to the format *JSONL Consolidated* (one of the formats expected by OCI Data Labeling).

5. Dataset Labeling: The labeling team, a key part of our process, will adapt the prelabeled dataset to our case study specification by manually labeling missing medical entities, i.e., dataset enrichment. We will review the labeling to ensure data labeling quality and consistency.

6. Dataset Creation: We will export the final labelled dataset from OCI Data Labeling to our Training Datasets Bucket. Then, we will save the final training dataset in the Hugging Face library format, a popular format for storing NLP datasets.

Our training dataset creation flow is sufficient for our case study, but it could be improved by implementing some data best practices for real-life projects. This includes automating the flow through data pipelines that automate the process from data collection to training dataset creation. Additionally, we should use data versioning tools to manage changes and experiment with different dataset versions. Documenting and keeping detailed records of data sources, collection methods, preprocessing decisions, and version histories is also important. Following these steps can ensure a more efficient and effective data flow.

Before we delve into the various stages of our data life cycle that we need to follow to create a training dataset from an existing one, it's essential that we have a well-defined set of requirements. These requirements will help us customize the preexisting Healthcare NER dataset to meet all our business needs. Therefore, our first focus should be outlining the requirements before proceeding with the rest of the process.

Framing the Problem (Step 1)

For our MedTALN Inc.'s running example, we need to create a training dataset with a comprehensive set of annotated textual examples from the medical domain in French. The examples should contain medical entities such as conditions, medications, symptoms, and procedures.

To simplify the process of creating the training dataset, the NLP consultant chose to use, as a starting point, an existing dataset that has already been curated for a similar purpose: Named Entity Recognition (NER) task for the French language, ideally in the healthcare domain.

The following are some assumptions for the labeling stage of our dataset life cycle:

- OCI Data Labeling Service (DLS) will be used as the annotation tool.
- Python notebooks shall be used to format the dataset in a format compatible with the DLS's import process.
- The labeling team is proficient with DLS tool.
- Detailed and clear annotation standards and guidelines and a thorough annotation review process shall be defined.

In summary, adhering to these requirements throughout the data preparation phase will ensure the development of a high-quality Healthcare NER training dataset.

Dataset Selection (Step 2)

Leveraging prelabeled datasets is not a straightforward task. NLP practitioners need to efficiently and cost-effectively search, evaluate, enrich, and adapt these datasets to prepare high-quality training datasets for downstream NLP tasks.

To identify the appropriate prelabeled dataset that meets our case study requirements, the NLP consultant has established the following selection criteria:

- Task specificity: The dataset must be specifically curated for the NLP task of Named Entity Recognition (NER).
- Domain Specificity: Ideally, the dataset should consist of examples from the medical domain, including annotations for medical entities such as conditions, medications, symptoms, and procedures.
- Language Specificity: The dataset must contain a significant number of French examples.
- Open Source and Accessibility: Preferred datasets should have an open source license and be publicly available from trusted sources such as Hugging Face[4] or Kaggle.[5]
- Data Quality, Relevance, and Cleanliness: Dataset's examples and annotations must exhibit an acceptable level of cleanliness, relevancy, and accuracy upon spot-checking.

---

[4] For more on Hugging Face, visit the website https://huggingface.co/
[5] For more details on Kaggle, visit the website https://www.kaggle.com/

CHAPTER 5 DATASET PREPARATION

By adhering to these criteria, the NLP consultant ensures the selection process of a prelabeled dataset is well documented and objective, resulting in a dataset that is both highly relevant and suitable for the specific NLP task at hand.

Selecting Datasets on Hugging Face

The dataset selection process begins with an in-depth search on the Hugging Face Hub, where we iteratively identify potential candidate datasets. Once these candidates are identified, we leverage Python code to thoroughly explore and evaluate them, ultimately determining the optimal dataset that will serve as the foundation for efficient and effective training dataset preparation.

When searching for datasets for Natural Language Processing (NLP), the Hugging Face Hub is a valuable repository. This platform hosts diverse community-curated datasets for NLP tasks in over 100 languages, including translation, Named Entity Recognition, and sentiment analysis. Furthermore, some datasets offer a **Dataset Viewer**, as illustrated in Figure 5-5, a user-friendly interface for data exploration. To search or explore the publicly available datasets, we can use the search box on the primary datasets page or within the top navigation to locate datasets. The search results may be filtered by language, task, and license for optimal results.

*Figure 5-5. Hugging Face **Dataset Viewer***

183

CHAPTER 5   DATASET PREPARATION

When embarking on the journey of finding a dataset (as illustrated in Figure 5-6), the multitude of results can be daunting, even after applying filters such as the NLP task, language, or size. However, by exploring the data with the ***Dataset Viewer***, if available, and looking for several other key features, you can gain a deeper understanding of the dataset and make an informed decision about which dataset can be a good candidate for your needs.

At the outset, it's crucial to examine carefully the dataset summary. This concise overview usually provides valuable insights into the dataset's purpose, creation process, intended use, supported tasks and languages, and size. It can help us easily distinguish irrelevant datasets and determine if they align with our selection criteria.

Equally important is comprehending the dataset's curation process. Understanding what motivated its creation and the major choices made in its assembly, as well as knowing the dataset curators, their affiliations, and any funding information, can instill confidence in the dataset's reliability and credibility.

The source of the data and the collection process are also important factors to consider. We should know where the data came from, such as news text, social media posts, or translated sentences, and the data selection or filtering criteria. It is also important to know whether the data was produced by humans or machine-generated and the people or systems who originally created it. If data was collected from other Hugging Face preexisting datasets, knowing the details of the original Hugging Face dataset can be helpful.

If the dataset contains annotations, it's important to understand the annotation process and any tools used. Knowing the amount of data annotated and the annotation guidelines provided to the annotators can help us assess the quality of the annotations. If available, knowing interannotator statistics, which measure the level of agreement between different annotators, can also be useful. We can also look for information on any validation processes used to ensure the quality of the annotations.

Finally, it's important to check the dataset's licensing information and ensure we understand and agree with its terms of use. This is crucial as it determines how we can use the dataset, whether it's for commercial or noncommercial purposes, and if we can modify or redistribute the dataset.

Data scientists use exploratory data analysis (EDA) to understand the dataset's characteristics. EDA is an iterative process that helps answer questions about the dataset, such as whether any data wrangling or transformation is needed to solve the NLP problem. By analyzing the dataset, NLP experts can assess its quality and usability. For example, suppose the NLP problem is token classification. In that case, the expert can use EDA to analyze the dataset's text length, the label count for each class, and the distribution of labels across the dataset classes. This information can help decide whether to truncate or delete short or lengthy examples from the dataset or balance the dataset by writing some code.

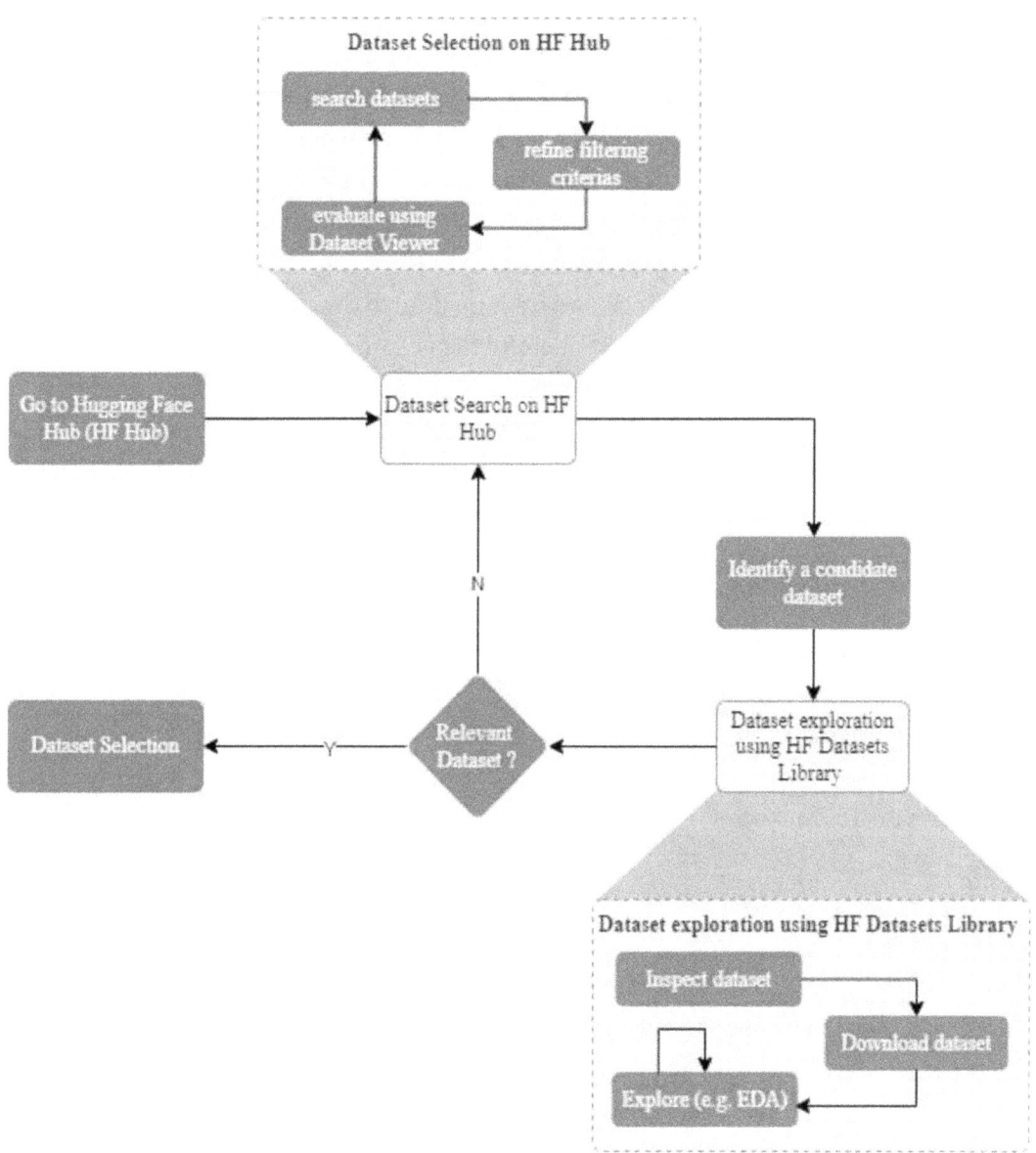

Figure 5-6. Dataset selection workflow using Hugging Face platform

Even if the ***Dataset Viewer*** is unavailable on the Hugging Face Hub, the NLP expert can use Python notebooks to explore and analyze a potential candidate dataset using the Hugging Face datasets.[6] This allows the expert to delve into the dataset's features and structure, like its text examples and labels, as well as its splits for training, validation, and testing.

The Hugging Face Datasets library offers a programmatic way to interact with datasets from the Hub, making integrating datasets into NLP projects easy. The library's streaming functionality ensures efficient data access, even when dealing with large datasets. We can effortlessly access and experiment with publicly available datasets with just a single line of code. This simplified approach to dataset integration can significantly enhance your workflow and boost productivity.

When working with datasets, it can take a long time to download them before you can even inspect them. To save time, it's helpful to get some general information about the dataset before downloading it. This information is stored in DatasetInfo and includes details such as the dataset description, features, and size. To inspect a dataset's attributes without committing to downloading it, as shown in the Listing 5-1, use the load_dataset_builder() function to load a dataset builder.

Listing 5-1. Inspect dataset

```
from datasets import load_dataset_builder
ds_builder = load_dataset_builder("conll2003")
# Inspect dataset
print(ds_builder.info.splits["train"].num_examples)
ds_builder.info.features
```

out

```
14041
{'id': Value(dtype='string', id=None),
 'tokens': Sequence(feature=Value(dtype='string', id=None), length=-1,
   id=None),
```

---

[6] For more information about Hugging Face Datasets library, please visit the website at https://huggingface.co/docs/datasets/index

```
'pos_tags': Sequence(feature=ClassLabel(names=['"', "''", '#', '$', '(',
')', ',', '.', ':', '``', 'CC', 'CD', 'DT', 'EX', 'FW', 'IN', 'JJ',
'JJR', 'JJS', 'LS', 'MD', 'NN', 'NNP', 'NNPS', 'NNS', 'NN|SYM', 'PDT',
'POS', 'PRP', 'PRP$', 'RB', 'RBR', 'RBS', 'RP', 'SYM', 'TO', 'UH', 'VB',
'VBD', 'VBG', 'VBN', 'VBP', 'VBZ', 'WDT', 'WP', 'WP$', 'WRB'], id=None),
length=-1, id=None),
'chunk_tags': Sequence(feature=ClassLabel(names=['O', 'B-ADJP', 'I-ADJP',
'B-ADVP', 'I-ADVP', 'B-CONJP', 'I-CONJP', 'B-INTJ', 'I-INTJ', 'B-LST',
'I-LST', 'B-NP', 'I-NP', 'B-PP', 'I-PP', 'B-PRT', 'I-PRT', 'B-SBAR',
'I-SBAR', 'B-UCP', 'I-UCP', 'B-VP', 'I-VP'], id=None), length=-1,
id=None),
'ner_tags': Sequence(feature=ClassLabel(names=['O', 'B-PER', 'I-PER',
'B-ORG', 'I-ORG', 'B-LOC', 'I-LOC', 'B-MISC', 'I-MISC'], id=None),
length=-1, id=None)}
```

This is a dataset of contains a train split with 14041 examples.

Candidate Healthcare NER Dataset

The CoNLL-2003 dataset,[7] a pivotal resource in the field of Named Entity Recognition (NER), serves as a benchmark for evaluating NER models. Initially crafted for English and German languages, its structure and annotation scheme have set a standard, inspiring the creation of datasets for additional languages. This dataset, introduced by Erik F. Tjong Kim Sang and Fien De Meulder in 2003, is notable for its comprehensive annotation of entities into four categories: locations (LOC), organizations (ORG), persons (PER), and miscellaneous (MISC).

As NER continues to evolve, the principles embodied in the CoNLL-2003 dataset remain a cornerstone for advancing the field, guiding researchers and practitioners in the creation and refinement of models capable of accurately recognizing and classifying named entities across a spectrum of languages and contexts. To illustrate the structure of the CoNLL-2003 dataset, Figure 5-7 provides a representative example.

---

[7] The Hugging Face CoNLL-2003 dataset is available at https://huggingface.co/datasets/eriktks/conll2003

CHAPTER 5   DATASET PREPARATION

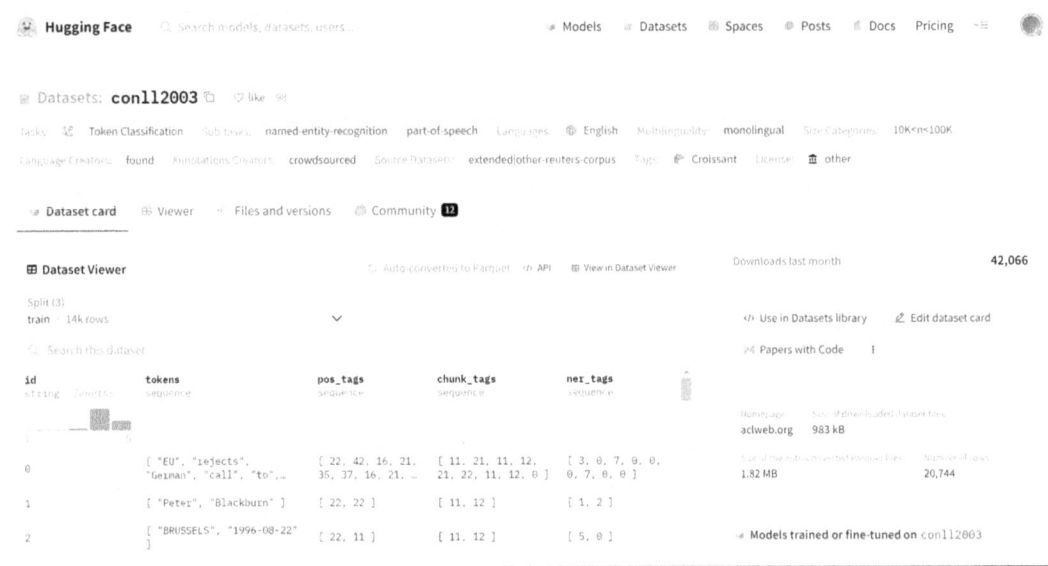

Figure 5-7. CoNLL-2003 dataset page on Hugging Face

Building on the foundation laid by the CoNLL-2003 dataset, several datasets have been developed for French language NER, adhering to its structure and annotation guidelines. These datasets are crucial for training and evaluating NER models tailored to French, reflecting the linguistic nuances and entity categories specific to the language.

- WikiNER[8]: Developed by Nothman et al. in 2013, WikiNER provides a substantial dataset for French, comprising 120,682 training instances and 13,410 test instances. It categorizes entities into locations (LOC), organizations (ORG), persons (PER), and miscellaneous (MISC), serving as a versatile resource for model training.

- Wikiann[9]: Initiated by Rahimi et al. in 2019, based on work by Pan, Xiaoman, et al., Wikiann further enriches the French NER dataset landscape. It offers 20,000 training instances, 10,000 validation instances, and 10,000 test instances, maintaining the same entity categories as WikiNER and promoting detailed model evaluation.

---

[8] WikiNER homepage: https://metatext.io/datasets/wikiner and Hugging Face Hub page: https://huggingface.co/datasets/Jean-Baptiste/wikiner_fr
[9] Wikiann Hugging Face Hub page: https://huggingface.co/datasets/wikiann

CHAPTER 5   DATASET PREPARATION

- MultiNERD[10]: Introduced by Tedeschi and Navigli in 2022, MultiNERD expands the scope of French NER datasets with 140,880 training instances, 17,610 validation instances, and 17,695 test instances. It broadens the spectrum of entity categories, including not only the standard ones but also animals (ANIM), biological terms (BIO), celestial entities (CEL), diseases (DIS), events (EVE), and more, thus offering a richer and more diverse dataset for comprehensive model training.

- MultiCoNER v2[11]: Proposed by Fetahu et al. in 2023, MultiCoNER v2 aligns with the French part of the dataset, presenting 120,682 training instances and 13,410 test instances. It classifies entities into detailed categories such as location, creative work, group, person, product, and medical, each with subcategories, thereby providing a nuanced and extensive dataset for French NER tasks.

- Pii-masking-200k[12]: This dataset, introduced by ai4Privacy in 2023, focuses specifically on Personally Identifiable Information (PII). This dataset is designed to address the pressing need for models that can accurately identify and handle sensitive personal information, ensuring privacy and compliance with data protection regulations.

- MedicalNER_Fr[13]: This open source dataset was published on Hugging Face by typica.ai in 2024 for educational purposes only. It has been specifically designed to assist in training Named Entity Recognition (NER) models for the French language within the medical and healthcare domain. The dataset is derived from MultiCoNER[11] Dataset (listed above).

---

[10] MultiNERD dataset Hugging Face Hub page: https://huggingface.co/datasets/tner/multinerd

[11] Multilingual Complex Named Entity Recognition Version 2 (MultiCoNER v2) dataset can be found at Hugging Face Hub: https://huggingface.co/datasets/MultiCoNER/multiconer_v2

[12] Pii-masking-200k dataset Hugging Face Hub page: https://huggingface.co/datasets/ai4privacy/pii-masking-200k

[13] TypicaAI/MedicalNER_Fr dataset can be found at Hugging Face: https://huggingface.co/datasets/TypicaAI/MedicalNER_Fr

CHAPTER 5   DATASET PREPARATION

While the datasets listed above represent a significant portion of resources available for French NER tasks, it's important to note that this list is not exhaustive. There exist other datasets tailored to the French language that might not have been mentioned here.

Amid the vast array of available datasets, each with its unique advantages and limitations, establishing clear selection criteria becomes paramount to align with the specific goals of our Healthcare NER project. The presence of datasets varying in quality necessitates a discerning approach; lower-quality datasets could complicate the training with unnecessary noise, demanding significant cleaning and preprocessing. Conversely, datasets overly concentrated on a particular domain, despite their value for specific applications, might not offer the comprehensive coverage required for broader NER tasks.

Upon careful evaluation by the NLP Consultant, typica.ai's "MedicalNER_Fr" dataset (typica.ai, 2024) was chosen as the dataset to begin creating the case study dataset (Figure 5-8). The following points outline the critical reasons and the underlying logic that, in adherence to the predefined selection criteria, led our consultant to prefer this specific dataset for our initiative:

- The Dataset's Focus on Healthcare: The dataset is explicitly focused on the healthcare domain, eliminating the need for additional costly healthcare domain-specific annotation and labeling.

- French Language Support: This dataset focuses on the French language, eliminating the need for additional filtering or preprocessing to isolate French data from a multilingual corpus.

- Open Source License and Availability: The dataset is open source and publicly available via the Hugging Face Hub, facilitating easy access and integration into our training pipeline. The open source license further underscores its suitability for academic and commercial projects alike.

- Comprehensive Entity Annotations: The dataset stands out for its rich annotation, including generic entities (e.g., person, location, organization) as well as medical entities, surpassing many alternative datasets.

- Overall Good Quality: Preliminary quality assessments of the dataset revealed a good degree of cleanliness and correctness. The dataset exhibited strong performance across key metrics of relevance, consistency, uniformity, and comprehensiveness.

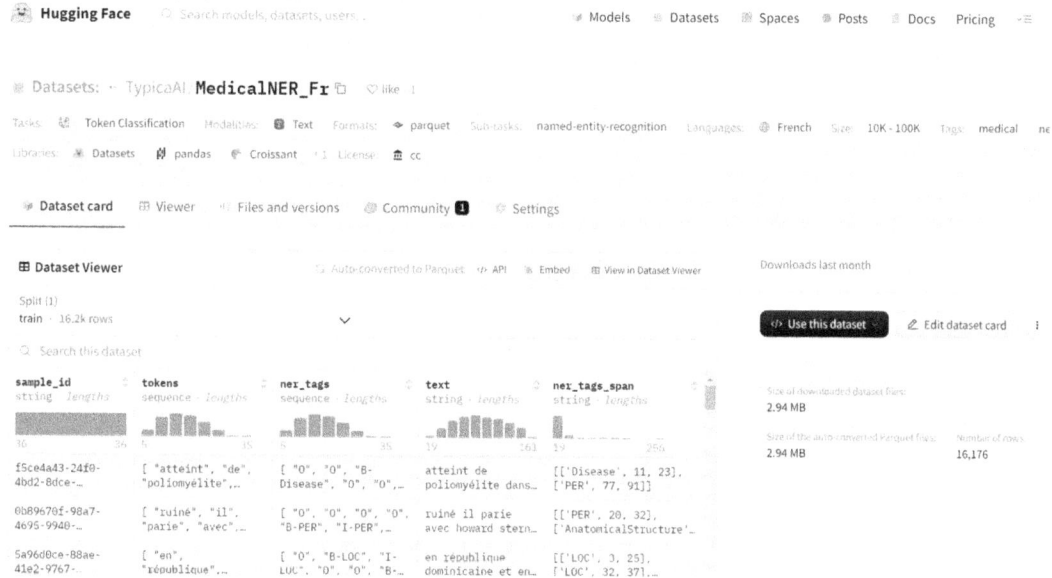

Figure 5-8. *The selected "TypicaAI/MedicalNER_Fr" dataset, on the Hugging Face Hub*

The combination of these factors makes the TypicaAI/MedicalNER_Fr dataset the optimal choice for our Healthcare NER project. It provides a solid foundation for the creation of a cost-effective training dataset, aligning well with the objectives of our case study.

Training Dataset Preparation

We have found a prelabeled dataset called "MedicalNER_Fr" (typica.ai, 2024). This dataset will be our starting point for creating our training dataset.

Our NLP consultant will assist us in acquiring, processing, annotating, and finalizing the training dataset.

CHAPTER 5   DATASET PREPARATION

We can review the key steps in preparing our Healthcare NER training dataset with the help of Figure 5-9's workflow diagram. Let's go through these steps together.

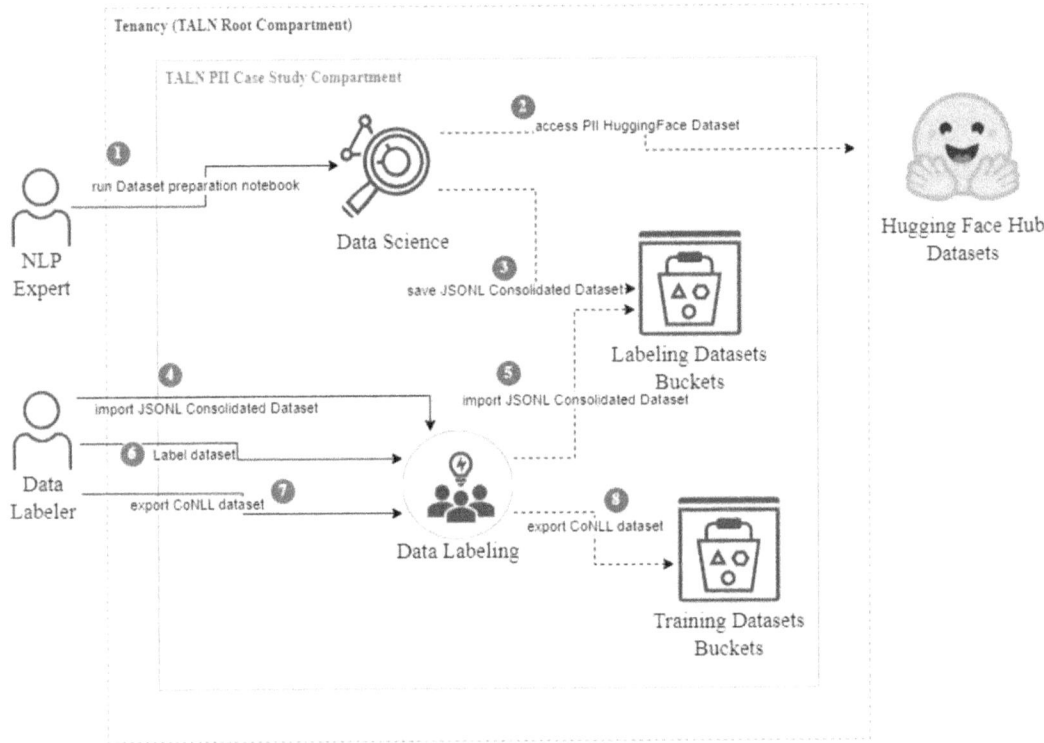

Figure 5-9. High-level architecture for dataset creation

1. Data Collection: The workflow begins by accessing the Hugging Face TypicaAI/MedicalNER_Fr dataset. This dataset is fetched using Python code within an OCI Data Science Jupyter notebook.

2. Data Wrangling: The retrieved *Hugging Face* dataset is processed and then transformed into the *JSONL Consolidated* format. This is a json-based format suitable for the subsequent data labeling process. Once transformed, the new dataset is saved to a Labeling Datasets Bucket.

3. Data Labeling: The transformed dataset, now in *JSONL Consolidated* format, is subsequently imported into the *OCI Data Labeling* Service (DLS) for annotation. Our labeling team enriches this dataset by identifying and labeling the missing medical annotations relevant to our case study.

4. Export Annotated Data: After the data labeling is completed, the annotated dataset is exported back to Training Datasets Bucket. This dataset now contains the original data along with the annotations provided by the labeling team.

Note During dataset enrichment, it is possible to add new medical entities that might be relevant to the Healthcare NER task. However, for simplicity, in this case study, we will focus only on adding missing annotations for existing entities rather than introducing new medical entities.

To create our Healthcare NER model training dataset, we need to retrieve, transform, label, and export the annotated data to CoNLL format. The OCI Data Labeling Service (DLS) can help improve the data quality by providing new labels or entities tailored to specific project requirements. This process results in a high-quality dataset that can train our Healthcare NER model.

John Doe, our NLP consultant, will play a dual role as a member of the data labeling and data science teams.[14] He will use OCI Data Science Notebooks and OCI Data Labeling Datasets to prepare the Data Labeling Service (DLS) Dataset and annotate its records. He will provide step-by-step guidance through Python notebooks and DLS procedures throughout the process. As a result, he will share his expertise in preparing datasets to train NLP models for downstream tasks.

Dataset Collection and Wrangling (Steps 3 and 4)

Data wrangling is the process of taking "raw" or "found" data and transforming it into something that can be used to generate insight and meaning (McGregor, 2021). Data wrangling is a crucial step in preparing datasets for Natural Language Processing (NLP). It involves converting unstructured data into a structured format that is essential for building successful NLP models. By performing effective data wrangling, practitioners ensure that the data is clean, consistent, and easy to understand, which improves the predictive capabilities of machine learning algorithms. Practitioners use techniques like

---

[14] Please note that, for the purpose of this case study, the terms "data labelers" and "data scientists" refer to our fictional NLP consultant, John Doe, who is the only member of these teams.

data munging and cleaning to develop robust NLP datasets. By meticulously preparing the data, they lay a solid foundation for maximizing the accuracy and efficiency of their models.

As part of our case study, we need to process the Hugging Face dataset by cleaning, balancing, and transforming it.

- Cleaning involves removing examples that are irrelevant to our case study, such as those that don't feature the entity of interest.

- Balancing ensures that the labeled entities are distributed evenly across all selected dataset examples.

- In the Transform step, we will convert the dataset to the JSONL Consolidated format, which is one of the formats supported by OCI Data Labeling. This format is suitable for the subsequent data labeling process.

Once the transformation is complete, the new dataset will be saved to a Labeling Datasets Bucket.

Before proceeding with the data wrangling process, it's important to understand the significance of a well-balanced dataset for training a classification model (e.g., token classification[15]). A balanced dataset refers to a dataset where the number of samples for each class is roughly equal. This is important because imbalanced datasets can lead to biased models that perform poorly on minority classes.

Balancing the dataset is a crucial step in our data wrangling process. As our Healthcare NER model resolves a classification problem, it performs token classification. However, the dataset may be imbalanced, meaning that some classes (e.g., first name, last name) have a large portion of the training dataset, while others are underrepresented (e.g., address, city). This creates a problem as the model may learn to predict the majority classes to achieve high accuracy, ignoring the underrepresented classes, which are equally or more important in real-world scenarios.

We need to balance the dataset to prevent bias toward one class, making it easier to train the model. Balancing ensures that the model does not favor the majority class simply because it contains more data. By ensuring that the dataset is balanced, we can improve the performance of our classification model and ensure that it is able to accurately classify all classes.

---

[15] Token classification is also known as Named Entity Recognition (NER).

CHAPTER 5 DATASET PREPARATION

There are several ways to deal with imbalanced data (Lewis Tunstall, 2022), including

- Randomly oversample the minority class.
- Randomly undersample the majority class.
- Gather more labeled data from the underrepresented classes.

Now, it is time to open the Python notebook prepared by our NLP consultant, John Doe, and execute the code for the dataset preparation.

Dataset Preparation Notebook

This notebook aims to assist you in preparing the dataset for the labeling phase using the *OCI Data Labeling Service (DLS)*. This notebook includes steps for downloading the dataset, cleaning it, balancing it, and transforming it to *JSONL Consolidated* format.

Go to your CPU-based notebook session; from JupyterLab file browser, open the notebook *prepare_dataset.ipynb* under the folder (see Figure 5-10):

/repos/john-doe-typica-ai/nlp-on-oci.git

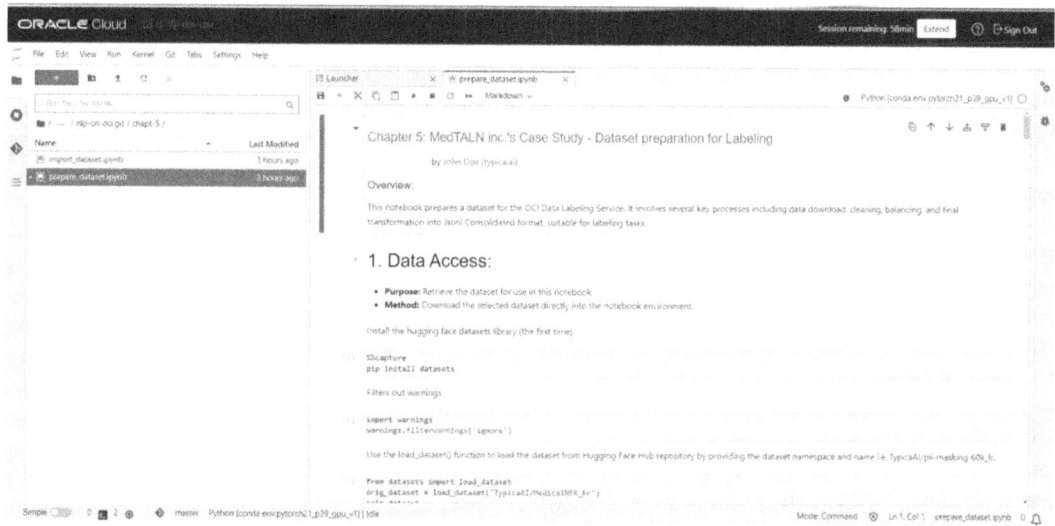

Figure 5-10. *Notebook for dataset preparation*

195

CHAPTER 5   DATASET PREPARATION

Loading

It starts by importing necessary libraries: load_dataset from the Hugging Face's Datasets library for data loading. Since the Datasets library is missing from our installed PyTorch conda env., we need to ensure that it is installed it using pip:

```
%%capture
pip install datasets
```

Note We use the cell magic %%capture to suppress the install output (irrelevant for understanding the notebook).

Filter all the warnings occurring during the execution of this notebook to be ignored (irrelevant for understanding the notebook):

```
import warnings
warnings.filterwarnings('ignore')
```

Use the load_dataset() function to load the dataset from its Hugging Face hub repository (by providing the repository namespace and dataset name, i.e., TypicaAI/MedicalNER_Fr; see Listing 5-2).

Listing 5-2. Prepare dataset notebook: data access

```
from datasets import load_dataset
orig_dataset = load_dataset("TypicaAI/MedicalNER_Fr")
orig_dataset
```

   Output

```
Downloading readme: 100%
10.6k/10.6k [00:00<00:00, 1.06MB/s]
Downloading data: 100%
2.94M/2.94M [00:00<00:00, 4.17MB/s]
Generating train split:
16176/0 [00:00<00:00, 214652.91 examples/s]
DatasetDict({
```

CHAPTER 5 DATASET PREPARATION

```
    train: Dataset({
        features: ['sample_id', 'tokens', 'ner_tags', 'text',
        'ner_tags_span'],
        num_rows: 16176
    })
})
```

Strip all the columns from the downloaded dataset except columns text ner_tags_span (Listing 5-3 and Listing 5-4).

Listing 5-3. Prepare dataset notebook: data access

```
from datasets import Dataset, DatasetDict

# Assuming cleaned_dataset['train'] is a dataset object
# Select only the desired columns
streamlined_dataset = orig_dataset['train'].remove_columns([col for col in orig_dataset['train'].column_names if col not in ['text', 'ner_tags_span']])

# Create a new dataset with only these columns
streamlined_dataset = DatasetDict({
    'train': streamlined_dataset
})

# Print the columns to verify
streamlined_dataset
```

Output

```
DatasetDict({
    train: Dataset({
        features: ['text', 'ner_tags_span'],
        num_rows: 16176
    })
})
```

Chapter 5 Dataset Preparation

Listing 5-4. Displaying the First Example of the Dataset

```
streamlined_dataset['train'][0]
```

Output

```
{'text': 'atteint de poliomyélite dans son enfance il devient fan de blues en écoutant big joe turner .', 'ner_tags_span': "[['Disease', 11, 23], ['PER', 77, 91]]"}
```

Wrangling Steps

This phase ensures the dataset is ready for labeling, involving two main tasks: cleaning and balancing the dataset (see Listings 5-5, 5-6, and 5-7).

– Cleaning: This function is useful for standardizing NER labels in a dataset by consolidating certain nonmedical entities types under a single label, "MISC."

Listing 5-5. Prepare dataset notebook: wrangling data cell

```python
import ast  # Import abstract syntax trees module to safely parse strings into lists

# Define the labels that need to be renamed to 'MISC'
labels_to_map = {'LOC', 'PER', 'PROD', 'CW', 'ORG', 'GRP'}

def rename_labels_to_misc(example):
    # Convert the string representation of list into an actual list
    ner_tags_span = ast.literal_eval(example['ner_tags_span'])

    # Rename labels to 'MISC'
    renamed_ner_tags_span = [
        ['MISC', start, end] if label in labels_to_map else [label, start, end]
        for label, start, end in ner_tags_span
    ]
```

CHAPTER 5   DATASET PREPARATION

```
# Update the example with the renamed labels
example['ner_tags_span'] = str(renamed_ner_tags_span)

return example
```

The Hugging Face "datasets" library "map" function applies "rename_labels_to_misc" across the dataset.

Listing 5-6. Clean dataset

```
# Apply the function across the dataset using map
cleaned_dataset = streamlined_dataset.map(rename_labels_to_misc,
batched=False)
cleaned_dataset
```

Output

```
Map: 100%
16176/16176 [00:00<00:00, 18957.06 examples/s]
DatasetDict({
    train: Dataset({
        features: ['text', 'ner_tags_span'],
        num_rows: 16176
    })
})
```

The Hugging Face "datasets" library "filter" is used to remove example with text length inf. to 50 characters.

Listing 5-7. Filter dataset

```
# Define the minimum text length (e.g., 10 characters)
min_text_length = 50
# remove examples with text lenght < min_text_length
filtered_dataset = cleaned_dataset.filter(lambda example:
len(example['text']) >= min_text_length)
```

CHAPTER 5 DATASET PREPARATION

```
orig_ds_count = len(orig_dataset['train'])
filtered_ds_count = len(filtered_dataset['train'])

print(f"Original number of examples: {orig_ds_count}")
print(f"Number of examples (after filtering): {filtered_ds_count}")
```

Output

```
Filter: 100%
16176/16176 [00:00<00:00, 163731.51 examples/s]
Original number of examples: 16176
Number of examples (after filtering): 14722
```

Plot the dataset size before and after the cleaning (Listing 5-8).

Listing 5-8. Plotting the Dataset Size After Cleaning

```
import matplotlib.pyplot as plt

# Data setup
categories = ['Original', 'After Cleaning']
values = [orig_ds_count, filtered_ds_count]

# Create bar plot
plt.figure(figsize=(8, 5))
plt.bar(categories, values, color=['blue', 'green'])
plt.title('Comparison of Dataset Sizes')
plt.ylabel('Number of Examples')
plt.show()
```

Output (Figure 5-11)

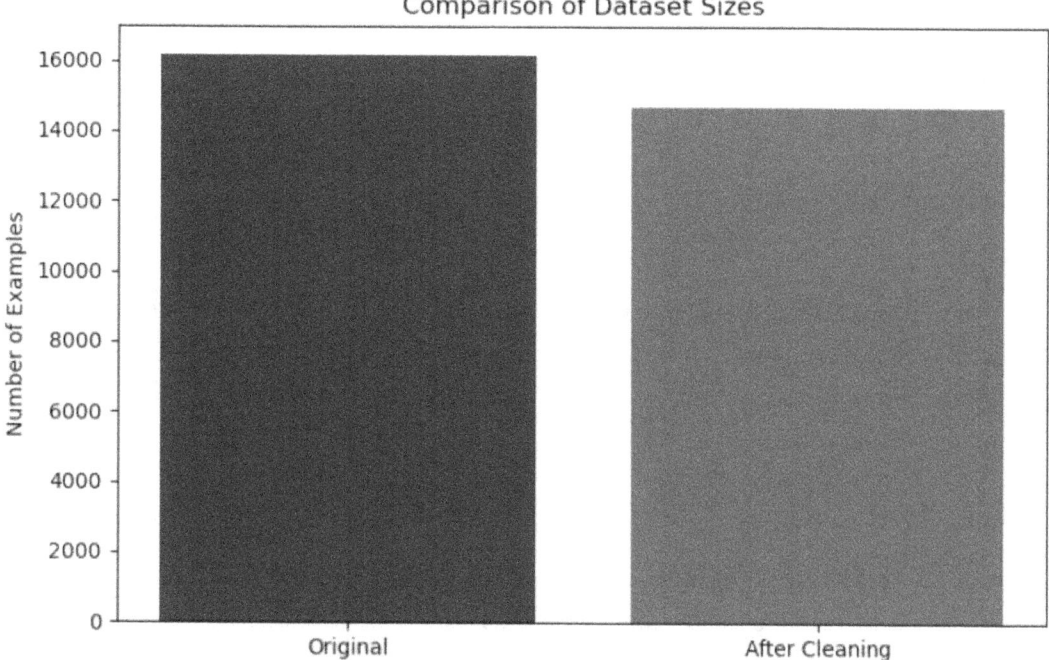

Figure 5-11. *Dataset size before and after cleaning*

- Balancing: Adjust the dataset to ensure all entities are equally represented, using undersampling techniques as necessary (detailed in Listings 5-9 to 5-15).

Listing 5-9. Defining Functions

```
from collections import Counter
import ast

def extract_labels(ner_tags_span):
    """Helper function to parse and filter labels from ner_tags_span."""
    return [label[0] for label in ast.literal_eval(ner_tags_span)]

def count_labels(dataset):
    """Counts occurrences of each label in the dataset's ner_tags_span
    field, excluding 'O'."""
```

```
# Use a list comprehension to gather all labels across the dataset and
  count them
label_counts = Counter(
    label for example in dataset['train']
    for label in extract_labels(example['ner_tags_span'])
)
return label_counts
```

Output

Define "plot_distributions" function. This function is designed to visually compare the distributions of a dataset's labels before and after the balancing step. It uses the matplotlib library to create a bar chart, which displays the counts of each label side by side for easy comparison.

Listing 5-10. Defining plot distributions function

```
def plot_distributions(before_counts, after_counts, process="Balancing"):
    # Sort labels to ensure they match up in the plot
    labels = sorted(before_counts.keys())
    before_values = [before_counts[label] for label in labels]
    after_values = None
    if after_counts:
      after_values = [after_counts[label] for label in labels]

    x = range(len(labels))  # Label location on x-axis

    # Create the bar plot
    plt.figure(figsize=(10, 5))
    plt.bar(x, before_values, width=0.4, label=f'Before {process}',
    color='b', align='center')
    if after_values:
      plt.bar(x, after_values, width=0.4, label=f'After {process}',
      color='r', align='edge')
```

```
# Add some text for labels, title and custom x-axis tick labels, etc.
plt.xlabel('Labels')
plt.ylabel('Counts')
plt.title(f'Label Distribution Before and After {process}')
plt.xticks(x, labels, rotation='vertical')
plt.legend()

# Show the plot
plt.tight_layout()
plt.show()
```

Output

Calculates and prints the label distribution to assess the dataset initial balance.

Listing 5-11. Calculating Label Distribution

```
# count labels for the cleaned dataset
label_counts = count_labels(filtered_dataset)
print("Label Counts:", label_counts)
```

Output

```
Label Counts: Counter({'Disease': 4291, 'AnatomicalStructure': 4246,
'Medication/Vaccine': 3952, 'MedicalProcedure': 2930, 'Symptom': 1577,
'MISC': 1550})
```

Plot the label distribution before balancing.

CHAPTER 5   DATASET PREPARATION

Listing 5-12. Cell 14

```
plot_distributions(label_counts, None)   # Plot the label counts
```

Output (Figure 5-12)

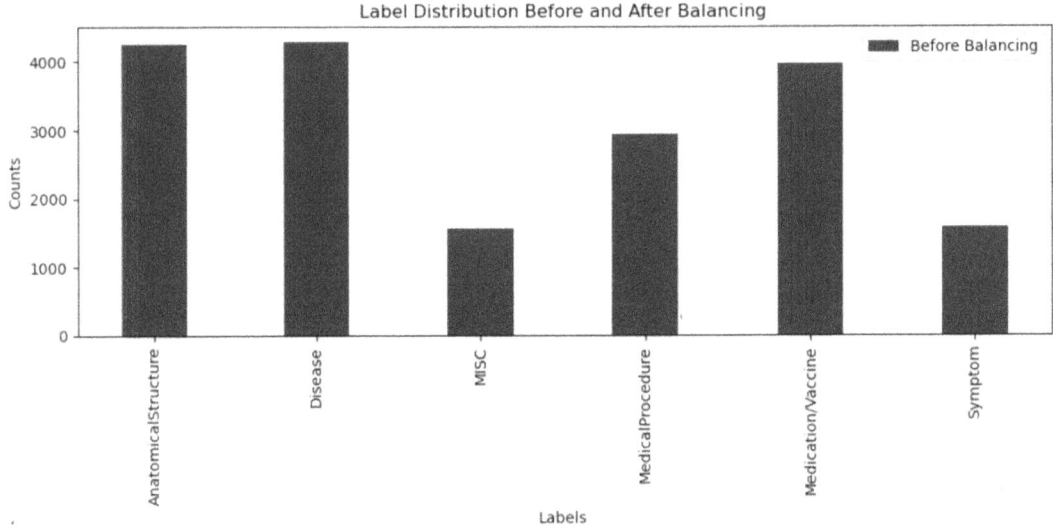

Figure 5-12. *Dataset before balancing*

We notice that the dataset's classes are imbalanced.

The "undersample_label" function applies the "random undersampling" approach to adjust the frequency of a specified label within a dataset to prevent overrepresentation.

Listing 5-13. Undersample function

```
import random
from datasets import Dataset, DatasetDict

def extract_labels(ner_tags_span):
    """
    Extracts the labels from the ner_tags_span field.
    Assumes ner_tags_span is a string representation of a list of [label,
    start, end] triplets.
    """
    tags = ast.literal_eval(ner_tags_span)
    labels = [tag[0] for tag in tags if tag[0] != 'O']
    return labels
```

204

CHAPTER 5 DATASET PREPARATION

```python
def undersample_label(dataset, label_to_undersample, target_count, seed):
    """Reduces the occurrences of a specified label to a target count."""
    examples_with_label = [ex for ex in dataset if label_to_undersample in
        extract_labels(ex['ner_tags_span'])]
    examples_without_label = [ex for ex in dataset if label_to_undersample
        not in extract_labels(ex['ner_tags_span'])]

    random.seed(seed)  # Set the random seed for reproducibility

    if len(examples_with_label) > target_count:
        examples_with_label = random.sample(examples_with_label,
            target_count)

    new_dataset = examples_with_label + examples_without_label
    return new_dataset

# The target count based on the 'Symptom' label
target_count = 2000

# Start with the filtered dataset
balanced_ds_list = filtered_dataset['train']

# Loop over labels with count > 2000 and apply undersampling
for label, count in label_counts.items():
    if count > target_count:
        print(f"Undersampling label '{label}' from {count} to "
            f"{target_count}")
        balanced_ds_list = undersample_label(balanced_ds_list, label,
            target_count, seed=42)

# Convert the balanced dataset to a Hugging Face dataset
balanced_dataset = DatasetDict({
    'train': Dataset.from_dict({
        'text': [ex['text'] for ex in balanced_ds_list],
        'ner_tags_span': [ex['ner_tags_span'] for ex in balanced_ds_list]
    })
})

print("Balancing complete. Dataset ready for training.")
```

Chapter 5 Dataset Preparation

Output

```
Undersampling label 'Disease' from 4291 to 2000
Undersampling label 'AnatomicalStructure' from 4246 to 2000
Undersampling label 'Medication/Vaccine' from 3952 to 2000
Undersampling label 'MedicalProcedure' from 2930 to 2000
Balancing complete. Dataset ready for training.
```

After undersampling is applied to labels with count > 2000, a new balanced dataset is created.

Note In this step, to balance the dataset, we applied the "random undersampling" approach to the majority class. This method reduces the number of examples in the overrepresented classes to match a chosen threshold. This strategy effectively balances class distribution, minimizing bias toward more frequently occurring labels. However, it does mean discarding a significant portion of the data from the overrepresented classes, which could potentially remove valuable information.

Recount and plot the label distribution after balancing.

Listing 5-14. Recalculate the new label counts after balancing dataset

```
# Recalculate the new label counts
new_label_counts = count_labels(balanced_dataset)
print("Label Counts:", new_label_counts)
```

Output

```
Label Counts: Counter({'Medication/Vaccine': 2312, 'AnatomicalStructure': 2236, 'MedicalProcedure': 2135, 'Disease': 2079, 'Symptom': 1465, 'MISC': 975})
```

Listing 5-15. Plot the new label counts after balancing dataset

```
# plot the new label distribution
plot_distributions(label_counts, new_label_counts)
# Plot the label new counts
```

CHAPTER 5 DATASET PREPARATION

Output (Figure 5-13)

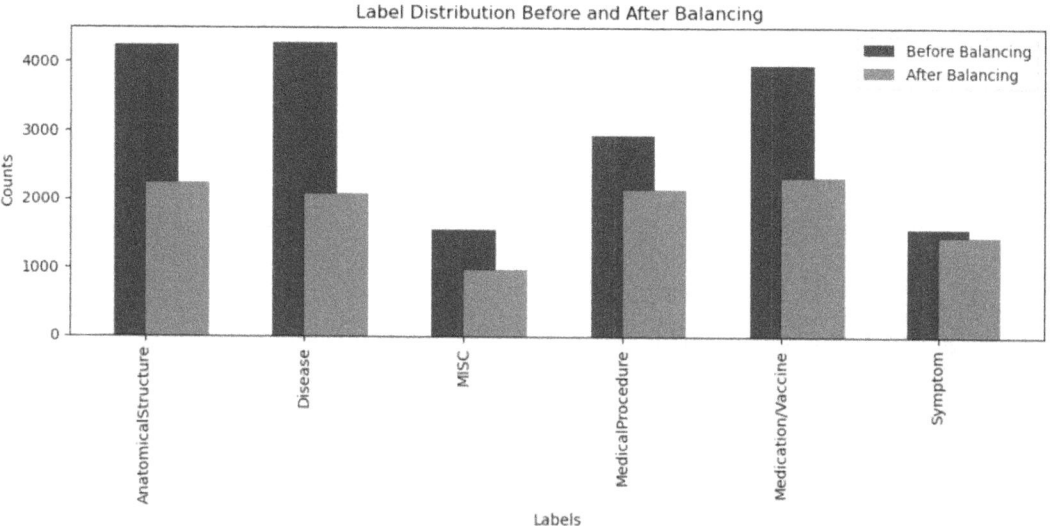

Figure 5-13. *Dataset before and after balancing*

Now we are ready to move to the dataset transformation step (provided in Listings 5-16 to 5-21).

- Transforming: Converts the balanced dataset to an OCI Data Labeling Dataset (JSONL Consolidated format).

Now that we have balanced our dataset, let's transform it into a suitable format for OCI Data Labeling JSONL Consolidated and save it to our Labeling Datasets Bucket.

The code snippet, based on our balanced dataset, creates in our *Labeling Datasets Bucket* the files for our future DLS Dataset to be imported:

- Metadata file (JSONL Consolidated format), which includes the new dataset details and annotations for each example

- Record files, which are text files where each one represents one example from the balanced dataset

Tip The default service limit for DLS record count is 10,000. If we want to avoid the record count limit issue, we need to reduce the size of our balanced dataset. In a real-world situation, if the number of records we need to create exceeds the number of available records, you must file a regular Customer Account Management (CAM) ticket to request an increase in the limit.

207

CHAPTER 5 DATASET PREPARATION

Listing 5-16. Calculate the new dataset size after balancing

```
len(balanced_dataset["train"])
```

Output

9052

Downsize the dataset to 9,000 examples by randomly selecting a subset of 9,000 examples.

Listing 5-17. Downsizing dataset

```
from datasets import DatasetDict

# Set the seed for reproducibility
seed = 42
new_size= 9000

# Shuffle the train dataset using the specified seed
shuffled_train_dataset = balanced_dataset["train"].shuffle(seed=seed)

# Select the first 3000 examples from the shuffled train dataset
healthcare_ner_dataset_v1 = shuffled_train_dataset.select(range(new_size))

# Convert the sampled dataset back to a DatasetDict if needed
healthcare_ner_dataset_v1 = DatasetDict({
    "train": healthcare_ner_dataset_v1
})
healthcare_ner_dataset_v1
```

Output

```
DatasetDict({
    train: Dataset({
        features: ['text', 'ner_tags_span'],
        num_rows: 9000
    })
})
```

Revalidate the label distribution after the downsizing of the balanced dataset.

CHAPTER 5   DATASET PREPARATION

Listing 5-18. Plot label distribution for balanced and downsized datasets

```
# Recalculate the new label counts
healthcare_ner_dataset_v1_label_counts = count_labels(healthcare_ner_
dataset_v1)
print("Label Counts:", healthcare_ner_dataset_v1_label_counts)
```

Output

```
Label Counts: Counter({'Medication/Vaccine': 2298, 'AnatomicalStructure':
2219, 'MedicalProcedure': 2122, 'Disease': 2069, 'Symptom': 1459,
'MISC': 969})
```

Listing 5-19. Plot label distribution for balanced and downsized datasets

```
# Plot the balanced label count vs. downsized label counts
plot_distributions(new_label_counts, healthcare_ner_dataset_v1_label_
counts, "Down-sizing")
```

Output (Figure 5-14)

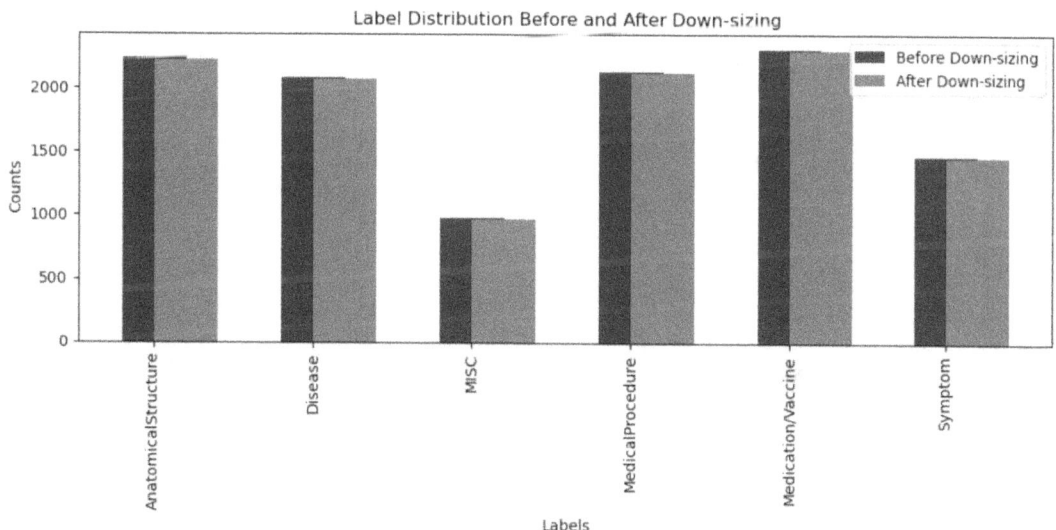

Figure 5-14. Plot label distribution for balanced and downsized datasets

The downsized dataset is balanced enough for our case study.

At this stage, we can save the downsized dataset to the training-datasets-bkt as a restore point for this dataset or for later use (if needed; see Figure 5-15).

209

CHAPTER 5 DATASET PREPARATION

Listing 5-20. Save dataset

```
healthcare_ner_dataset_v1.save_to_disk('/home/datascience/buckets/training-
datasets-bkt/healthcare_ner_dataset_v1.0.0')
```

Output

```
Saving the dataset (1/1 shards): 100%
9000/9000 [00:00<00:00, 151055.97 examples/s]
```

Note Saving the dataset to the bucket is now as easy as saving it to the local file system, thanks to the notebook session storage mount feature.

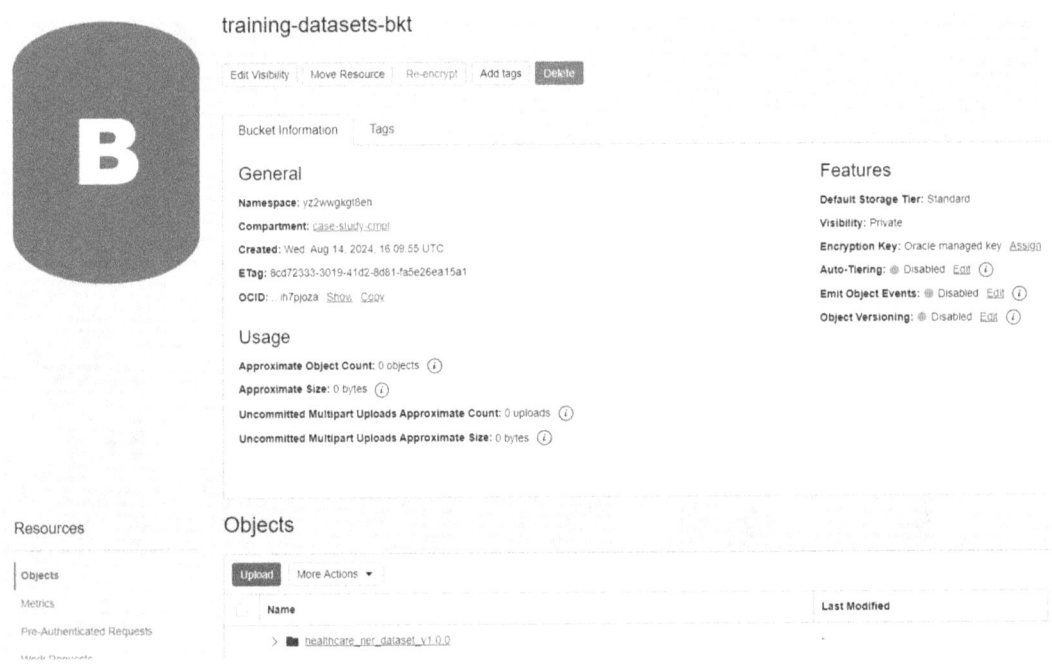

Figure 5-15. *Prelabeled dataset restore point in bucket training-datasets-bkt after cleaning, balancing, and downsizing*

Let's create the DLS Dataset in JSONL Consolidated format.

CHAPTER 5   DATASET PREPARATION

Listing 5-21. prepare_dataset notebook: transforming

```
import oci
from datasets import load_dataset
import json
import io
import os
import tempfile
from tqdm import tqdm

# Initialize OCI Object Storage Client with notebook session's resource
principal
signer = oci.auth.signers.get_resource_principals_signer()
object_storage_client = oci.object_storage.ObjectStorageClient(config={},
signer=signer)

#set dataset display name
dataset_display_name = "healthcare_ner_dataset_v1.0.0"
dataset_description = "Healthcare NER Dataset for the Case Study",

# Initialize Object Storage bucket infos
namespace = object_storage_client.get_namespace().data
bucket_name = "labelling-datasets-bkt"

# Base folder in the bucket
base_folder = f"{dataset_display_name}/"

labels_set = set()
annotations_list = []

for idx, item in enumerate(tqdm(healthcare_ner_dataset_v1["train"],
                        desc="processing dataset records",
                        total=len(healthcare_ner_dataset_v1["train"])
                        )
                    ):
    text = item['text']
    ner_tags_span = eval(item['ner_tags_span'])
# Convert string to list if necessary
```

CHAPTER 5   DATASET PREPARATION

```python
    # Prepare annotations for this row
    entities = []
    for label,start, end in ner_tags_span:
        label = label.replace("/","")
        labels_set.add(label)  # Add to the set of unique labels
        entities.append({
            "entityType": "TEXTSELECTION",
            "labels": [{"label_name": label}],
            "textSpan": {"offset": start, "length": end - start}
        })

    if len(entities) > 0:
        file_name = f"rec-{idx}.txt"
        annotations_list.append({
            "sourceDetails": {"path": file_name},
            "annotations": [{"entities": entities}]
        })

        # Upload the text to OCI bucket
        record_filename = f"{base_folder}{file_name}"
        record_body = io.BytesIO(text.encode('utf-8'))

        #with open(temp_file_path, 'rb') as f:
        object_storage_client.put_object(namespace,
                                        bucket_name,
                                        record_filename,
                                        record_body, #f, #
                                        content_type='text/plain'
                                        )
# Prepare the dataset metadata, i.e., JSONL Metadata
dataset_details = {
    "displayName": dataset_display_name, # "Healthcare NER Dataset v1.0.0",
    "description": "Healthcare NER Case Study - Dataset ready for adding
    annotations using OCI Data Labeling Service.",
    "labelsSet": [{"name": label} for label in labels_set],
```

CHAPTER 5 DATASET PREPARATION

```
    "annotationFormat": "ENTITY_EXTRACTION",
    "datasetFormatDetails": {"formatType": "TEXT"}
}

# Metadata and annotations as JSONL string
jsonl_data = json.dumps(dataset_details) + '\n' + '\n'.join(json.
dumps(annotation) for annotation in annotations_list)

# Write the metadata and annotations to a JSONL file
jsonl_data = io.BytesIO(jsonl_data.encode('utf-8'))

#with open(temp_file_path, 'rb') as f:
object_storage_client.put_object(namespace,
                                 bucket_name,
                                 f"{base_folder}dataset_metadata.jsonl",
                                 jsonl_data, #f, #
                                 content_type='application/json')
print(f'The Dataset {dataset_details["displayName"]} was created
successfully in the bucket {bucket_name}')
```

Output

```
processing dataset records: 100%|██████████████| 9000/9000
[05:34<00:00, 26.91it/s]
The Dataset healthcare_ner_dataset_v1.0.0 was created successfully in the
bucket labelling-datasets-bkt
```

Tip I faced issues while importing a dataset due to misleading "quota limit errors" while using put_object to save a JSONL metadata object in the bucket. To resolve this, I used a temporary file handler as the object body parameter in put_object to create the metadata file in the labeling bucket.

213

CHAPTER 5 DATASET PREPARATION

We can confirm, as illustrated in Figure 5-16, that the number of objects (Approximate Object Count) in our labeling bucket matches the count in the JSONL Consolidated format: 10,000 record files, plus 1 metadata file.

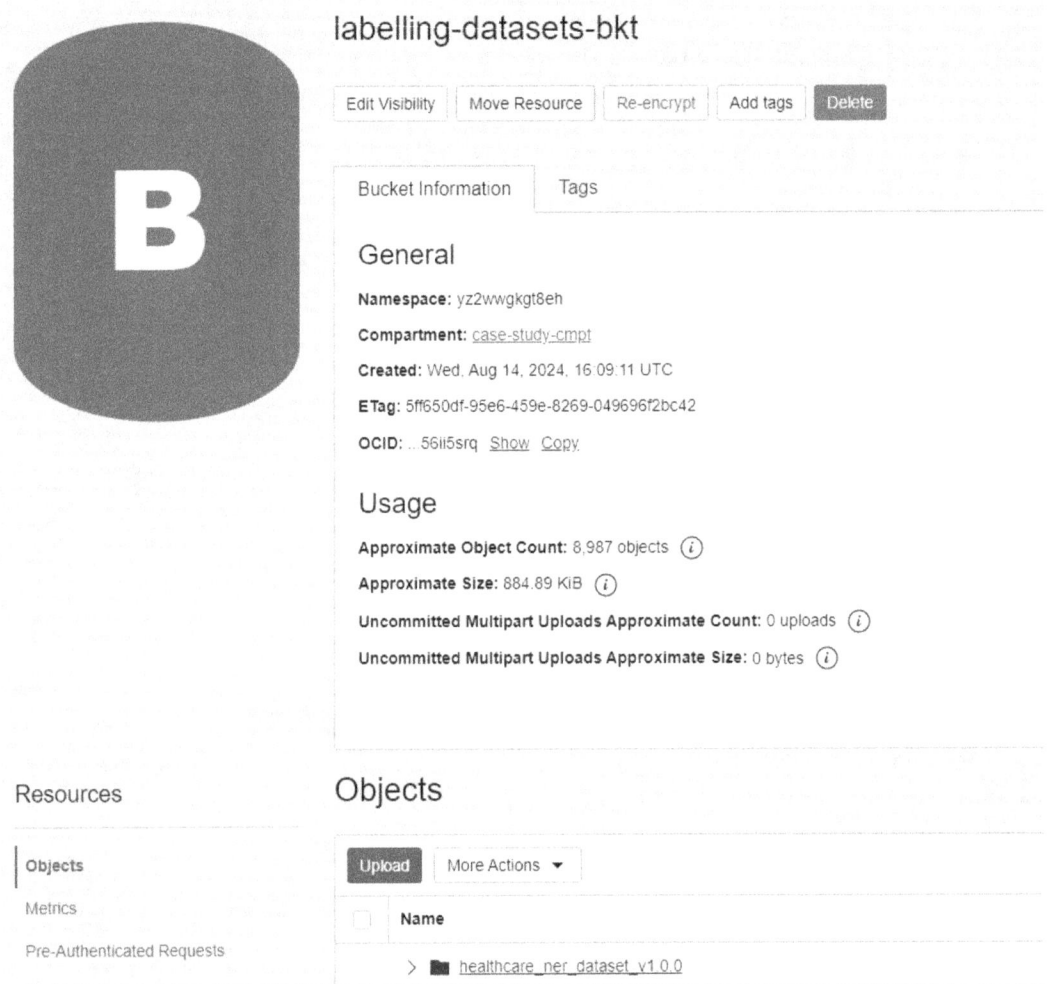

Figure 5-16. Object count in Labeling Datasets Bucket

Here, we can see the dataset's files (the dataset metadata as JSONL and records as txt files; see Figure 5-17).

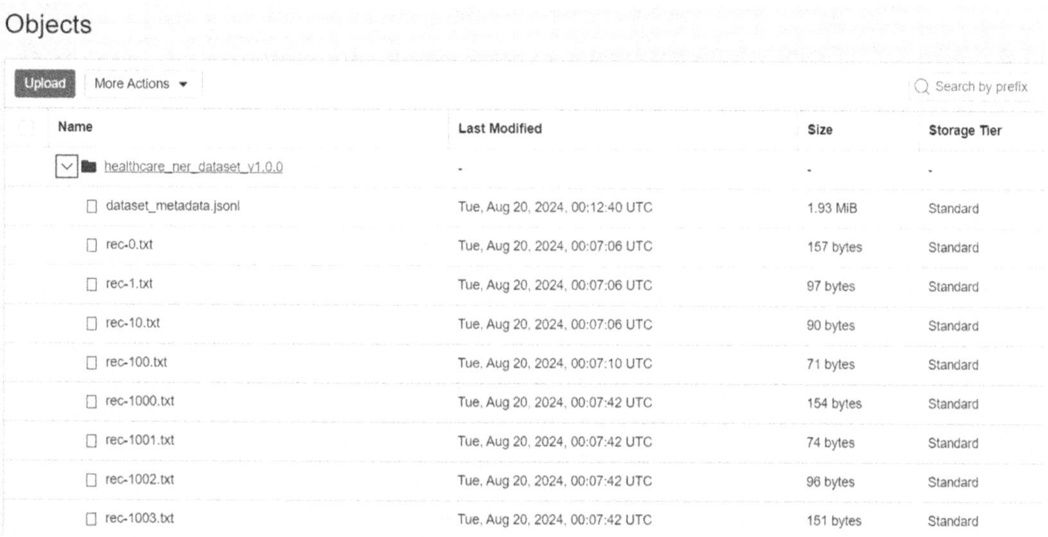

CHAPTER 5   DATASET PREPARATION

Figure 5-17. Healthcare NER dataset v1.0.0 files in JSONL Consolidated format saved to Labeling Datasets Bucket

---

Tip   The NLP consultant did not mount the Labeling Datasets Bucket in the notebook sessions to prevent accidental changes or loss of the data labelers' work, ensuring data security and avoiding unnecessary costs.

Dataset Labeling (Step 5)

Now that the dataset has been collected and transformed into a DLS-compatible format, we are ready to begin the labeling process, specifically adapting and enriching the preexisting dataset. The first step is to import the dataset into a OCI Data Labeling Service (DLS) as a DLS Dataset.

OCI Data Labeling Service (DLS)

The OCI Data Labeling service (DLS) is a tool that enables AI professionals to hand label or annotate datasets. DLS Datasets consist of two primary components: data records and metadata. The metadata contains details about annotations and basic information about the dataset. Data records are representations of data, such as a piece of text in the context of NLP projects. Labels, which are strings of text, become annotations when associated with a data record. ("About Data Labeling—Oracle").

> **Note** To avoid any confusion about the term "datasets," we have decided to provide more context and clarify its usage throughout the book chapters. Initially, we used the term "dataset" as a generic term to describe the input data required for training an NLP model. Later, we discussed Hugging Face Hub Datasets and the Hugging Face Datasets library. Currently, in the context of the Data Labeling service (DLS), when we refer to "datasets," we are specifically talking about the core resource within DLS which is DLS Datasets.[16]

OCI Data Labeling provides public APIs for data labeling. These APIs can expedite the data labeling by enabling automated data annotation at scale. By utilizing DLS APIs and Python SDK, dataset creation, record creation, and annotation can be automated.

From a cost standpoint, DLS can significantly reduce the expenses incurred in data annotation. As we discussed previously, DLS has a competitive pricing policy that offers the first 1000 records free of cost, making the data labeling process even more cost-effective.

In the upcoming sections, our NLP consultant, John Doe, will guide us through the process of annotating our training dataset using OCI Data Labeling Datasets.

Dataset Import

This section is designed to guide you through the straightforward process of importing a preannotated dataset already created in the Data Labeling JSONL Consolidated format.

At the time of writing, OCI does not offer an API for importing datasets. To overcome this limitation, our NLP consultant developed a notebook that automates the dataset import process.

---

[16] To learn more about DLS Datasets, please visit https://docs.oracle.com/en-us/iaas/Content/data-labeling/using/datasets.htm

CHAPTER 5 DATASET PREPARATION

> **Note** Developing a Jupyter notebook to import datasets through Python code instead of using the DLS UI is a deliberate choice made for this book to demonstrate to readers how to effectively use DLS APIs.

Dataset Import Notebook

This notebook aims to assist you in importing the dataset into a *DLS Dataset*.

Before starting the import process, we need to add a new policy to our data science policies as shown in Figure 5-18. This new policy allows us to create and manage DLS Datasets from notebook sessions.

Figure 5-18. New policy for data science policies

Listing 5-22. New data science policy

```
allow dynamic-group data-science-dyn-grp to manage data-labeling-family in
compartment case-study-cmpt
```

CHAPTER 5 DATASET PREPARATION

After adding the new policy, open the CPU-based notebook session. Then, from JupyterLab file browser, open the notebook *import_dataset.ipynb* under the folder (Figure 5-19):

/repos/john-doe-typica-ai/nlp-on-oci.git/chapt-5/

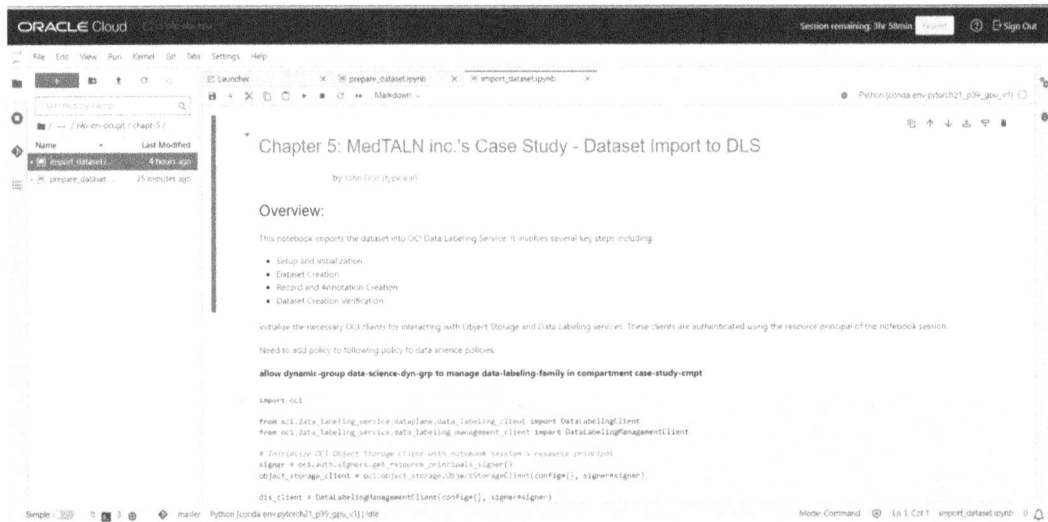

Figure 5-19. *Notebook for dataset import*

Initialization

Initialize the necessary OCI clients for interacting with Object Storage and Data Labeling Services (Listing 5-23). These clients are authenticated using the resource principal of the notebook session.

Listing 5-23. Initialization

```
import oci

from oci.data_labeling_service_dataplane.data_labeling_client import DataLabelingClient
from oci.data_labeling_service.data_labeling_management_client import DataLabelingManagementClient
```

```
# Initialize OCI Object Storage Client with notebook session's resource
principal
signer = oci.auth.signers.get_resource_principals_signer()
object_storage_client = oci.object_storage.ObjectStorageClient(config={},
signer=signer)

dls_client = DataLabelingManagementClient(config={}, signer=signer)
dls_dp_client = DataLabelingClient(config={}, signer=signer)
```

Output

Function to Create a Dataset in OCI Data Labeling Service: This function automates the creation of a dataset in OCI's Data Labeling Service. It takes in various parameters, including compartment details, object storage information (where the data is stored), and labels, and combines them to define and create a dataset (Listing 5-24).

Listing 5-24. Function to create a dataset in OCI Data Labeling Service

```
from oci.data_labeling_service.models import ObjectStorageSourceDetails
from oci.data_labeling_service.models import DatasetFormatDetails
from oci.data_labeling_service.models import LabelSet
from oci.data_labeling_service.models import Label
from oci.data_labeling_service.models import CreateDatasetDetails
from oci.data_labeling_service.data_labeling_management_client import
DataLabelingManagementClient

def create_dataset(compartment_id,
                   namespace,
                   bucket,
                   prefix,
                   ds_display_name,
                   ds_description,
                   ds_format_type,
                   ds_annotation_format,
                   ds_labels):

    # Create the Dataset Source Details object
    dataset_source_details_obj = ObjectStorageSourceDetails(namespace=names
    pace, bucket=bucket, prefix=prefix)
```

```python
# Create the Dataset Format Details object
dataset_format_details_obj = DatasetFormatDetails(format_type=ds_
format_type)

# Create the LabelSet object from the list of labels
label_set_obj = LabelSet(
    items=[oci.data_labeling_service.models.Label(name=label) for label
    in ds_labels]
)

# Create the Dataset Details object
create_dataset_obj = CreateDatasetDetails(display_name=ds_display_name,
                    description=ds_description,
                    compartment_id=compartment_id, annotation_
                    format=ds_annotation_format,
                    dataset_source_details=dataset_source_details_obj,
                    dataset_format_details=dataset_format_details_obj,
                    label_set=label_set_obj)

# Create the dataset and handle exceptions
try:
  response = dls_client.create_dataset(create_dataset_details=create_
           dataset_obj)
  #print(response)
except Exception as error:
  response = error

return response
```

Output

Function to Create a Record in OCI Data Labeling Service: This function facilitates the creation of a record within a dataset in OCI's Data Labeling Service. It constructs the necessary details from the provided parameters and interacts with the Data Labeling Service to register the record (Listing 5-25).

Listing 5-25. Function to create a record in OCI Data Labeling Service

```
from oci.data_labeling_service_dataplane.models import
CreateObjectStorageSourceDetails
from oci.data_labeling_service_dataplane.models import CreateRecordDetails

def create_ds_rec(compartment_id, dataset_id, prefix, rec_name):
    relative_path = rec_name
    name = rec_name

    source_details_obj = CreateObjectStorageSourceDetails(relative_
                        path=relative_path)

    create_record_obj = CreateRecordDetails(name=name,
                        dataset_id=dataset_id,
                        compartment_id=compartment_id,
                        source_details=source_details_obj)
    try:
        response = dls_dp_client.create_record(create_record_details=create_
                    record_obj)
        #print(response.data)
        response = response

    except Exception as error:
        response = error

    return response
```

Output

Function to Add Annotations to a Record in OCI Data Labeling Service: This function adds text selection annotations to an existing record in OCI's Data Labeling Service. It processes a list of annotations, creating entities that specify the label, text offset, and length for each annotated segment (Listing 5-26).

CHAPTER 5   DATASET PREPARATION

Listing 5-26. Function to add annotations to a record in OCI Data Labeling Service

```
from oci.data_labeling_service_dataplane.models import Label
from oci.data_labeling_service_dataplane.models import TextSelectionEntity
from oci.data_labeling_service_dataplane.models import CreateAnnotationDetails

def add_rec_annotation(record_id, annotations_list):

    entity_type = "TEXTSELECTION"

    # Initialize an empty list to store the entities
    entities_obj = []

    for ent_obj in annotations_list:

        # Extract label, offset, and length
        label = ent_obj["labels"][0]["label_name"]
        offset = ent_obj["textSpan"]["offset"]
        length = ent_obj["textSpan"]["length"]
        # Create the labels_obj with the label
        labels_obj = [oci.data_labeling_service_dataplane.models.
                    Label(label=label)]

        # Create the text_span_obj with offset and length
        span_obj = oci.data_labeling_service_dataplane.models.
                TextSpan(length=length, offset=offset)

        # Create the TextSelectionEntity and add it to the
          entities_obj list
        entity = TextSelectionEntity(entity_type=entity_type,
                labels=labels_obj, text_span=span_obj)
        entities_obj.append(entity)

    # entities_obj now contains the desired list of TextSelectionEntity objects
    #print(entities_obj)
    create_annotation_details_obj = CreateAnnotationDetails(record_
                                id=record_id, compartment_
                                id=compartment_id,
                                entities=entities_obj)
```

CHAPTER 5 DATASET PREPARATION

```
    try:
        response = dls_dp_client.create_annotation(create_annotation_
                    details=create_annotation_details_obj)
        #print(response.data)
    except Exception as error:
        response = error
```

Output

Dataset Import

Creation of the Dataset with its annotated records.

Load Dataset Metadata from OCI Object Storage: This code block retrieves and processes the metadata for a dataset stored in OCI Object Storage. The metadata is then used to create and label our dataset in the Data Labeling Service (Listing 5-27).

Listing 5-27. Function to add annotations to a record in OCI Data Labeling Service

```
import json
import os

#compartment where to create the dataset
compartment_id = os.environ['NB_SESSION_COMPARTMENT_OCID']
# Object Storage namespace
namespace = object_storage_client.get_namespace().data
# Dataset Object Storage bucket
bucket_name = "labelling-datasets-bkt"

# Dataset name
ds_name = "healthcare_ner_dataset_v1.0.0"
# Dataset metadata file name (Jsonl Consolidated created in prepare_dataset notebook)
ds_metadata_jsonl_fname = "dataset_metadata.jsonl"

prefix = f"{ds_name}/" # Dataset folder in Object Storage bucket
object_name = f"{prefix}{ds_metadata_jsonl_fname}"
```

```
metadata_jsonl = object_storage_client.get_object(
                namespace,
                bucket_name,
                object_name)
print(f"Dataset metadata file {object_name} loaded")

#load jsonl
metadata_jsonl_obj = [json.loads(jline) for jline in metadata_jsonl.data.
                        content.decode('utf-8').splitlines()]
```

Output

```
Dataset metadata file healthcare_ner_dataset_v1.0.0/dataset_metadata.
jsonl loaded
```

Extract Metadata and Create Dataset in OCI Data Labeling Service: This code block extracts the dataset metadata from a JSONL file and then initiates the creation of a dataset in OCI's Data Labeling Service using the metadata extracted. This process ensures that the dataset is correctly created and ready for further operations, such as record creation (Listing 5-28).

Tip The dataset creation process is asynchronous, so the code waits until the newly created dataset reaches the "ACTIVE" state. This is done using *oci.wait_until*, which periodically checks the dataset's status.

Listing 5-28. Function to add annotations to a record in OCI Data Labeling Service

```
ds_display_name = metadata_jsonl_obj[0]['displayName']
ds_description = metadata_jsonl_obj[0]['description']
ds_annotation_format = metadata_jsonl_obj[0]['annotationFormat']
ds_format_type = metadata_jsonl_obj[0]["datasetFormatDetails"]
['formatType']
ds_labels = [label['name'] for label in metadata_jsonl_obj[0]['labelsSet']]

#print(metadata_jsonl_obj)
```

CHAPTER 5 DATASET PREPARATION

```python
print(f"Start the creation of the dataset {ds_display_name} ...")
#print(ds_annotation_format)
#print(ds_format_type)
#print(ds_labels)

ds_resp = create_dataset(compartment_id,
        namespace,
        bucket_name,
        prefix,
        ds_display_name,
        ds_description,
        ds_format_type,
        ds_annotation_format,
        ds_labels)

if ds_resp.status == 201: #status created
    # Extract the dataset's OCID (unique identifier)
    dataset_id = ds_resp.data.id
    print(f"Dataset named {ds_display_name} created successfully.\nDataset OCID: {dataset_id}")

    # Retrieve opc-request-id from the response headers (optional for
      logging)
    opc_request_id = ds_resp.headers.get("opc-request-id")
    print(f"OPC Request ID: {opc_request_id}")

    # Wait until the dataset reaches the 'ACTIVE' lifecycle state
    print(f"Wait for the dataset {ds_display_name} to be in ACTIVE status...")

    get_dataset_response = dls_client.get_dataset(dataset_id)

    oci.wait_until(
        dls_client,
        get_dataset_response,
```

CHAPTER 5 DATASET PREPARATION

```
        evaluate_response=lambda r: r.data.lifecycle_state == 'ACTIVE',
        max_wait_seconds=60,  # Maximum wait time in seconds
        max_interval_seconds=3   # Check every 30 seconds
    )
    print(f"Dataset {ds_display_name} is now ACTIVE. You can start creating records.")
```

Output

```
Start the creation of the dataset healthcare_ner_dataset_v1.0.0 ...
Dataset named healthcare_ner_dataset_v1.0.0 created successfully.
Dataset OCID: ocid1.datalabelingdataset.oc1.ca-toronto-1.
amaaaaaa3hvgr2qan3yenas7wktkowma6gsjyvfe72ac2xr2pe76iwvkljaq
OPC Request ID: 64B04E4F37A447BDA6C6FB89E0C244BE/19B00E41E5115CDA21C0D
00805AFE1E2/999AB8CB07D840AFC03CA3A1B7375B1E
Wait for the dataset healthcare_ner_dataset_v1.0.0 to be in ACTIVE status...
Dataset healthcare_ner_dataset_v1.0.0 is now ACTIVE. You can start creating records.
```

Create Records and Annotations for Dataset in OCI Data Labeling Service: This code block handles the creation of records and their corresponding annotations for the newly created dataset in OCI's Data Labeling Service. After the dataset is successfully created, the code iterates through each record in the metadata, creating records in the dataset. For each created record, associated annotations are added by looping through the entities defined in the metadata (Listing 5-29).

Listing 5-29. Create records and annotations for dataset in OCI Data Labeling Service

```
import json
from tqdm import tqdm

#loop on records in metadata and create annotated records in the dataset
for idx, json_obj in enumerate(tqdm(metadata_jsonl_obj[1:],
                                    desc="Importing dataset records",
```

```
                        total=len(metadata_jsonl_obj[1:])
                    )
                ):
    rec_name = json_obj["sourceDetails"]["path"]
    #print(f'create record {idx} record name : {rec_name}')
    rec_resp =  create_ds_rec(compartment_id, dataset_id, prefix, rec_name)

    if rec_resp.status==200:
      record_id = rec_resp.data.id

      for annot_obj in json_obj["annotations"]:
        annotations_list = annot_obj["entities"]
        annot_resp = add_rec_annotation(record_id, annotations_list)
        #print(annot_resp)
```

Output

```
Importing dataset records: 100%|████████████████| 9000/9000
[1:15:05<00:00, 2.00it/s]
```

Now that the dataset import process is complete, the labeling team can begin exploring and correcting the dataset annotations using the DLS Dataset UI.

Dataset Labeling

The annotators should check the **Dataset Details** page to confirm the successful import process. On this page, they will find sections such as Dataset Information, Labeling Instructions, Labels, and Tags. The Dataset Information section is particularly important as it provides crucial details about the dataset, including its name, description, and source. Most importantly, it also shows the count of labeled records for the dataset. As shown in Figure 5-21, for our dataset, we can see that all 9000 records are labeled, indicating that there were no errors during the import.

Here are the steps for the labeling team to review and correct the dataset's annotations using the DLS Dataset UI.

1. Open the navigation menu, and click ***Analytics and AI***. Under ***Machine Learning***, click ***Data Labeling***.

2. Click ***Datasets*** (Figure 5-20).

CHAPTER 5   DATASET PREPARATION

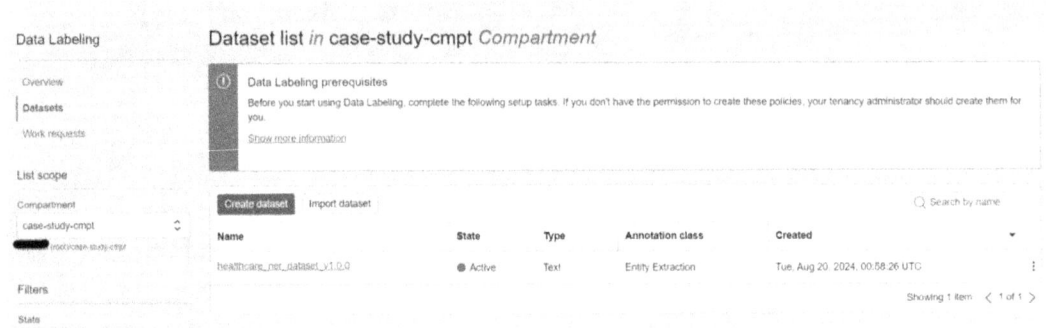

Figure 5-20. Dataset list

3. Then, click the case study dataset, i.e., healthcare_ner_dataset_v1.0.0.

Figure 5-21. Dataset Details page

The annotators can explore and edit data records through the **Details** section, using **Data records** or **Gallery view**.

In this step, leveraging OCI Data Labeling features, we enrich our imported dataset with the missing medical-related annotations.

Below are the steps to annotate the dataset for medical-related labels (Oracle, 2023):

1. In the **Dataset Details** page:

    - If Data records is selected, there are two ways to get to the **Add labels** page:

228

- Click the name of the text you want to label.
- Click the *Action* icon for the text you want to label, and click *Edit Label*.

2. In the *Add labels* page (Figure 5-22):

   a. Under Label, select the label for the part of the text you want to label. If there is only one label in the dataset, the label is automatically selected.

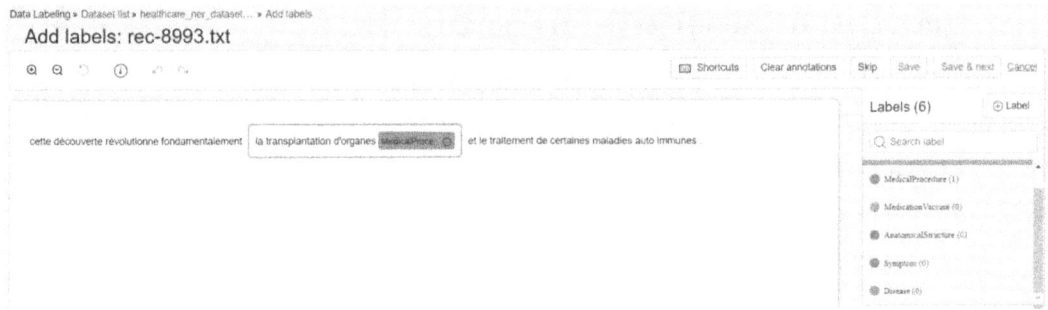

Figure 5-22. Record before manual annotation

   b. Highlight the part of text to be labeled (Figure 5-23).

   c. Repeat Steps 2.a and 2.b until you have added all the labels you wanted to add to the text.

Figure 5-23. Record after manual annotation

Caution For our case study, we do not use nested NER.

CHAPTER 5   DATASET PREPARATION

3. Click **Save & next** to save the changes and label the next item of text, or click **Save** to save the changes. If you don't want to label an item of text, click **Skip** to move to the next item of text. If you're at the last item of text, click **Save & done** to save the changes and return to the **Dataset Details** page.

4. (Optional) Click **Cancel** to return to the **Dataset Details** page.

5. If you have added a label to a record, but not saved it, you're prompted to confirm that you want to cancel.

When labeling medical entities, annotators can refer to the annotation guidelines provided by the labeling team lead if they need help with how to annotate a particular entity. They can access the labeling instructions by clicking the **Labeling instructions** icon, as shown in Figure 5-24.

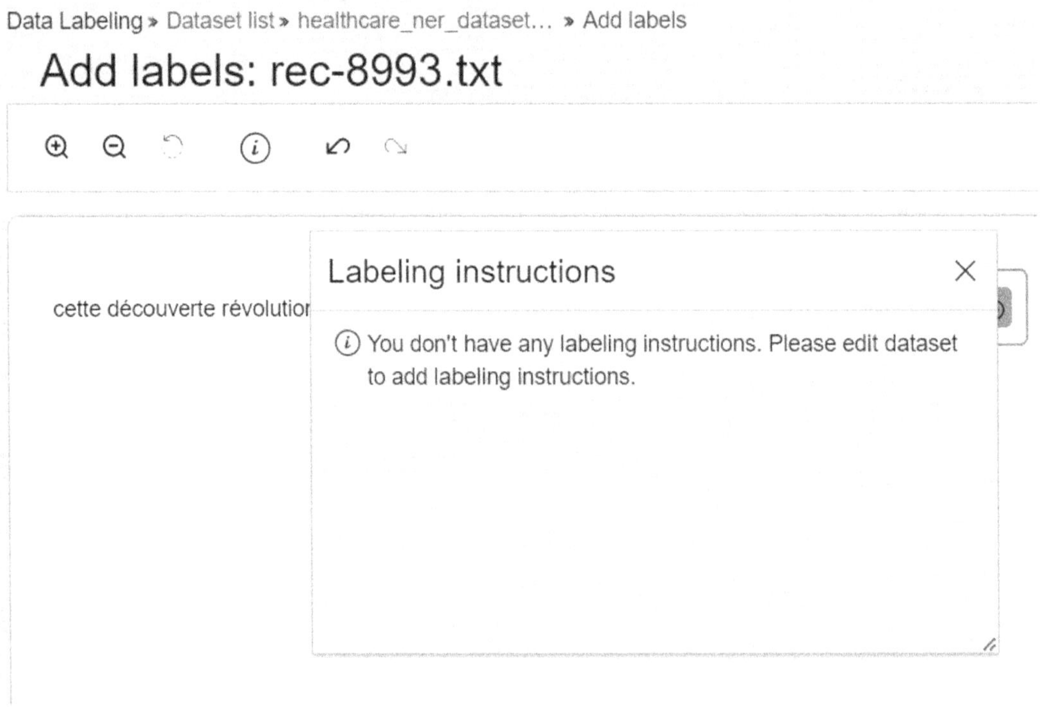

Data Labeling » Dataset list » healthcare_ner_dataset... » Add labels

Add labels: rec-8993.txt

cette découverte révolution

Labeling instructions   ×

ⓘ You don't have any labeling instructions. Please edit dataset to add labeling instructions.

Figure 5-24. Labeling instructions window

> **Tip** The dataset field *Labeling instructions* (optional) should include clear instructions and guidelines for the labeling team.

At this stage, we will assume that the labeling team has completed the annotation process for our dataset, and it is now ready for the training phase. However, before proceeding, it is crucial to validate the annotations to ensure they meet the project's requirements.

Quality Assurance (QA)

There are different methods for performing dataset quality assurance (QA), and for our small dataset, we will be using the human-in-the-loop validation approach. While human subjectivity can introduce labeling errors, it is important to understand that when guided by clear guidelines, human judgment plays a crucial role in the validation process.

The human-in-the-loop[17] (HITL) validation method (Munro, 2021), which integrates human judgment into the validation process, is particularly valuable for verifying the accuracy of annotations provided by automated systems or less experienced annotators. Humans review and rectify annotations as needed, typically using a user interface for validation tasks. This method is highly recommended for final quality assurance to address subtleties that machines may overlook. This approach is extensively discussed in Robert Munro's work, *Human-in-the-Loop Machine Learning*.

Another popular method is spot-checking, which involves randomly selecting a sample of annotations for expert annotators to review in detail. This method can be used as an initial quality check or on an ongoing basis to monitor the quality of annotations throughout the data collection process. Automated scripts can be used to check for common annotation errors or inconsistencies, such as overlapping, missing, or labels that do not conform to predefined rules or schemas.

---

[17] Human-in-the-loop (HITL) is a term used in various contexts. One such context is machine learning, which refers to humans assisting computers in making correct decisions, such as when validating training datasets.

CHAPTER 5   DATASET PREPARATION

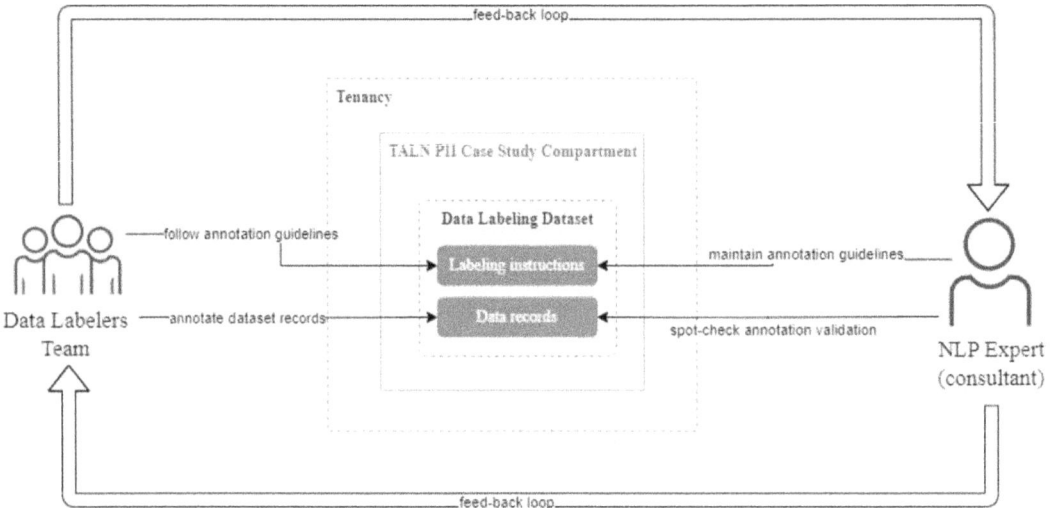

Figure 5-25. HITL spot-check annotation validation

As outlined in Figure 5-25, for our dataset QA, we will combine HITL validation with spot-checking. Our NLP expert, who deeply understands the data labeling process, will perform some spot-checking of our dataset records particularly for medical-related annotations, using the OCI Data Labeling features. This can be done by following these steps:

1. Open the navigation menu and navigate to **Analytics and AI ➤ Machine Learning ➤ Data Labeling**.

2. Click **Datasets**, and then click the name of the dataset you want to edit, i.e., healthcare_ner_dataset_v1.0.0.

3. On the **Dataset Details** page, enter the **Add labels** page, to start spot-checking of the dataset records (Figure 5-26).

    a. Click the **Action** icon for the record you want to validate, and click **Edit Label**.

    b. Navigate between records by clicking the **Skip** button (Ctrl+K).

CHAPTER 5 DATASET PREPARATION

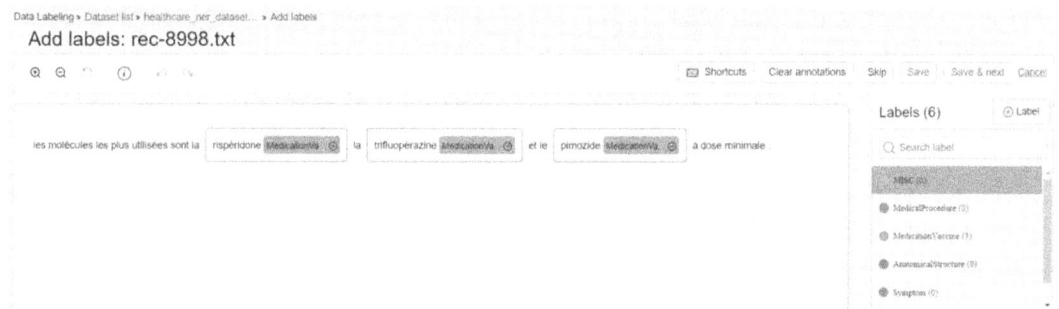

Figure 5-26. *Dataset record in edition mode in the **Add labels** page*

A robust dataset annotation review process involves several best practices to ensure data quality and relevance. Here's a concise outline:

- Annotation Guidelines: Create detailed guidelines with clear definitions and examples of each label or category used. These should be accessible to all annotators for reference.

- Using the **Labeling instructions** tab of DLS Datasets in the **Dataset Details** page, annotators can view any instructions entered by the labeling team lead (e.g., John Doe, our NLP consultant).

- Train annotators using a subset of the data, followed by calibration sessions to align their understanding and application of the guidelines.

- Iterative Review: Implement an iterative review system where another reviews annotations by one annotator to catch and correct errors or inconsistencies.

- Regularly conduct quality checks using metrics such as interannotator agreement (IAA) (Sowmya Vajjala, 2020) to assess annotation consistency and accuracy across the team.

- Establish feedback loops that allow annotators to discuss challenging cases or suggest changes to the annotation guidelines.

- Final Validation: Once the annotation is complete, perform a final spot-check review to randomly review a small sample of annotated data to ensure that it has been correctly labeled.

Data labeling validation is an ongoing process that necessitates collaboration, clear guidelines, and a balance between human judgment and automation. By adhering to these best practices, we can bolster the accuracy of our labeled datasets and thus enhance the performance of our NLP models.

Manually validating and correcting annotations is time-consuming, so for simplicity, we will assume the prelabeled dataset imported into DLS is good enough for our case study. We will proceed directly to the dataset creation step (export).

Dataset Creation (Step 6)

Our labeling team has completed the enrichment and validation of our Healthcare NER dataset.

They will now deliver it to the data scientists' team by exporting the labeled dataset in CoNLL format to the Training Datasets Bucket.

In ***Data Labeling***, we can export datasets to any Object Storage location in our tenancy. This enables us to maintain different versions of the same dataset or use it elsewhere, such as in the input phase for our NLP model training.

We can export our dataset to different file formats such as JSONL, JSONL Compact Plus Content, spaCy, and CoNLL V2003. We selected CoNLL V2003 as the output format for this case study.

Caution   It is important to keep in mind that when exporting text in the CoNLL format, recursive and overlapping entities can be ignored. While this may be acceptable in some cases, it is important to carefully consider whether this is appropriate for your specific use case.

Below are the steps to export our dataset:

1. Open the navigation menu and navigate to ***Analytics and AI ➤ Machine Learning ➤ Data Labeling***.

2. Click **Datasets**, and then click the name of the dataset you want to export, i.e., healthcare_ner_dataset_v1.0.0.

3. In the ***Dataset Details*** page, click **Export** to display the ***Export Dataset*** panel (Figure 5-27).

   Namespace is read-only and shows where the JSON files are stored.

4. For Bucket, select the Training Datasets Bucket, i.e., training-datasets-bkt.

   (Optional) To change the compartment where the Object Storage bucket resides, click ***Change Compartment***. Select the bucket from the list.

5. For the export file format, choose CoNLL V2003.

6. Change the default prefix value to change the version of the dataset from 1.0.0 to 1.10, i.e., healthcare_ner_dataset_v1.1.0/.

   The exported dataset files are stored starting with this path prefix.

7. Leave the checkbox ***Include unlabeled records to export*** unchecked.

Caution When working on a supervised learning task such as Named Entity Recognition (NER), it is important to remember that exporting all records, including those that are yet to be labeled, may not be a useful approach. It is recommended to export only the labeled records, as they are the ones that the model will learn from during the training phase. Exporting the unlabeled records can lead to a model that yields erroneous predictions.

CHAPTER 5   DATASET PREPARATION

8. Click **Export dataset**.

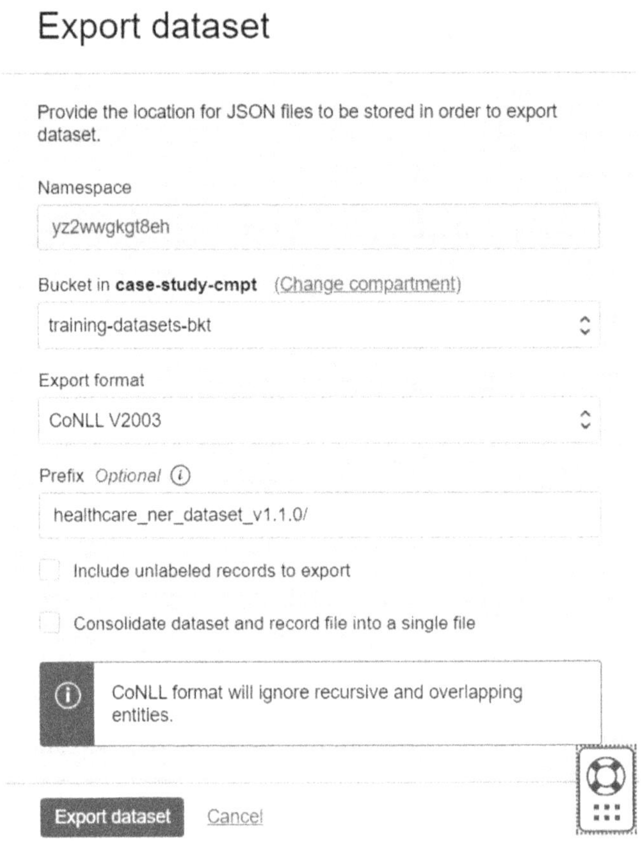

Figure 5-27. Dataset export dialog box

When the exporting process starts, the dataset status is changed to "Updating," and thus, the only action permitted becomes "View work requests." This action allows us to track the progress of the dataset export operation, i.e., **Snapshot Dataset**. Figure 5-28 illustrates this.

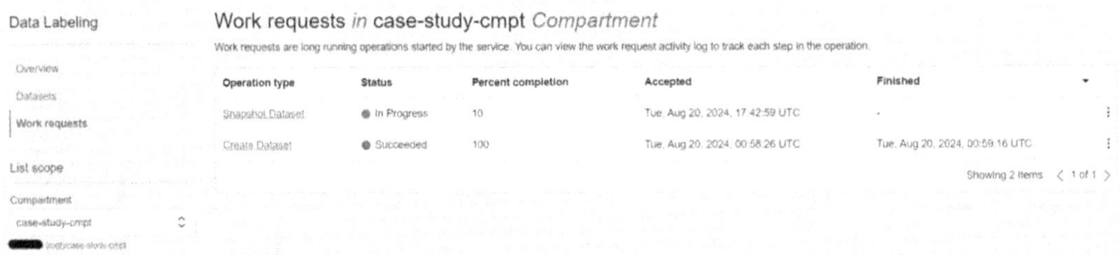

Figure 5-28. Dataset work requests

CHAPTER 5   DATASET PREPARATION

Additionally, you can view the full path of the CoNLL export in the **Snapshot path** field of the associated work request while it's being exported. As shown in Figure 5-29, the **Snapshot path** of our CoNLL file (replace yz2wwgkgt8eh with your Object storage namespace):

https://objectstorage.ca-toronto-1.oraclecloud.com/n/yz2wwgkgt8eh/b/training-datasets-bkt/o/healthcare_ner_dataset_v1.1.0/healthcare_ner_dataset_v1.0.0_1724175778995.conll

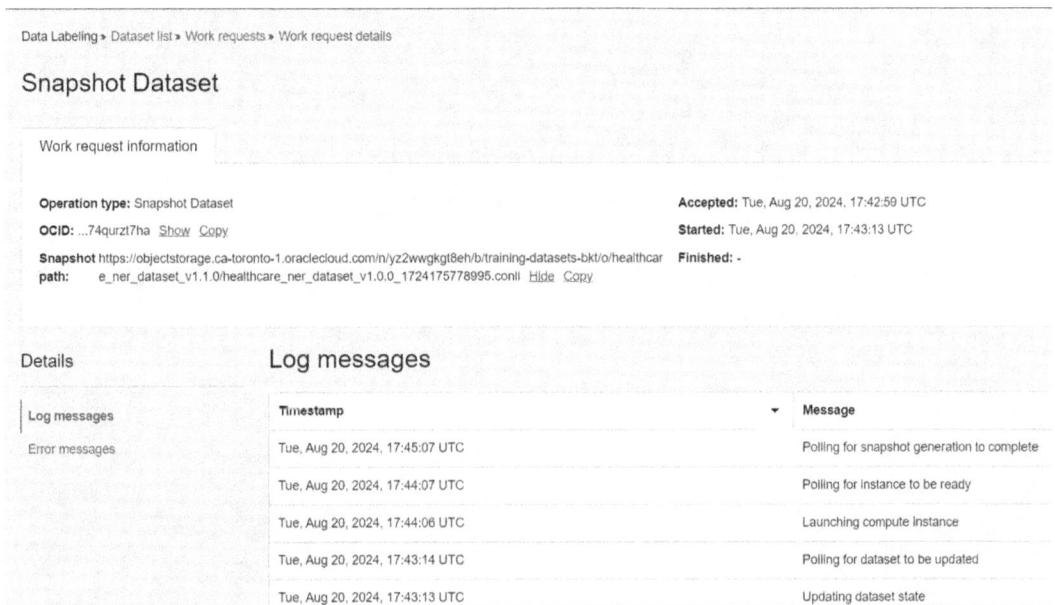

Figure 5-29. Dataset export Work request details page

Once the export is completed, the full path of the CoNLL export can be found on the **Dataset Details** page under the **Snapshot path** field, as shown in Figure 5-30.

CHAPTER 5   DATASET PREPARATION

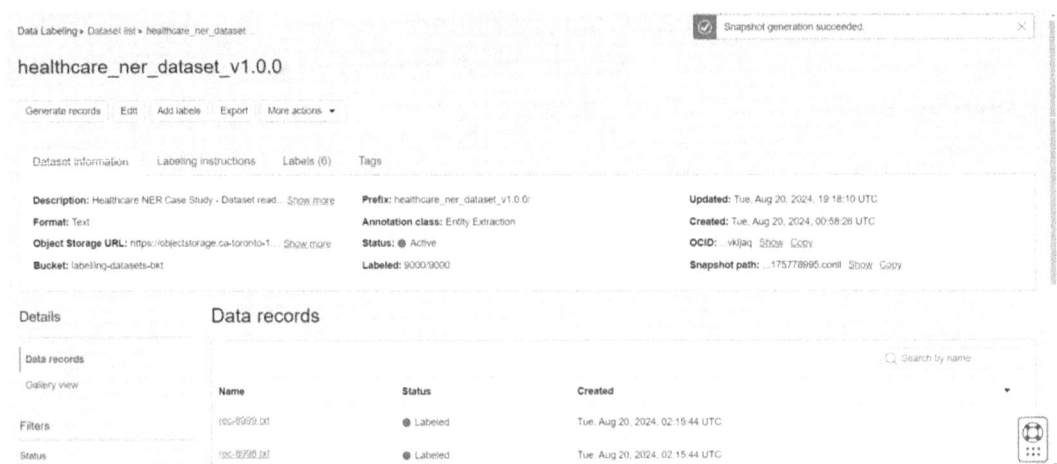

Figure 5-30. Dataset Details page with Snapshot path

To check if the exported dataset healthcare_ner_dataset_v1.1.0, including the CoNLL file exists, go to the Training Datasets Bucket, as shown in Figure 5-31.

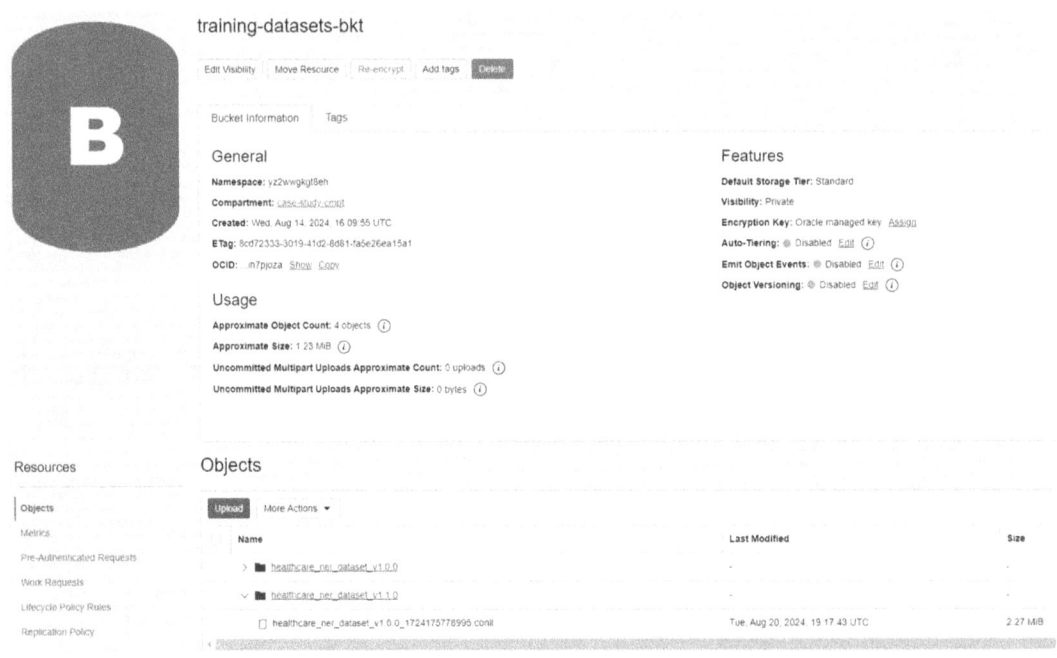

Figure 5-31. Exported dataset with CoNLL file in Object Storage Bucket

CHAPTER 5   DATASET PREPARATION

During the training phase, the data scientists will use the exported CoNLL file to train a Healthcare NER model using the Hugging Face Datasets library.[18]

Additional Notes

This section provides additional details about DLS that are helpful but optional to the main discussion of our case study's dataset preparation steps.

Dataset Import Using DLS UI

Another option for importing the dataset is to use the DLS UI's Import Dataset feature. This is possible because we have created DLS-compatible files in a consolidated JSONL format.

To proceed, you first need to locate the URL of the dataset's metadata file (JSONL). Here are the steps to copy this URL:

1. Go to the Labeling Datasets Bucket, i.e., labelling-datasets-bkt containing our dataset files.

2. Under Objects, click the folder that contains the dataset metadata file, i.e., healthcare_ner_dataset_v1.0.0.

3. Find the metadata file, i.e., healthcare_ner_dataset_v1.0.0/dataset_metadata.jsonl.

4. From the **Action** menu for the metadata file, select **View Object Details** (Figure 5-32).

5. Copy the value for URL path (URI) (replace yz2wwgkgt8eh with your Object storage namespace):

    https://objectstorage.ca-toronto-1.oraclecloud.com/n/yz2wwgkgt8eh/b/labelling-datasets-bkt/o/healthcare_ner_dataset_v1.0.0%2Fdataset_metadata.jsonl

---

[18] Hugging Face Transformers library and Hugging Face Datasets library.

CHAPTER 5   DATASET PREPARATION

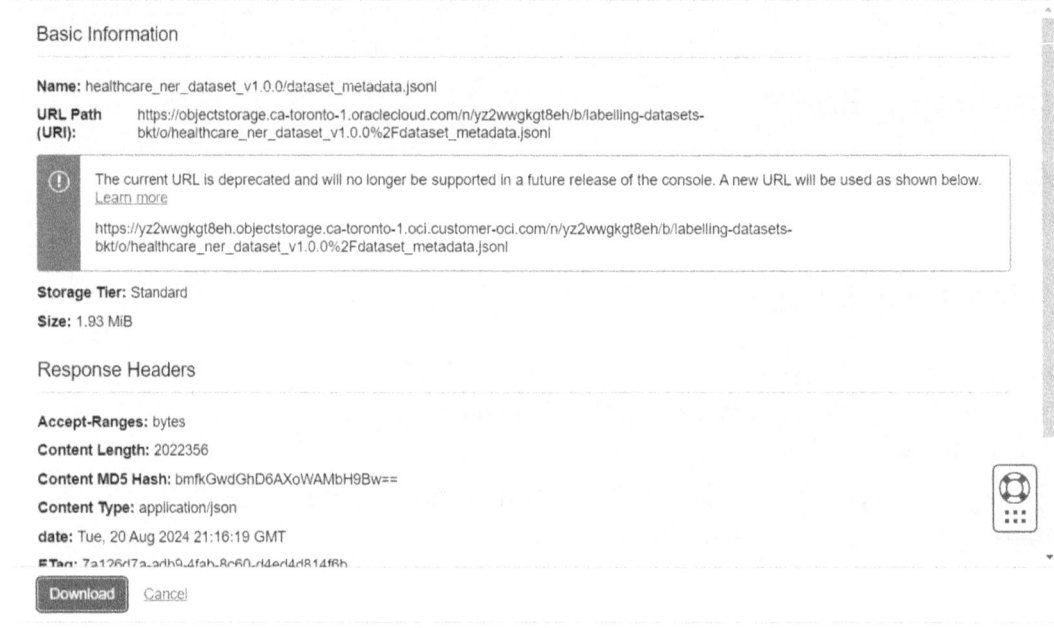

*Figure 5-32. Metadata JSONL file: **Object Details** dialog box from the **Bucket** page*

Follow these steps to import[19] a dataset into **Data Labeling**.

6. Open the navigation menu and click **Analytics and AI**. Under **Machine Learning**, click **Data Labeling**.

7. Click **Datasets**.

8. Click the **Import dataset** button (as shown in Figure 5-33).

---

[19] For more details about importing dataset-supported formats, refer to the documentation at https://docs.oracle.com/en-us/iaas/Content/data-labeling/using/datasets-import-about.htm

CHAPTER 5   DATASET PREPARATION

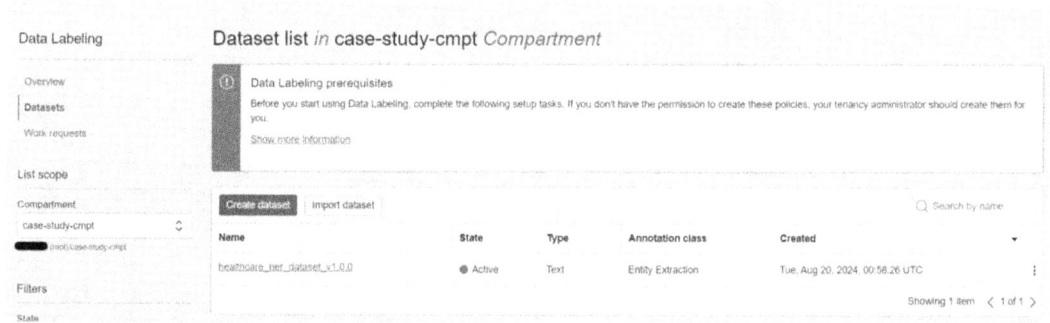

Figure 5-33. Dataset list page

9. To retrieve metadata and records that are already in Object Storage, click **Select from Object Storage**, and follow these steps:

   In Object Storage location, paste the metadata file URL (copied earlier) in the field: **Enter the Object Storage URL for your metadata file**.

   For example (replace yz2wwgkgt8eh with your Object storage namespace):

   https://objectstorage.ca-toronto-1.oraclecloud.com/n/
   yz2wwgkgt8eh/b/labelling-datasets-bkt/o/healthcare_ner_
   dataset_v1.0.0%2Fdataset_metadata.jsonl

   When you have specified it, Namespace, Bucket, Prefix, and Object fields are all populated from the URL.

10. Under **File location**, make sure that the check box *A record is present in the same metadata path* is selected (as shown in Figure 5-34).

241

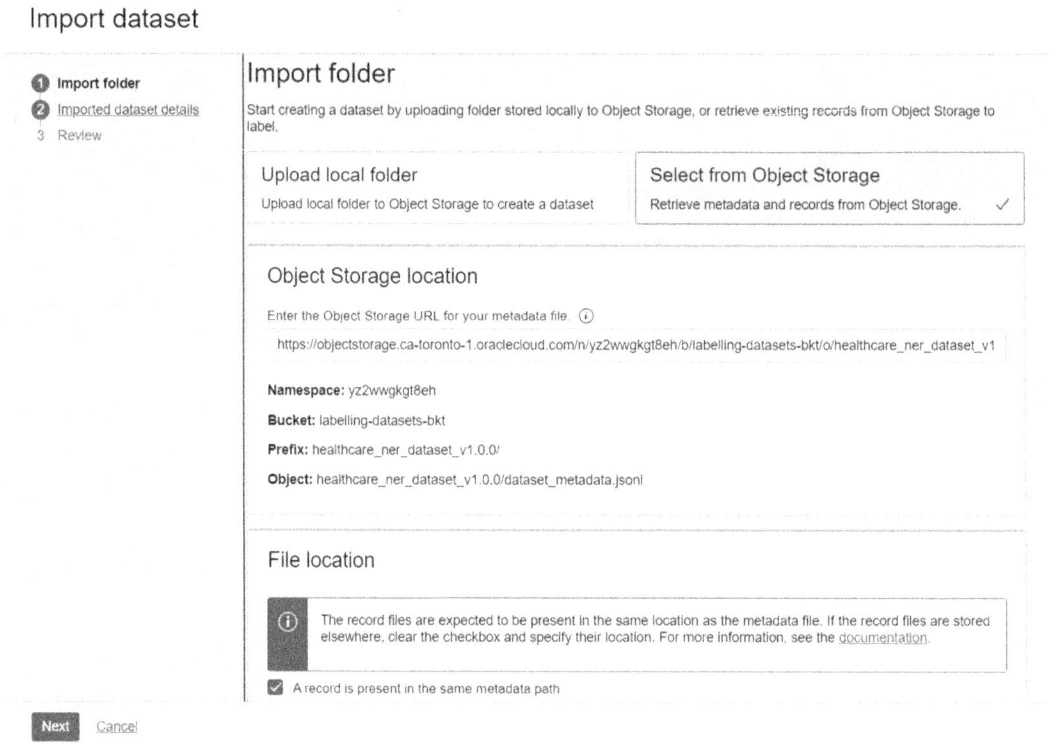

CHAPTER 5   DATASET PREPARATION

*Figure 5-34. Dataset import process: **Import folder** step*

11. Click **Next**.

12. On the **Add dataset details** page (Figure 5-35), the fields are populated from the metadata file, except for the field **Import format**.

    a. Make sure *JSONL Consolidated* is the option selected for the field **Import format**.

CHAPTER 5   DATASET PREPARATION

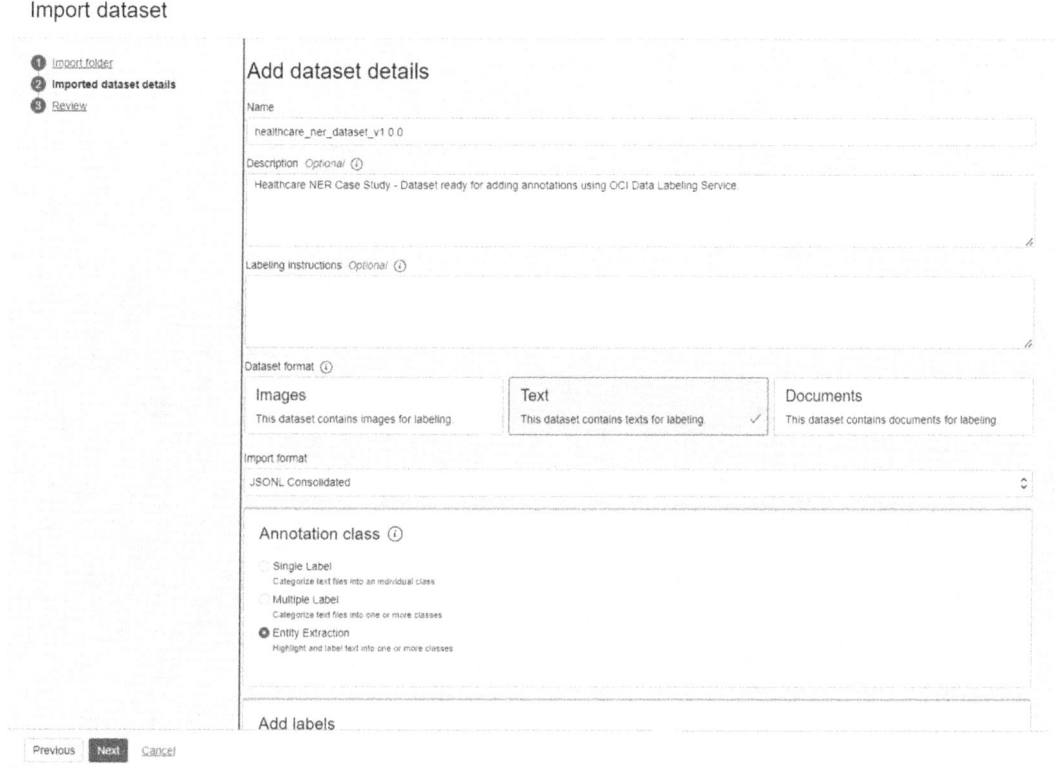

Figure 5-35. *Dataset import process:* ***Add dataset details*** *step*

13. Click ***Next***.

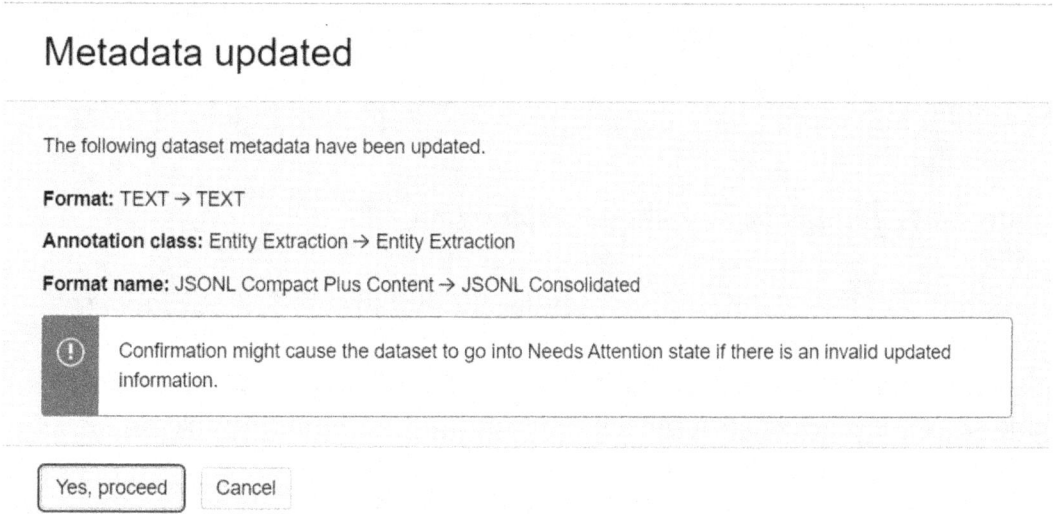

Figure 5-36. *Modifying metadata format*

CHAPTER 5 DATASET PREPARATION

14. *Click **Yes, proceed*** (Figure 5-36).

15. Click **Next**.

16. On the ***Review*** page, as shown in Figure 5-37, verify the information that you entered. If the dataset details need editing, click ***Edit***. If you need to go back and change any values, click ***Edit***.

17. Click ***Import***.

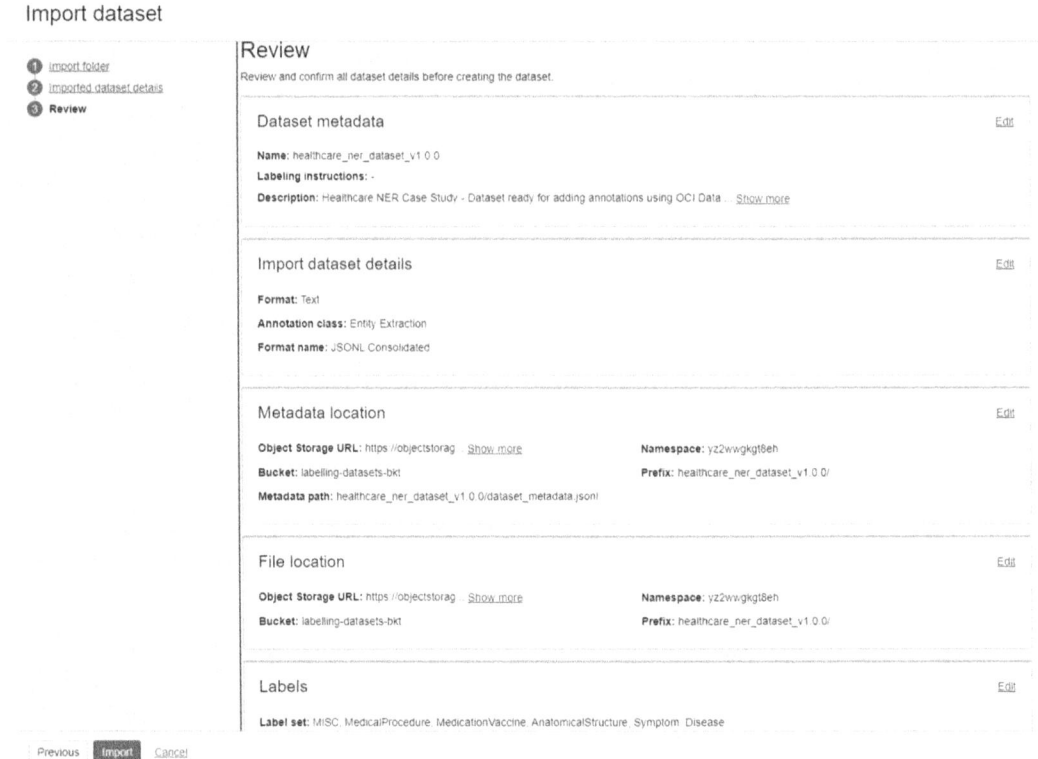

***Figure 5-37.** Dataset import process: **Review** step*

When a dataset is imported, records are generated automatically. During this process, the dataset state changes to "Updating." Once the records are generated, the files used to create them will appear on the ***Dataset Details*** page. However, it's important to keep in mind that generating records can take a considerable amount of time, especially for larger datasets.

The dataset is marked as "ACTIVE" only when the import process is complete, and all the records are created successfully.

CHAPTER 5 DATASET PREPARATION

Record Count Limit

The OCI Data Labeling Service (DLS) has a limit on the number of records that can be created, as illustrated in Figure 5-38. By default, the service limit for record count is set to 10,000.

When creating or importing records for a dataset, it's crucial to ensure that the total number of records across all DLS Datasets does not exceed the available record count to avoid issues related to the record count limit. If the number of records requiring labeling surpasses the available limit, you will need to file a Customer Account Management (CAM) ticket to request an increase in the record count limit.

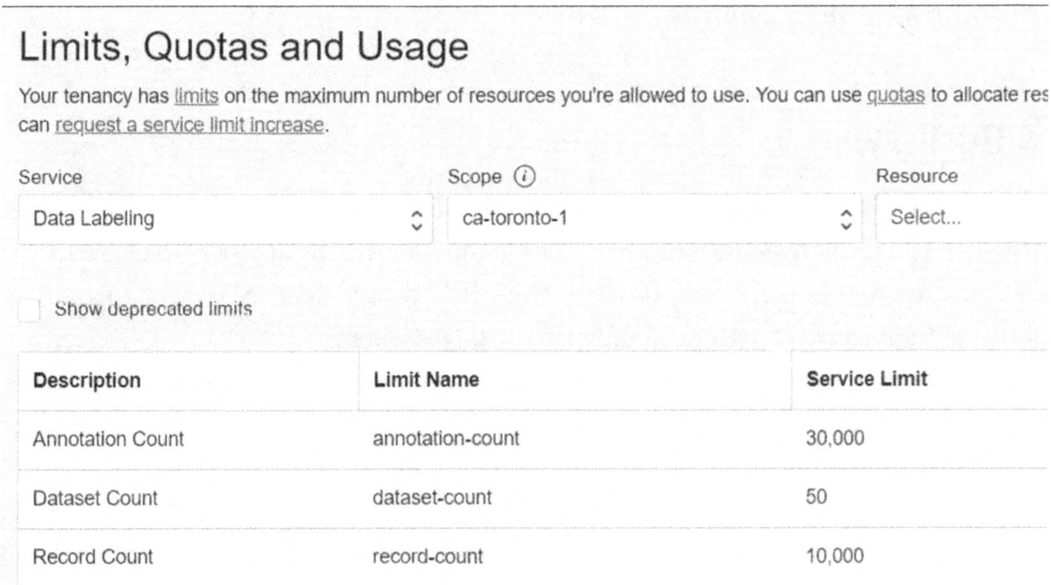

Figure 5-38. Default DLS record count service limit

Our tenancy administrator and the NLP consultant discussed increasing the DLS record count limit, leading to a request that was approved by Oracle, raising the limit to 30,000. Fortunately, due to this proactive action, we still have ample capacity for additional datasets, even after importing our 9,000-record dataset, as shown in Figure 5-39.

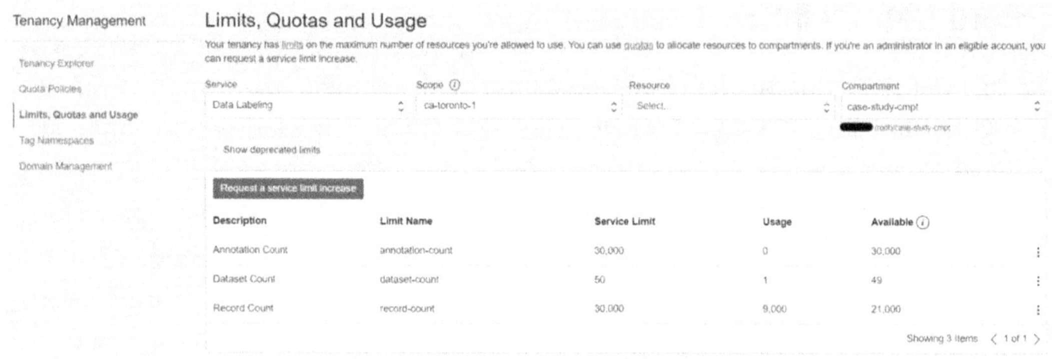

CHAPTER 5   DATASET PREPARATION

Figure 5-39.   DLS record count service limit after increasing the limit and importing the training dataset

Summary

This chapter, while based on the fictional narrative of MedTALN Inc., accurately reflects the real-world challenges and solutions encountered in NLP projects at typica.ai. The strategies and best practices demonstrated by John Doe, the fictional consultant, are grounded in practical experience from actual NLP projects on OCI.

A key focus of the chapter is on building robust yet cost-effective training datasets. This is achieved by leveraging community-curated datasets from reputable sources like Hugging Face and utilizing efficient annotation tools like the OCI Data Labeling Service. The chapter emphasizes the importance of these approaches, particularly for non-English languages, where finding and securing specialized annotators can be both costly and time-consuming, with variable annotation quality. Domain-specific data, especially in medical fields and languages such as French, German, Arabic, or Hindi, requires specialized expertise that is not always readily available.

The chapter also demonstrates how to use Python code within OCI Data Science Notebook Sessions to collect, clean, and import datasets using OCI Data Labeling APIs. This hands-on approach provides readers with practical steps to streamline the dataset building process.

Overall, this chapter serves as a practical guide for preparing training datasets across various NLP tasks, domains, and languages. It offers a detailed methodology and actionable insights to help NLP practitioners and professionals efficiently navigate the dataset building process, empowering them to address the complexities of the NLP landscape with greater confidence.

References

Bowne-Anderson, H. (2020). *The unreasonable importance of data preparation. Your models are only as good as your data.* Retrieved from O'Reilly's Radar: https://www.oreilly.com/radar/the-unreasonable-importance-of-data-preparation/

Joe Reis, M. H. (2022). *Fundamentals of Data Engineering.* O'Reilly Media, Inc.

Lewis Tunstall, L. v. (2022). *Natural Language Processing with Transformers, Revised Edition.* O'Reilly Media, Inc.

McGregor, S. E. (2021). *Practical Python Data Wrangling and Data Quality.* O'Reilly Media, Inc.

Munro, R. (2021). *Human-in-the-Loop Machine Learning.* Manning Publications

Oracle. (2023). *Data Labeling.* Retrieved from docs.oracle.com: https://docs.oracle.com/en-us/iaas/Content/data-labeling/using/home.htm

Sowmya Vajjala, B. M. (2020). *Practical Natural Language Processing.* O'Reilly Media, Inc.

typica.ai. (2024). MedicalNER_Fr: Named Entity Recognition Dataset for the French language in the medical and healthcare domain. https://huggingface.co/datasets/TypicaAI/MedicalNER_Fr

CHAPTER 6

Model Fine-Tuning

This chapter focuses on the process of fine-tuning a pretrained model for healthcare Named Entity Recognition (NER). This chapter provides an in-depth exploration of training the healthcare NER model using OCI's Data Science platform and Hugging Face tools. It covers the fine-tuning process, performance evaluation, and best practices that contribute to creating robust and cost-effective NLP models.

Preliminaries

This section delves into the fundamentals of language models (LMs) and their evolution, with a focus on their pivotal role in transfer learning—a key approach in building our Healthcare Named Entity Recognition (NER) model.

Transfer learning allows us to fine-tune pretrained models that have already absorbed vast amounts of linguistic and healthcare-specific knowledge. This approach streamlines the process of creating an effective and cost-efficient Healthcare NER model.

By the end of this section, you will have a solid understanding of the core concepts behind fine-tuning, preparing you for the practical steps of building our Healthcare NER model using OCI Data Science Notebooks and Hugging Face libraries.

Language Models (LMs)

As we discussed in Chapter 1, language models (LMs) are a foundational concept in NLP. Their importance becomes even more pronounced when we consider fine-tuning pretrained language models (PLMs) for task-specific applications, such as our case study.

CHAPTER 6 MODEL FINE-TUNING

Evolution of LMs

The concept of "language models" has long been central to Natural Language Processing (NLP), representing the tools and algorithms used to understand, predict, and generate human language. Initially, language models were relatively simple, based on statistical methods that could only capture basic patterns in text. However, the term "language models" as we understand it today—referring to sophisticated systems capable of understanding and generating natural language in a manner similar to how humans do—began to take shape in the early 2000s.

As illustrated in Figure 6-1, the evolution of language models (LMs) in NLP has been marked by a series of pivotal milestones, each building on the innovations of its predecessors. Below is an outline of this progression.

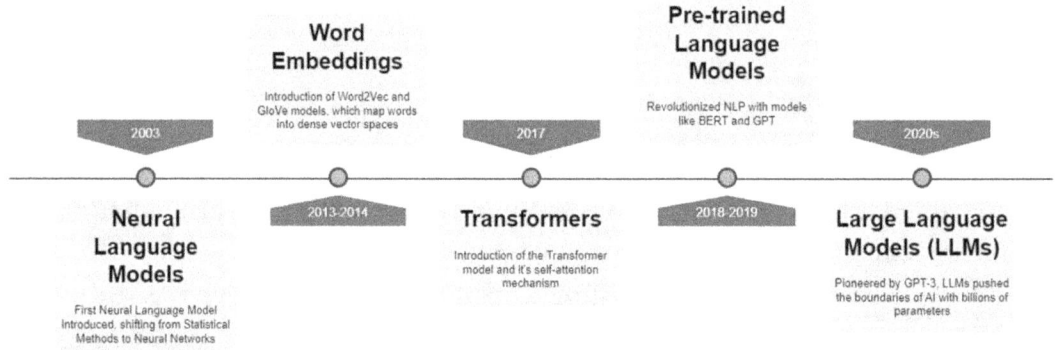

Figure 6-1. *Significant milestones in the evolution of language models*

Neural Language Models (2003)

The first crucial milestone came in 2003 with the introduction of **neural language models** by Yoshua Bengio (Bengio, Ducharme, Vincent, & Jauvin, 2003) and his colleagues. This was the first time neural networks were applied to language modeling, moving beyond the limitations of statistical N-gram models. Bengio's model introduced the concept of representing words in a continuous vector space, allowing for the capture of more complex relationships between words. This approach laid the foundation for the use of deep learning in NLP, setting the stage for the future of language models.

This milestone is crucial because it introduced key ideas that underpin today's language models: the use of neural networks for processing language and the representation of words as dense vectors in a semantic space. These concepts have

CHAPTER 6   MODEL FINE-TUNING

since been expanded and refined, leading to the creation of even more powerful models that can handle a broader range of language tasks with greater accuracy and versatility.

The evolution from these early neural language models to the sophisticated Transformer-based models of today represents a rapid advancement in our ability to create systems that can process and generate human language. The work of Bengio and his team not only changed the trajectory of NLP research but also set the stage for the development of the powerful language models we rely on today, such as BERT, GPT, and other Transformer-based architectures.

Word Embeddings: Word2Vec and GloVe (2013–2014)

A decade later, the development of **word embeddings** through models like **Word2Vec** (Tomas Mikolov, 2013) and **GloVe** (Jeffrey Pennington, 2014) represented a significant leap forward. These models transformed the way words were represented by embedding them in a dense vector space, where words with similar meanings were positioned closer together. This breakthrough allowed language models to understand and capture the semantic relationships between words more effectively, improving performance across a wide range of NLP tasks and providing the foundational representations that would be crucial for subsequent advances.

This development was crucial for several reasons:

- Semantic Understanding: Word embeddings allowed language models to understand the meaning of words in a much deeper way than previous methods. This was a significant leap from the earlier bag-of-words or N-gram models, which struggled to capture semantic relationships between words.

- Improved Generalization: These embeddings enabled language models to generalize better across tasks by using the learned word representations, which improved performance in tasks like sentiment analysis, Named Entity Recognition, and machine translation.

- Foundation for Advanced Models: Word embeddings laid the foundation for subsequent developments in NLP. They became a fundamental building block for more advanced models, including those that used deep learning and Transformers.

Word embeddings paved the way for the even more powerful and versatile Transformer-based models.

Transformers (2017)

While **Transformers** (Ashish Vaswani, 2017) were the next major milestone and arguably the most transformative, Word2Vec and GloVe were essential stepping stones. They introduced the idea of learning word representations in a way that could be effectively utilized by deep learning models, including Transformers. Without the groundwork laid by word embeddings, the leap to Transformers might not have been as impactful or feasible.

Transformers replaced the sequential processing of RNNs and LSTMs with a self-attention mechanism that allowed for parallel processing of text, leading to significantly better performance on a wide range of tasks. This architecture enabled models to capture complex dependencies in text more efficiently and laid the groundwork for the development of pretrained language models. Transformers quickly became the standard architecture for modern NLP, underpinning almost all of today's advanced language models.

Pretrained Language Models (2018–2019)

Building on the Transformer architecture, pretrained language models (PLMs) like BERT (Jacob Devlin, 2018) and GPT (Radford, 2018) represented a new paradigm in NLP. These models were pretrained on massive corpora of text in an unsupervised manner and then fine-tuned on specific tasks, allowing for remarkable performance with minimal task-specific data. BERT introduced bidirectional training, enabling a deep understanding of context, while GPT focused on text generation. PLMs revolutionized how NLP tasks were approached, making it possible to achieve state-of-the-art results across a variety of applications with much less training data.

Large Language Models (LLMs) (2020s)

The 2020s have been characterized by the rise of large language models (LLMs), with models like GPT-3 pushing the boundaries of what is possible in NLP. With billions of parameters, LLMs demonstrated the ability to generate highly coherent, contextually relevant text and perform complex tasks with little to no fine-tuning. These models have become the pinnacle of NLP, enabling sophisticated applications such as conversational

AI, content generation, and more. LLMs continue to redefine the capabilities of language models, driving innovation and setting new benchmarks for performance and versatility in the field.

Acronyms

In the rapidly evolving field of Natural Language Processing (NLP), several acronyms are frequently used to describe the different types of Transformer-based pretrained language models. Understanding these acronyms is essential for grasping the nuances of the various models and their applications.

- PLM (Pretrained Language Model): This term broadly refers to models that are pretrained on large corpora of text before being fine-tuned for specific downstream tasks. PLMs have become the cornerstone of modern NLP applications, enabling tasks like text classification, sentiment analysis, and more.

- LLM (Large Language Model): LLMs are a subset of PLMs distinguished by their large scale, typically involving billions of parameters. These models, such as GPT-4, are known for their ability to generate coherent and contextually relevant text, making them highly versatile and effective across a wide range of NLP tasks, including text generation, translation, and conversation.

- MLM (Masked Language Model): MLM refers to a type of PLM where the model is trained to predict masked words within a sentence, used in models like BERT. These models learn deep contextual representations of language, which are useful for tasks like Named Entity Recognition (NER) and text classification.

- GLM (Generative Language Model): GLM is a term occasionally used to describe models focused on text generation, such as GPT (Generative Pretrained Transformer). However, these models are more commonly referred to by their specific architectures, like GPT, rather than the generic GLM term. These models excel at generating humanlike text and are widely used in applications like chatbots and content creation.

- SLM (Small Language Model): Though not a widely standardized acronym, SLM is used descriptively to refer to smaller-scale language models. These models have fewer parameters and are optimized for efficiency, making them suitable for applications where computational resources are limited or where quick, cost-effective processing is needed.

These acronyms represent key concepts in the realm of Transformer-based pretrained language models. It is increasingly important to be familiar with these terms to understand and take advantage of the latest developments in NLP.

Taxonomy of Pretrained Language Models

Understanding this taxonomy helps in selecting the right model for specific NLP tasks, as well as grasping the underlying principles that differentiate various models such as BERT, GPT, and T5.

Below are the main categories in the taxonomy of pretrained language models, focusing on aspects such as language coverage, model size, and training data:

- Language Coverage: Pretrained language models vary widely in the languages they support. Monolingual models, such as **BERT** for English, **CamemBERT** for French, and **AraBERT** for Arabic, are tailored to capture the nuances and specificities of a single language, making them highly effective for language-specific tasks. These models are trained on large corpora of text in their respective languages, allowing them to understand and generate text with high accuracy. On the other hand, multilingual models like **mBERT** and XLM-R are trained on multiple languages, making them versatile for cross-lingual applications. Although they may not capture the subtleties of each language as well as monolingual models, they are invaluable for applications that need to support multiple languages simultaneously, especially in diverse linguistic regions like the Maghreb, where multilingual capabilities are essential.

- Domain Specialization: Domain-specific language models are tailored to understand and process text within a particular field. **BioBERT** is a prime example, focusing on biomedical literature to aid in tasks like entity recognition and relation extraction in the medical domain. Similarly, **SciBERT** is designed for processing scientific texts, making it useful for researchers who need to analyze academic papers and scholarly articles. **FinBERT**, on the other hand, is specialized for the financial sector, enabling tasks like sentiment analysis and trend prediction in financial documents. These models are pretrained on large datasets relevant to their respective domains, which allows them to outperform general-purpose models in specialized tasks. For startups and enterprises in the Maghreb focusing on specific industries, such as healthcare or finance, these domain-specific models can provide a significant advantage.

- Openness (Open vs. Proprietary): The openness of a language model refers to whether it is freely available for use and modification or is restricted by proprietary access. Open models like BERT, RoBERTa, and GPT-2 are widely accessible, with both their code and pretrained weights available for public use. These models have spurred innovation by allowing researchers and developers to build upon them, customize them, and apply them to a wide range of tasks. On the other hand, proprietary models like **GPT-3**, **Claude**, and **Cohere** are controlled by their developers, typically accessed via APIs. While these models often offer cutting-edge performance, their restricted access can limit how they are used, particularly for small businesses and startups that may not afford the costs associated with proprietary APIs. In regions like the Maghreb, where resources can be limited, open models often provide the best balance between cost and performance.

- Model Size: The size of a Transformer model, typically measured in the number of parameters, significantly impacts its capabilities and resource requirements. Small models like **DistilBERT** and **ALBERT** are designed to be lightweight and efficient, offering faster inference times and lower computational costs while sacrificing some accuracy compared to larger models. These models are particularly useful for applications with resource constraints, such as mobile applications or real-time systems. In contrast, large models like **BERT-large** and **GPT-3** contain hundreds of millions to billions of parameters, allowing them to perform complex tasks with high accuracy but at the cost of requiring substantial computational resources. These large models are typically used for cutting-edge applications where performance is critical, such as advanced AI research or large-scale industrial applications. For startups in the Maghreb, the choice between small and large models will depend on the specific needs of their applications and the resources available to them.

- Architecture Variants: Transformer models can be categorized based on their architecture into **encoder-only**, **decoder-only**, and **encoder–decoder** models, each serving different purposes. Encoder-only models like BERT are primarily used for understanding tasks such as text classification, Named Entity Recognition, and sentiment analysis. These models are powerful at generating meaningful embeddings of input text, making them ideal for tasks requiring deep comprehension of the text. Decoder-only models, such as GPT, are focused on text generation, excelling in tasks where producing coherent and contextually relevant text is essential, like in dialogue systems and creative writing. Encoder–decoder models like T5 and BART combine the strengths of both architectures, making them versatile for tasks that require both understanding and generation, such as summarization, translation, and question answering.

- Use Case Focus: Different pretrained models are optimized for specific use cases, which makes them particularly suited for certain tasks. For instance, conversational agents, models like DialoGPT are specifically optimized for dialogue, enabling them to generate more natural and contextually appropriate responses in conversations.

Healthcare-Specific Pretrained Language Models

In the rapidly evolving field of Natural Language Processing (NLP), the need for specialized models tailored to specific domains, such as healthcare, has become increasingly apparent. For tasks like Named Entity Recognition (NER) in the healthcare domain, especially when dealing with non-English languages such as French, leveraging domain-specific small pretrained language models offers a promising approach to achieving high-quality results while maintaining cost-effectiveness.

This section explores the benefits and strategies of utilizing such models, particularly in the context of building a Healthcare NER model for the French language.

Why Domain-Specific Models for Healthcare

Healthcare is a highly specialized field with unique terminologies, abbreviations, and contextual nuances that are not commonly found in general language corpora. Therefore, general-purpose language models, even those pretrained on large datasets, often fall short when applied to healthcare-related tasks. Domain-specific pretrained models, particularly those trained on healthcare-related text, are better equipped to handle the intricacies of medical language, leading to more accurate and reliable NER results.

In the French language, this challenge is compounded by the relative scarcity of large-scale annotated healthcare datasets compared to English. As a result, the use of small, domain-specific pretrained models becomes even more crucial for tasks like Healthcare NER. These models can provide a more precise understanding of medical entities and relationships, ultimately leading to better performance in downstream tasks.

The advantages of small Healthcare-specific pretrained language models are as follows:

- Cost-Effectiveness: Large pretrained models, while powerful, are often expensive to train and deploy due to their size and complexity. They require significant computational resources, which can drive up costs, particularly in environments where budgets are constrained. Small pretrained language models, on the other hand, are more efficient in terms of resource usage. They can be fine-tuned more quickly, reducing the time and computational power needed for training, which directly translates to cost savings.

- Specialization: Small models that are pretrained on domain-specific corpora, such as healthcare texts in French, offer the advantage of being more specialized. While they may not have the broad coverage of larger models, their training on domain-specific data allows them to excel in specific tasks. For instance, a small model trained on French medical texts would be more adept at recognizing and classifying entities like drug names, medical conditions, and procedures compared to a general-purpose model.

- Adaptability and Fine-Tuning: Another significant advantage of small pretrained models is their adaptability. These models can be fine-tuned on smaller, task-specific datasets, such as a curated NER dataset for the French healthcare domain, with greater ease than larger models. The fine-tuning process is less resource-intensive, which allows for quicker iterations and more experimentation, enabling developers to optimize the model's performance for the specific task at hand.

- Balancing Model Quality and Cost-Effectiveness: In the context of building a Healthcare NER model for the French language, leveraging domain-specific small pretrained language models strikes an effective balance between model quality and cost-effectiveness.

Despite their smaller size, domain-specific models often outperform larger general-purpose models when applied to specialized tasks. For instance, a small French healthcare model pretrained on medical journals, clinical notes, and healthcare-related articles will likely achieve higher accuracy in NER tasks than a larger model trained on a general corpus. This targeted performance is crucial in healthcare, where accuracy and reliability are paramount.

Furthermore, the reduced complexity of small models also facilitates their deployment in various environments, including cloud-based services and edge devices. This flexibility is particularly beneficial in healthcare settings, where data privacy and real-time processing are often critical. Small models can be deployed closer to where the data is generated, enabling quicker and more secure processing of sensitive healthcare information.

Leveraging domain-specific, small pretrained language models for the French healthcare domain appears to offer a compelling strategy for building a high-quality, cost-effective Healthcare NER model for our case study.

Why Open Pretrained Models

With free access to state-of-the-art models, platforms like Hugging Face help accelerate NLP adoption. Models published on Hugging Face are mostly open source, meaning they are freely available to anyone, eliminating the need to invest in proprietary software or licenses. This makes advanced NLP technology accessible without upfront costs.

CHAPTER 6 MODEL FINE-TUNING

In addition to the inherent advantages of pretrained models, open source models from Hugging Face provide additional benefits.

- Access to State-of-the-Art Models: Hugging Face provides access to a wide range of state-of-the-art models trained on large, diverse datasets. Leveraging these models allows us to benefit from cutting-edge research without the associated development and training costs.

- Community and Support: Hugging Face has a large community and extensive documentation, which reduces the cost associated with troubleshooting, experimenting, and developing custom models. This community-driven support accelerates development and reduces the need for in-house expertise.

Basically, using pretrained models from Hugging Face is a smart choice because it saves on computational resources, time, and data collection while giving us access to top-notch models that can be adjusted for specific tasks with minimal extra effort.

Cost-Saving Strategies for the Training Phase

One of the biggest challenges in deploying NLP solutions is finding the right balance between cost-effectiveness, functionality, and performance. Training NLP models can be particularly expensive due to the high demand for GPU resources, especially when training is only needed intermittently.

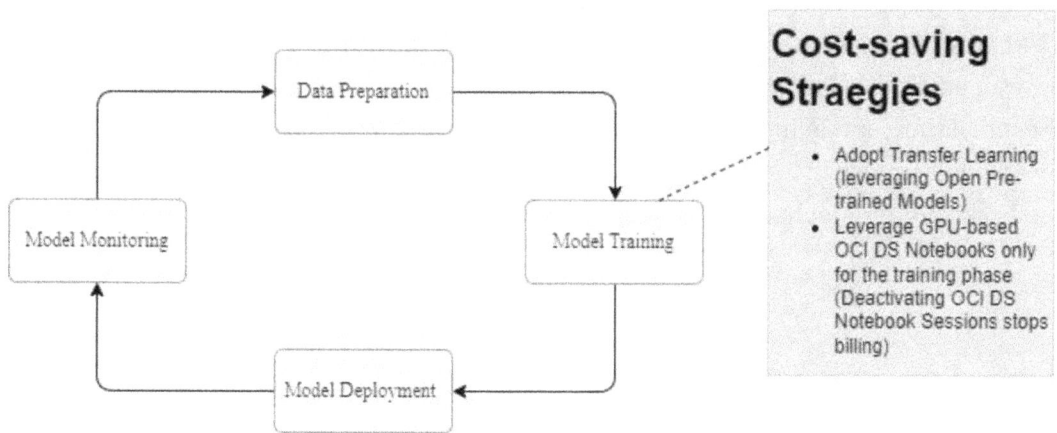

Figure 6-2. Cost-savings strategies for the training phase

As depicted in Figure 6-2, the following cost-saving strategies will be applied during the model training phase and will be explained in more detail in the upcoming sections:

- Adopting a Transfer Learning Approach: We utilized healthcare-specific pretrained models for the French language available on the Hugging Face Hub. This approach allows us to leverage existing models, reducing the need for extensive, costly training from scratch.

- Leveraging OCI Data Science Notebooks: By using OCI's GPU-based Data Science Notebooks only during the training phase and deactivating them during idle times, we minimized costs. OCI's billing stops when notebooks are deactivated, which enhances cost efficiency.

To keep expenses low, we decided to use open pretrained models available on Hugging Face for our case study. This strategy provides access to high-quality models without direct costs and fosters innovation and customization through the open source community.

Through this case study, you will see how strategic decisions, such as leveraging OCI's features and external resources like the Hugging Face Hub, can significantly reduce the financial barriers to implementing advanced NLP projects. By capitalizing on OCI's capabilities and embracing transfer learning, it's possible to achieve substantial cost savings in traditionally expensive areas.

Transfer Learning–Based Fine-Tuning Workflow

The following text provides a high-level overview of our workflow for fine-tuning pretrained models for Healthcare NER. This workflow outlines the sequential steps from the initial problem definition to the final evaluation and ranking of fine-tuned models. By following this structured process, we ensure that each step of the training phase is methodically approached, resulting in high-performance Healthcare NER models.

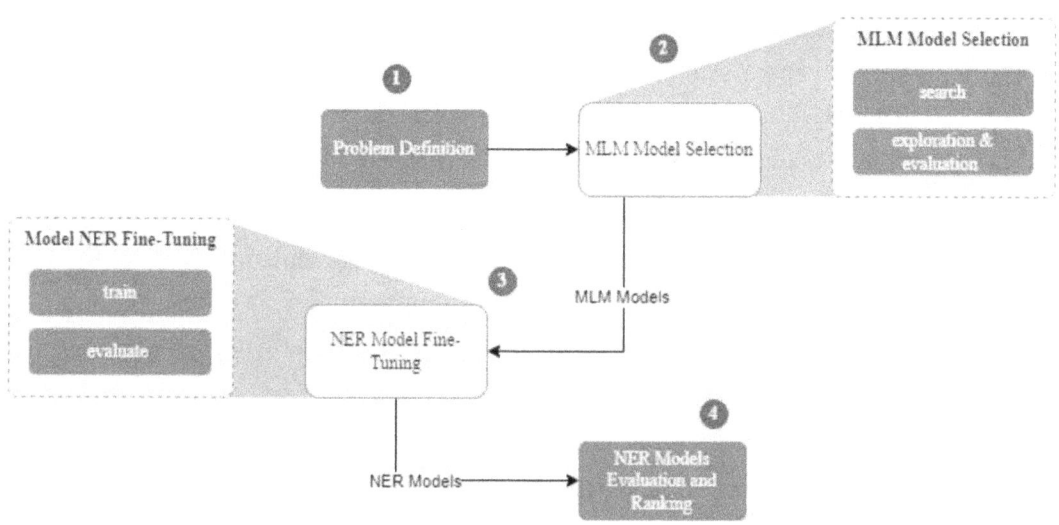

Figure 6-3. *High-level flow for fine-tuning a pretrained language model MLM of NER*

Figure 6-3 illustrates the key steps of the workflow, which include defining selection criteria for pretrained models, choosing appropriate pretrained models, fine-tuning them for NER, and evaluating and ranking the models based on their performance. Each phase of the workflow builds upon the previous one, ensuring a logical progression from identifying the problem to deploying the model, with a strong emphasis on cost-effectiveness.

Key steps in the training process are as follows:

1. Problem Definition: The initial phase of the workflow involves problem definition. During this phase, we define the criteria for selecting Masked Language Models (MLM) that are suitable for our case study, such as those that support the French language for the healthcare domain.

2. MLM Model Selection: In this phase, our main focus is to choose the best Masked Language Modeling (MLM) models from the Hugging Face Hub that can be fine-tuned for our healthcare NER task. This phase consists of two important steps:

    a. Search: We start by searching the Hugging Face Hub for candidate MLM models that support the French language and have been pretrained on healthcare data.

b. Exploration and Evaluation: Once potential models are identified, they are further explored and evaluated to determine their suitability for fine-tuning. This involves assessing how well these models can predict masked tokens related to the healthcare domain. Models that perform well in this step are selected for the next phase.

3. NER Model Fine-Tuning: After selecting the most promising MLM models, the third phase involves fine-tuning these models to adapt them specifically for the NER task. This phase is divided into two substeps:

    a. Train: The selected models are fine-tuned on a healthcare-specific dataset where they learn to recognize and classify medical entities.

    b. Evaluate: After training, the models are evaluated on a validation set to assess their effectiveness. Key metrics such as precision, recall, and F1 score determine how well the models perform on the NER task.

4. NER Model Evaluation and Ranking: The last step of the workflow includes evaluating and ranking the fine-tuned NER models. During this phase, each model's performance is compared using the F1 measure. The evaluation results are then used to rank the models, which helps in selecting the best performing NER model. The winning healthcare NER task is tested against a small manually created dataset to visually and manually assess its predictions by a human expert.

To fine-tune our Healthcare NER models, John Doe, our NLP consultant, will use OCI Data Science Notebooks. He will provide step-by-step guidance through Python notebooks and Hugging Face libraries (Transformers and Datasets libraries) throughout the process.

Pretrained Model Selection

In this section, we outline the process of selecting pretrained Masked Language Modeling (MLM) models for fine-tuning, leveraging Hugging Face (HF).

Framing the Problem (Step 1)

In this phase, our NLP consultant has worked on establishing objective criteria for choosing Masked Language Modeling (MLM) models to fine-tune for our healthcare Named Entity Recognition (NER) task.

The criteria for selecting the MLM models are based on key factors such as the model's relevance to the healthcare domain, its support for the French language, and its architecture and size.

Table 6-1 summarizes the selection criteria for the pretrained MLM models.

Table 6-1. Pretrained MLM selection criteria

Criteria	Description	Example Value
Language	The model must support French.	fr
NLP Task	The model must be designed for the Masked Language Modeling task.	fill-mask
Library	The model must be implemented in the PyTorch library.	pytorch
Transformer Framework	The model must use the Hugging Face Transformers library.	transformers
Domain-Specific Tags	The model must be tagged with one or more relevant domain-specific tags.	healthcare, medical, clinical, biomedical, biology, life science
Single Language Support	The model must support only one language, specifically French. Ensuring it is specialized for the target language without other language influences.	

Additionally, as shown in Listing 6-1, the NLP consultant prepared a small dataset to evaluate the predictions of the selected MLM models. This evaluation shall provide insights into the models' effectiveness in predicting masked medical terms, allowing us to shortlist the top five models for the fine-tuning phase.

Listing 6-1. Tiny handmade dataset for MLM model selection

```
examples = [
    {
        "text": "Le medecin donne des {} en cas d'infections des voies respiratoires.",
```

```
            "expected_entities": [{'antibiotiques': 1}]
    },
    {
            "text": "Le médecin recommande des {} pour réduire l'inflammation
            dans les poumons.",
            "expected_entities": [{'corticoïdes': 1}, {'anti-inflammatoires':
            0.9}]
    },
    {
            "text": "Pour soulager les symptômes d'allergie, le médecin
            prescrit des {}.",
            "expected_entities": [{'antihistaminiques': 1}]
    },
    {
            "text": "Pour gérer le diabète, le médecin prescrit une {}.",
            "expected_entities": [{'insulinothérapie': 1}]
    },
    {
            "text": "Après une blessure musculaire, le patient doit suivre
            une {}.",
            "expected_entities": [{'physiothérapie': 1}, {'rééducation': 0.8}]
    },
    {
            "text": "En cas d'infection bactérienne, le médecin recommande
            une {}.",
            "expected_entities": [{'antibiothérapie': 1}]
    }
]
```

Tip Though tiny, this domain-specific test set has been carefully crafted to contain healthcare-related sentences with well-known medical terms masked (e.g., antibiotics). Pretrained MLM models that fail to accurately predict these masked medical terms are considered to perform poorly in the healthcare domain in French and, thus, discarded from our top five selection.

CHAPTER 6 MODEL FINE-TUNING

MLM Model Selection from Hugging Face (Step 2)

As depicted in Figure 6-4, the workflow of selecting pretrained Masked Language Modeling (MLM) involves searching and filtering models based on the defined selection criteria using HF APIs and evaluating programmatically their performance against a handcrafted tiny evaluation dataset. This systematic approach ensures that we identify the top five MLM models best suited for our healthcare NER task.

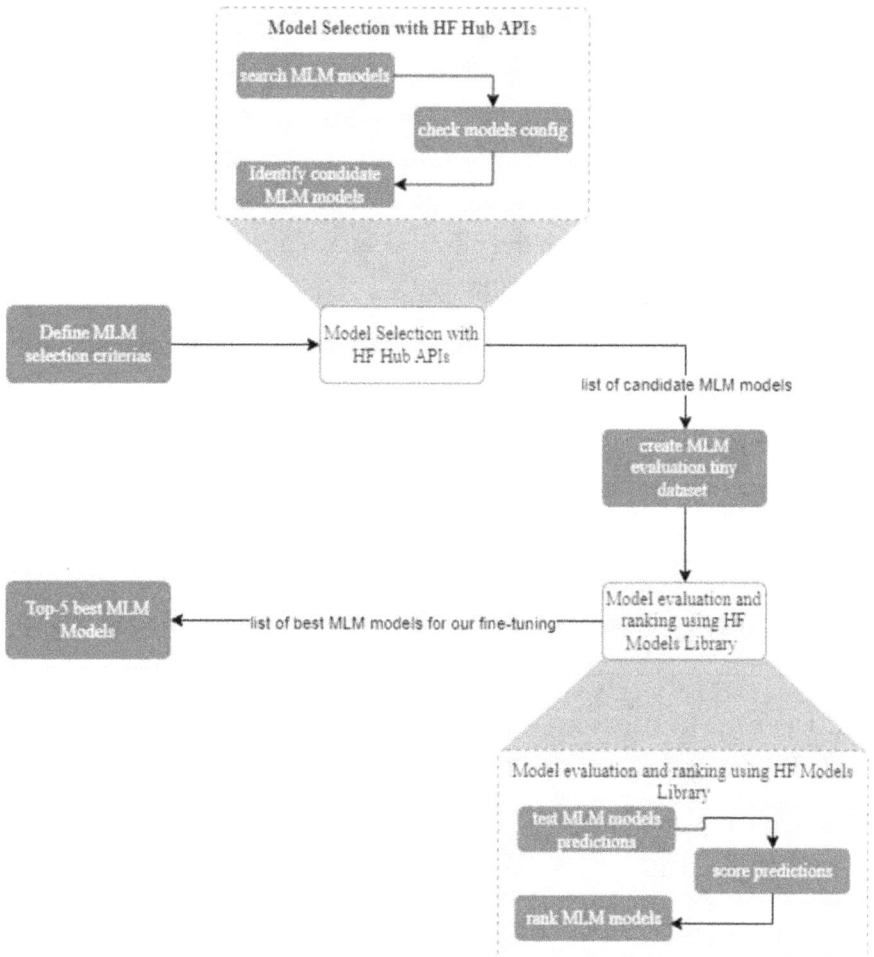

Figure 6-4. Pretrained MLM model selection workflow using Hugging Face

Pretrained Model Selection Notebook

This notebook guides you through the process of selecting a list of pretrained models that suitable for our case study, i.e., MLM supporting healthcare domain and the French language that can be fine-tuned into a Healthcare NER model.

This notebook will help us identify top-performing MLM models based on specific objective criteria. The notebook is structured into the following key steps:

1. Search for MLM Models: We begin by searching for candidate MLM models on the Hugging Face Hub that are suitable for our needs.

2. Evaluate and Rank Models: We evaluate the selected MLM models on a small, handcrafted dataset to determine which models best predict medical entities in the fill-mask task. The models are then ranked based on their performance.

For the training, we will need to open the GPU-based OCI Data Science Notebook Session.

1. Open the GPU-based notebook session:

    a. Go to **Analytics & AI ➤ Machine Learning ➤ Data Science ➤ Projects**.

    b. Open the our OCI DS project, i.e., cs-nlp-prj.

    c. If it is deactivated, activate the CPU-based OCI Data Science Notebook Session, i.e., cs-nlp-nbs-cpu.

    d. Open the CPU-based OCI Data Science Notebook Session, i.e., cs-nlp-nbs-cpu.

2. From the JupyterLab file browser, open the notebook ***select_mlm_models.ipynb*** under the folder (Figure 6-5):

    /repos/john-doe-typica-ai/nlp-on-oci.git/chapt-6

CHAPTER 6 MODEL FINE-TUNING

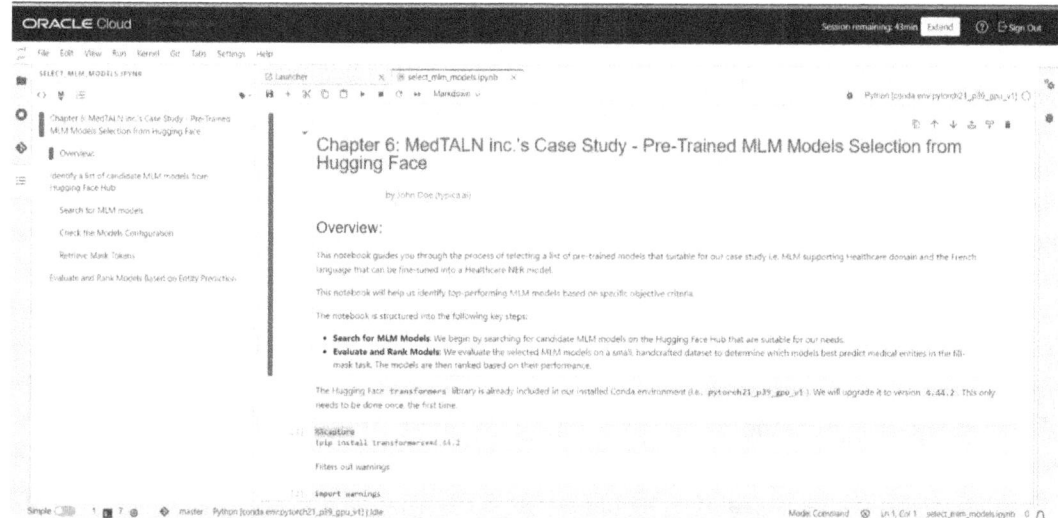

Figure 6-5. Notebook for MLM model selection

Identify a List of Candidate MLM Models from Hugging Face Hub

We will use the Hugging Face Transformers library to search and filter models programmatically (see Listings 6-2 through 6-5).

1. Search Models: Search for the MLM models supporting the French language.

2. Filter Found Models: Filter out the returned models to retain only models supporting only French (monolingual) and have at least one of the healthcare domain–related tags.

The Hugging Face "Transformers" library is already included in our installed conda environment (i.e., pytorch21_p39_gpu_v1). We will upgrade it to version "4.44.2." This only needs to be done once, the first time.

Listing 6-2. Installing Dependencies

```
%%capture
!pip install transformers==4.44.2
```

CHAPTER 6　MODEL FINE-TUNING

Search MLM Models

Listing 6-3. Searching Models

```
from huggingface_hub import list_models
# Fetch the list of models with the specified criteria
models = list_models(
    language ="fr", task="fill-mask", library = "pytorch", cardData = True
)

# List of tags to filter by
filter_tags = ["healthcare", "medical",  "clinical", "biomedical",
"biology", "life science"]

# Print the model IDs and some basic information
included_models = []
for model in models:
    if  len(model.card_data.language) == 1 and \
        model.card_data.library_name == 'transformers' and \
        any(tag in model.tags for tag in filter_tags):
      included_models.append(model.modelId)

included_models
```

　　Output

```
['Dr-BERT/DrBERT-4GB',
 'Dr-BERT/DrBERT-7GB',
 'Dr-BERT/DrBERT-4GB-CP-PubMedBERT',
 'almanach/camembert-bio-base',
 'Dr-BERT/DrBERT-7GB-Large',
 'abazoge/DrLongformer',
 'abazoge/DrBERT-4096',
 'PantagrueLLM/jargon-general-base',
 'PantagrueLLM/jargon-general-biomed',
 'PantagrueLLM/jargon-biomed-4096',
 'PantagrueLLM/jargon-multidomain-base',
 'PantagrueLLM/jargon-biomed',
 'PantagrueLLM/jargon-NACHOS',
 'PantagrueLLM/jargon-NACHOS-4096']
```

Check the Model Configuration

In this step, we validate that the selected models adhere to the architecture of the BERT base model, specifically with 12 hidden layers and 12 attention heads. This ensures consistency in terms of model size for our fine-tuned models.

Listing 6-4. Checking the Model's Configuration

```
from transformers import AutoConfig

# List of models to check
model_ids = included_models
# Initialize an empty dictionary to store model ID and their details
models_with_right_config = []

# Function to fetch the number of layers and attention heads
def get_model_details(model_id):
    try:
        # Load the model configuration
        config = AutoConfig.from_pretrained(model_id, trust_remote_
            code=False)

        # Get the number of layers and attention heads
        num_layers = config.num_hidden_layers
        num_heads = config.num_attention_heads

        return num_layers, num_heads
    except Exception as e:
        return f"Error retrieving config for {model_id}: {e}", None

# Iterate through the models and populate the dictionary with their details
for model_id in model_ids:
    details = get_model_details(model_id)
    if details[0] == 12 and details[1] == 12:
      models_with_right_config.append(model_id)

models_with_right_config
```

CHAPTER 6   MODEL FINE-TUNING

Output

```
['Dr-BERT/DrBERT-4GB',
 'Dr-BERT/DrBERT-7GB',
 'Dr-BERT/DrBERT-4GB-CP-PubMedBERT',
 'almanach/camembert-bio-base',
 'abazoge/DrLongformer',
 'abazoge/DrBERT-4096']
```

Retrieve Mask Tokens

In this step, we retrieve the mask tokens for each of the models that retained models so far. Using the "AutoTokenizer" from the Hugging Face Transformers library, we load the tokenizer associated with each model and extract its mask token.

We then validate that the mask token matches the expected tokens, specifically "[MASK]" or "<mask>", which are commonly used in models like BERT and its variants. Models that meet this criterion are stored in the "models_with_mask_tokens" dictionary for further processing.

Listing 6-5. Identifying Mask Tokens

```python
from transformers import AutoTokenizer

# Initialize an empty dictionary to store model ID and mask token
models_with_mask_tokens = {}

# Function to fetch the mask token using the tokenizer
def get_mask_token_via_tokenizer(model_id):
    try:
        # Load the tokenizer
        tokenizer = AutoTokenizer.from_pretrained(model_id)

        # Get the mask token
        return tokenizer.mask_token
    except Exception as e:
        return f"Error retrieving tokenizer for {model_id}: {e}"
```

CHAPTER 6 MODEL FINE-TUNING

```
# Iterate through the models and populate the dictionary with mask tokens
for model_id in models_with_right_config:
    mask_token = get_mask_token_via_tokenizer(model_id)
    if mask_token in ["[MASK]", "<mask>"]:
        models_with_mask_tokens[model_id] = mask_token

# Print the constructed dictionary
models_with_mask_tokens
```

Output

```
{'Dr-BERT/DrBERT-4GB': '<mask>',
 'Dr-BERT/DrBERT-7GB': '<mask>',
 'Dr-BERT/DrBERT-4GB-CP-PubMedBERT': '[MASK]',
 'almanach/camembert-bio-base': '<mask>',
 'abazoge/DrLongformer': '<mask>',
 'abazoge/DrBERT-4096': '<mask>'}
```

Evaluate and Rank Models Based on Entity Prediction

In this step, we evaluate a set of models to determine their effectiveness in predicting specific medical entities within masked sentences. Using the "fill-mask" pipeline from the Hugging Face Transformers library, each model is tested on a series of examples where key medical terms are masked (as detailed in Listing 6-6).

The models are scored based on how well their predictions match a combination of generic and specific expected entities. These scores are then aggregated to produce a cumulative score for each model. Finally, the models are ranked based on their cumulative scores, helping us identify the most effective model for our healthcare NER task.

Listing 6-6. Ranking Models

```
from transformers import pipeline

# Define the generic expected entities with their weights
generic_expected_entities = [
    {'médicaments': 0.3},
    {'traitements': 0.3},
    {'soins': 0.3},
```

CHAPTER 6   MODEL FINE-TUNING

```
        {'remèdes': 0.3},
        {'conseils': 0.1},
        {'indications': 0.1},
        {'instructions': 0.05},
        {'interventions': 0.05},
        {'compléments': 0.05}
]
# Define the examples and their specific expected entities
examples = [
    {
        "text": "Le medecin donne des {} en cas d'infections des voies
        respiratoires.",
        "expected_entities": [{'antibiotiques': 1}]
    },
    {
        "text": "Le médecin recommande des {} pour réduire l'inflammation
        dans les poumons.",
        "expected_entities": [{'corticoïdes': 1}, {'anti-
        inflammatoires': 0.9}]
    },
    {
        "text": "Pour soulager les symptômes d'allergie, le médecin
        prescrit des {}.",
        "expected_entities": [{'antihistaminiques': 1}]
    },
    {
        "text": "Pour gérer le diabète, le médecin prescrit une {}.",
        "expected_entities": [{'insulinothérapie': 1}]
    },
    {
        "text": "Après une blessure musculaire, le patient doit suivre
        une {}.",
        "expected_entities": [{'physiothérapie': 1}, {'rééducation': 0.8}]
    },
```

```
    {
        "text": "En cas d'infection bactérienne, le médecin recommande
        une {}.",
        "expected_entities": [{'antibiothérapie': 1}]
    }
]

models = models_with_mask_tokens

# Initialize a dictionary to store the cumulative scores for each model
model_scores = {model_name: 0 for model_name in models}

# Iterate over each model
for model_name, mask_token in models.items():
    print(f"Testing {model_name} ...")
    try:
        # Load the fill-mask pipeline for the current model
        fill_mask = pipeline("fill-mask", model=model_name, tokenizer=model_
                    name, trust_remote_code=False)

        # Iterate over each example
        for example in examples:
            # Prepare the example sentence with the correct mask token
            masked_example = example["text"].format(mask_token)
            specific_expected_entities = example["expected_entities"]

            # Combine generic and specific entities, giving priority to
            specific ones
            combined_expected_entities = {**{k: v for d in generic_expected_
                                            entities for k, v in d.items()},
                                        **{k: v for d in specific_expected_
                                            entities for k, v in d.items()}}

            # Get predictions
            results = fill_mask(masked_example)

            # Extract the top predicted tokens
            predicted_tokens = [result['token_str'] for result in results]
```

CHAPTER 6   MODEL FINE-TUNING

```
            # Calculate a score based on matching expected entities
            score = 0
            for entity, weight in combined_expected_entities.items():
                if entity in predicted_tokens:
                    score += weight

            # Add the score to the cumulative score for the model
            model_scores[model_name] += score

    except:
        print(f"Error in {model_name}")

# Rank models based on their cumulative scores
ranked_models = sorted(model_scores.items(), key=lambda item: item[1],
            reverse=True)

# Print the final ranking
print("\nModel Ranking based on Weighted Entity Match Scores (top-5): ")
for rank, (model_name, score) in enumerate(ranked_models, 1):
    #print only the top-5 models
    if rank <= 5:
      print(f"{rank}. {model_name}: Cumulative Score = {score}")
```

Output

```
Testing Dr-BERT/DrBERT-4GB ...
Testing Dr-BERT/DrBERT-7GB ...
Testing Dr-BERT/DrBERT-4GB-CP-PubMedBERT ...
Testing almanach/camembert-bio-base ...
Testing abazoge/DrLongformer ...
Testing abazoge/DrBERT-4096 ...

Model Ranking based on Weighted Entity Match Scores (top-5):
1. Dr-BERT/DrBERT-4GB: Cumulative Score = 3.65
2. abazoge/DrBERT-4096: Cumulative Score = 2.85
3. Dr-BERT/DrBERT-7GB: Cumulative Score = 2.75
4. almanach/camembert-bio-base: Cumulative Score = 2.5
5. Dr-BERT/DrBERT-4GB-CP-PubMedBERT: Cumulative Score = 0.2
```

Our NLP consultant, John Doe, has developed a systematic and methodical approach to selecting suitable pretrained language models. This approach has saved much time for MedTALN Inc.'s project by expediting the model selection process while being based on objective criteria. John's approach emphasizes the importance of a thoughtful, criteria-driven process in selecting NLP models for transfer learning.

Caution Our MLM model selection notebook interacts with Hugging Face's API in real time. Since pretrained models on the Hugging Face Hub are frequently updated or removed, the output of our selection notebook may change based on the current availability of models.

Healthcare NER Model Fine-Tuning

This section will walk you through the high-level workflow for fine-tuning a pretrained language model to build our case study healthcare Named Entity Recognition (NER) model (as shown in Figure 6-6). By fine-tuning a domain-specific pretrained language model, we can leverage transfer learning to build an accurate healthcare NER model without starting from scratch.

CHAPTER 6   MODEL FINE-TUNING

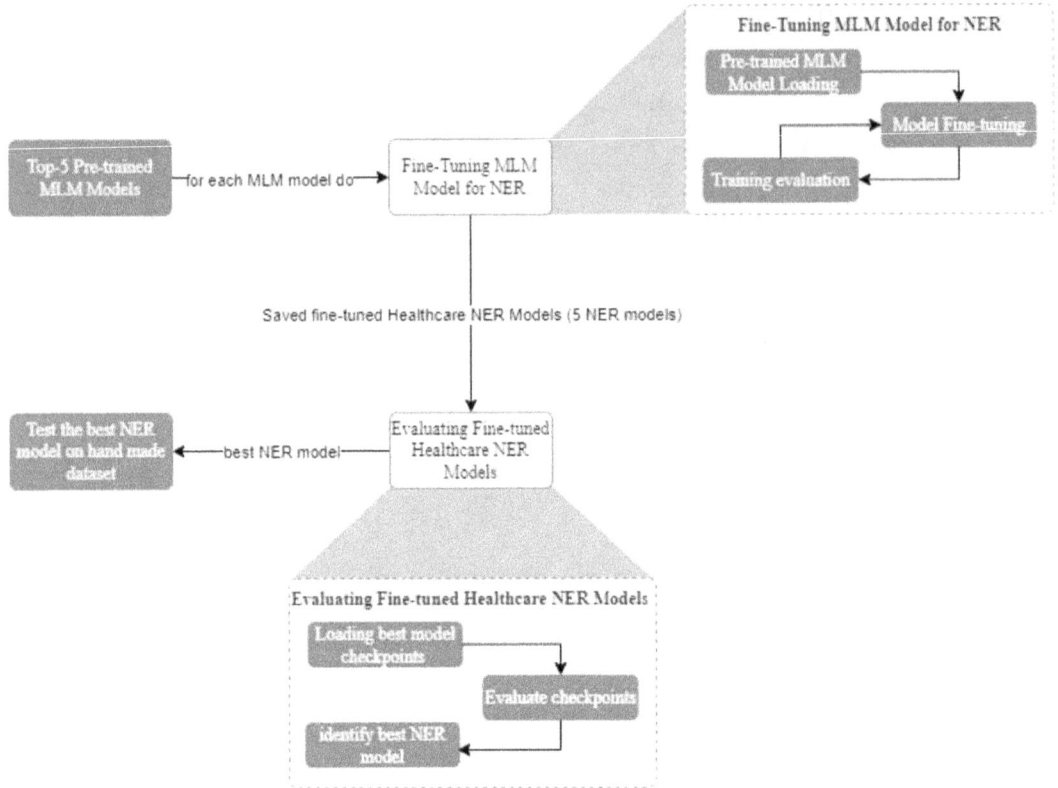

Figure 6-6. *Workflow for training and evaluating the Healthcare NER model*

Training Dataset Creation Notebook

1. Open the GPU-based notebook session:

   a. Go to **Analytics & AI ➤ Machine Learning ➤ Data Science ➤ Projects**.

   b. Open the our OCI DS project, i.e., cs-nlp-prj.

   c. If it is deactivated, activate the CPU-based OCI Data Science Notebook Session, i.e., cs-nlp-nbs-cpu.

   d. Open the CPU-based OCI Data Science Notebook Session, i.e., cs-nlp-nbs-cpu.

2. From the JupyterLab file browser, open the notebook *create_healthcare_ner_dataset.ipynb* under the folder (Figure 6-7; explained in Listings 6-7 to 6-11):

   /repos/john-doe-typica-ai/nlp-on-oci.git/chapt-6

CHAPTER 6   MODEL FINE-TUNING

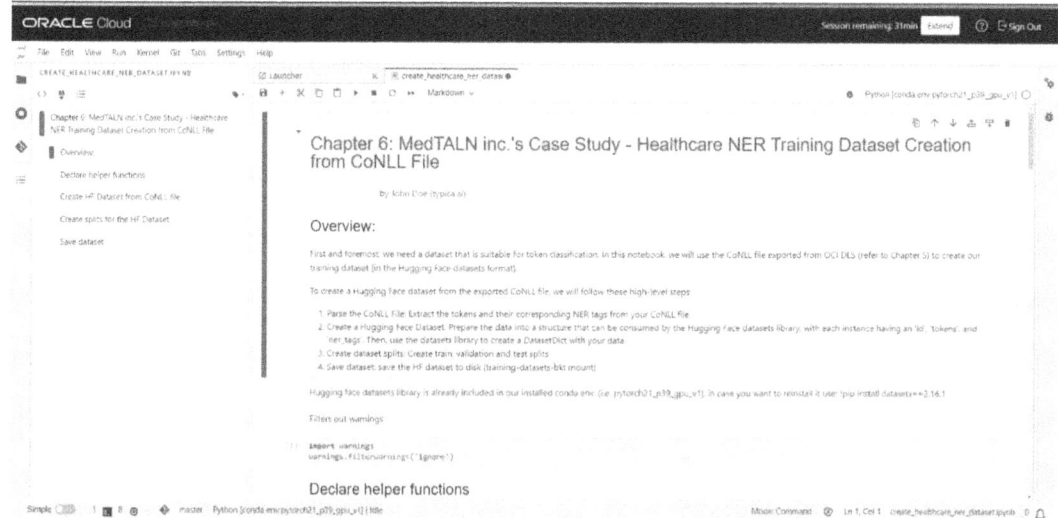

Figure 6-7. Dataset creation notebook

Declare Helper Functions

Listing 6-7. Declaring Functions

```
import re

def split_token(token, tag):
    """
    Splits tokens if they end with specific punctuation characters (.,;!?)
    and assigns
    'O' to the punctuation, leaving other tokens intact.
    """
    # Define the punctuations to split
    punctuations = ".,;!?"
    # Check if the token ends with a punctuation that should be split
    if token[-1] in punctuations:
        # Return the token without the last character and the punctuation
        # as separate tokens
        return [(token[:-1], tag), (token[-1], 'O')]
    else:
        # Return the token as is if it doesn't end with specified punctuation
        return [(token, tag)]
```

CHAPTER 6   MODEL FINE-TUNING

```python
# This function reads your .conll file and extracts sentences and their
NER tags.
def parse_conll_file(file_path):
    sentences = []
    current_sentence = []
    with open(file_path, 'r', encoding='utf-8') as file:
        for line in file:
            if line.startswith("-DOCSTART-") or line.strip() == "":
                if current_sentence:
                    sentences.append(current_sentence)
                    current_sentence = []
            else:
                parts = line.strip().split()
                token = parts[0]
                tag = parts[-1] if len(parts) > 1 else 'O'  # Default to
                        'O' if no tag is present
                # Split token if it contains punctuation
                current_sentence.extend(split_token(token, tag))
        if current_sentence:  # Add the last sentence if it exists
            sentences.append(current_sentence)
    return sentences

# This function extracts unique NER tags ensuring 'O' is first, and
prepares the data for the dataset creation.
def prepare_dataset(sentences):
    unique_tags = set()
    for sentence in sentences:
        for _, tag in sentence:
            #if tag not in excluded_tags:
            unique_tags.add(tag)

    # Ensure 'O' is first, then sort the rest of the tags
    unique_tags.discard('O')  # Remove 'O' to avoid duplication
    unique_tags = ['O'] + sorted(unique_tags)  # Prepend 'O' and sort
                the rest
```

278

CHAPTER 6 MODEL FINE-TUNING

```
    tag_to_id = {tag: id for id, tag in enumerate(unique_tags)}

    # Prepare data for Hugging Face Dataset
    data = {'id': [], 'tokens': [], 'ner_tags': []}
    for i, sentence in enumerate(sentences):
        tokens, tags = zip(*sentence)
        data['id'].append(str(i))
        data['tokens'].append(list(tokens))
        data['ner_tags'].append([tag_to_id.get(tag, tag_to_id['O']) for tag
        in tags ]) #if tag not in excluded_tags

    return data, unique_tags

from datasets import Dataset, DatasetDict, Features, ClassLabel,
Sequence, Value

# This function creates the dataset using the prepared data and unique
NER tags.
def create_hf_dataset(data, unique_tags):
    features= Features({
                'id': Value(dtype='string', id=None),
                'tokens': Sequence(feature=Value(dtype='string', id=None),
                length=-1, id=None),
                'ner_tags': Sequence(feature=ClassLabel(num_
                classes=len(unique_tags), names=unique_tags))
            })

    dataset = Dataset.from_dict(data, features=features)
    dataset_dict = DatasetDict({'train': dataset})
    return dataset_dict
```

Output

Create HF Dataset from CoNLL File

Parse the .conll file, prepare the data, and create the dataset.

CHAPTER 6   MODEL FINE-TUNING

Listing 6-8. Creating the dataset

```
# Set the CoNLL file path
file_path = "/home/datascience/buckets/training-datasets-bkt/healthcare_
ner_dataset_v1.1.0/healthcare_ner_dataset_v1.0.0_1724175778995.conll"

# Parse the .conll file
sentences = parse_conll_file(file_path)
# Prepare the dataset and extract unique NER tags
data, unique_tags = prepare_dataset(sentences)
# Create the Hugging Face dataset
dataset_dict = create_hf_dataset(data, unique_tags)

print("Dataset created successfully!")

dataset_dict
```

  Output

```
Dataset created successfully!
[9]:
DatasetDict({
    train: Dataset({
        features: ['id', 'tokens', 'ner_tags'],
        num_rows: 9000
    })
})
```

Inspect a row randomly.

Listing 6-9. Displaying a Random Row

```
dataset_dict["train"].shuffle(seed=42)[0]
```

  Output

```
{'id': '2015',
 'tokens': ['un',
  'vaccin',
  'vivant',
  'atténué',
```

280

```
    'est',
    'maintenant',
    'disponible',
    ' ',
    '.'],
 'ner_tags': [0, 5, 0, 0, 0, 0, 0, 0, 0]}
```

Create Splits for the HF Dataset

At this step, we create train, validation, and test splits.

Listing 6-10. Splitting the Dataset

```
from datasets import DatasetDict

ds_train_devtest = dataset_dict['train'].train_test_split(test_size=0.25, seed=42)
ds_devtest = ds_train_devtest['test'].train_test_split(test_size=0.25, seed=42)

healthcare_ner_dataset = DatasetDict({
    'train': ds_train_devtest['train'],
    'validation': ds_devtest['train'],
    'test': ds_devtest['test']
})

healthcare_ner_dataset
```

Output

```
DatasetDict({
    train: Dataset({
        features: ['id', 'tokens', 'ner_tags'],
        num_rows: 6750
    })
    validation: Dataset({
        features: ['id', 'tokens', 'ner_tags'],
        num_rows: 1687
    })
```

```
    test: Dataset({
        features: ['id', 'tokens', 'ner_tags'],
        num_rows: 563
    })
})
```

Save Dataset

We will save the Hugging Face dataset to the "training-datasets-bkt/healthcare_ner_dataset_v1.2.0" directory. The dataset is ready for Named Entity Recognition (NER) training.

The "v1.2" part of the version denotes that the dataset has been fully processed and is in a format suitable for NER tasks, making it easy to reference this specific state of the dataset in future training runs or evaluation tasks.

Listing 6-11. Saving the Dataset

```
healthcare_ner_dataset.save_to_disk("/home/datascience/buckets/training-datasets-bkt/healthcare_ner_dataset_v1.2.0")
```

   Output

```
Saving the dataset (1/1 shards): 100%
6750/6750 [00:00<00:00, 95833.62 examples/s]
Saving the dataset (1/1 shards): 100%
1687/1687 [00:00<00:00, 21889.73 examples/s]
Saving the dataset (1/1 shards): 100%
563/563 [00:00<00:00, 3193.01 examples/s]
```

Training Notebook

With our top five pretrained MLM models identified and the training dataset saved in the Hugging Face format, we can now proceed with fine-tuning these models to develop a set of Healthcare NER models. Our NLP consultant will guide us through each stage of the fine-tuning process.

To better understand the key steps involved, we can refer to the workflow diagram in Figure 6-8. Let's go through these steps together.

CHAPTER 6   MODEL FINE-TUNING

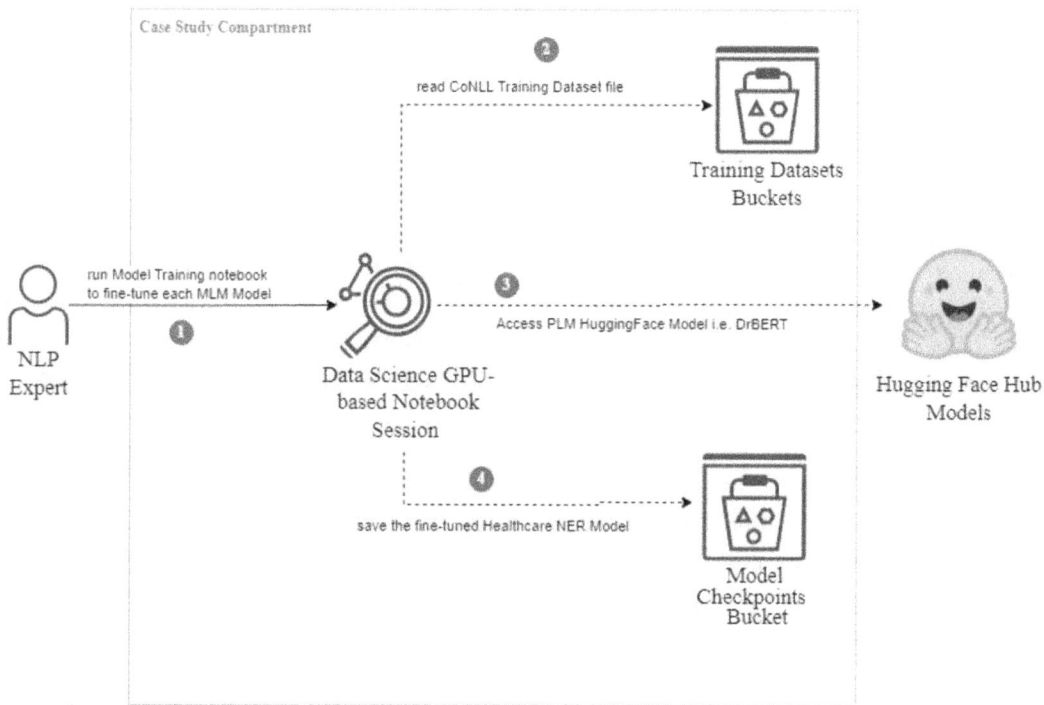

Figure 6-8. *High-level data flow interactions for model training*

1. Executing the Training Notebook: The workflow begins by executing the Python code of Model Training Jupyter notebook. This notebook is executed within the OCI Data Science GPU-based Notebook Session.

2. Labeled Dataset Access: The training process starts by retrieving the labeled dataset CoNLL file created during training dataset preparation process (refer to Chapter 5). This CoNLL file is retrieved from the Training Datasets Bucket which is the holding place of the training datasets once labeled and exported from OCI Data Labeling Service (DLS) in the CoNLL format.

CHAPTER 6   MODEL FINE-TUNING

3. Pretrained MLM Model Fine-Tuning: During the training phase, each of the top five MLM models (selected earlier) is loaded from Hugging Face and fine-tuned to produce distinct Healthcare NER models. Each resulting model is unique, based on the specific MLM used for fine-tuning.

4. Saving the Fine-Tuned Model Checkpoint: During the model training, checkpoints are saved at each training epoch to the Model Checkpoints Bucket (mount directory).

Caution As a temporary workaround for an obscure I/O error that occurs when the trainer writes directly to a bucket mount, we use a staging folder in the Data Science GPU-based Notebook Session's file system ("local_training_dir"). Once the trainer completes the training, we move the model artifacts to the Model Checkpoints Bucket (i.e., "/home/datascience/buckets/models-ckpt-bkt").

For the training, we will need to open the GPU-based OCI Data Science Notebook Session.

1. Open the GPU-based notebook session:

   i. Go to **Analytics & AI ➤** Machine Learning **➤ Data Science ➤ Projects**.

   ii. Open the our OCI DS project, i.e., cs-nlp-prj.

   iii. If it is deactivated, activate the GPU-based OCI Data Science Notebook Session, i.e., cs-nlp-nbs-gpu.

   iv. Open the GPU-based OCI Data Science Notebook Session, i.e., cs-nlp-nbs-gpu.

2. From the JupyterLab file browser, open the notebook ***train_healthcare_ner_model.ipynb*** under the folder (Figure 6-9; explained in Listings 6-12 to 6-23):

   /repos/john-doe-typica-ai/nlp-on-oci.git/chapt-6

CHAPTER 6   MODEL FINE-TUNING

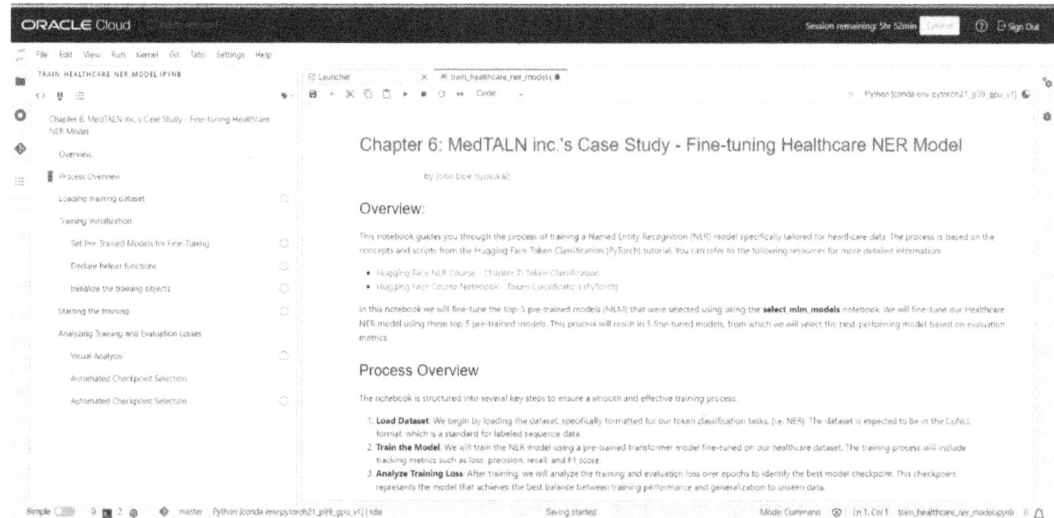

Figure 6-9. *Training notebook open in GPU-based OCI DS Notebook Session*

The Hugging Face "Transformers" library is already included in our installed conda environment (i.e., "pytorch21_p39_gpu_v1"). We will upgrade it to version "4.44.2." This only needs to be done once, the first time.

Listing 6-12. Installing Dependencies

```
%%capture
!pip install transformers==4.44.2
!pip install accelerate==0.33.0
!pip install seqeval==1.2.2
```

Loading Training Dataset

First, we need to load the Hugging Face dataset that was saved earlier to the "training-datasets-bkt/healthcare_ner_dataset_v1.2.0" directory (refer to notebook "create_hf_dataset.ipynb"). This dataset "v1.2" is ready for Named Entity Recognition (NER) training.

Listing 6-13. Loading Dataset

```
from datasets import load_from_disk

healthcare_ner_dataset = load_from_disk("/home/datascience/buckets/
training-datasets-bkt/healthcare_ner_dataset_v1.2.0")
healthcare_ner_dataset
```

285

CHAPTER 6   MODEL FINE-TUNING

Output

```
DatasetDict({
    train: Dataset({
        features: ['id', 'tokens', 'ner_tags'],
        num_rows: 6750
    })
    validation: Dataset({
        features: ['id', 'tokens', 'ner_tags'],
        num_rows: 1687
    })
    test: Dataset({
        features: ['id', 'tokens', 'ner_tags'],
        num_rows: 563
    })
})
```

Listing 6-14. Setting NER Feature Variable

```
ner_feature = healthcare_ner_dataset["train"].features["ner_tags"]
ner_feature
```

Output

```
Sequence(feature=ClassLabel(names=['O', 'B-AnatomicalStructure',
'B-Disease', 'B-MISC', 'B-MedicalProcedure', 'B-MedicationVaccine',
'B-Symptom', 'I-AnatomicalStructure', 'I-Disease', 'I-MISC',
'I-MedicalProcedure', 'I-MedicationVaccine', 'I-Symptom'], id=None),
length=-1, id=None)
```

Listing 6-15. Setting NER labels Variable

```
label_names = ner_feature.feature.names
label_names
```

Output

```
['O',
 'B-AnatomicalStructure',
 'B-Disease',
 'B-MISC',
 'B-MedicalProcedure',
 'B-MedicationVaccine',
 'B-Symptom',
 'I-AnatomicalStructure',
 'I-Disease',
 'I-MISC',
 'I-MedicalProcedure',
 'I-MedicationVaccine',
 'I-Symptom']
```

Training Initialization

Set Pretrained Models for Fine-Tuning

In the following cell, we initialize the pretrained MLM model that that will be fine-tuned using our Healthcare NER dataset. The top five models, chosen based on their Weighted Entity Match Scores, are as follows:

Top five MLM models:

- Dr-BERT/DrBERT-4GB

- abazoge/DrBERT-4096

- Dr-BERT/DrBERT-7GB

- almanach/camembert-bio-base

- Dr-BERT/DrBERT-4GB-CP-PubMedBERT

CHAPTER 6 MODEL FINE-TUNING

Listing 6-16. Initializing Pretrained Model

```
from transformers import AutoTokenizer

model_checkpoint = "Dr-BERT/DrBERT-4GB"
tokenizer = AutoTokenizer.from_pretrained(model_checkpoint)
```

Output

Declare Helper Functions

Listing 6-17. Defining Helper Functions

```
import evaluate
import numpy as np

def align_labels_with_tokens(labels, word_ids):
    new_labels = []
    current_word = None
    for word_id in word_ids:
        if word_id != current_word:
            # Start of a new word!
            current_word = word_id
            label = -100 if word_id is None else labels[word_id]
            new_labels.append(label)
        elif word_id is None:
            # Special token
            new_labels.append(-100)
        else:
            # Same word as previous token
            label = labels[word_id]
            # If the label is B-XXX we change it to I-XXX
            if label % 2 == 1:
                label += 1
            new_labels.append(label)

    return new_labels
```

```python
# This function tokenize the dataset and align labels with tokens
def tokenize_and_align_labels(examples):
    tokenized_inputs = tokenizer(
        examples["tokens"], truncation=True, is_split_into_words=True
    )
    all_labels = examples["ner_tags"]
    new_labels = []
    for i, labels in enumerate(all_labels):
        word_ids = tokenized_inputs.word_ids(i)
        new_labels.append(align_labels_with_tokens(labels, word_ids))

    tokenized_inputs["labels"] = new_labels
    return tokenized_inputs

#This function compute eveluation metrics

metric = evaluate.load("seqeval")

def compute_metrics(eval_preds):
    logits, labels = eval_preds
    predictions = np.argmax(logits, axis=-1)

    # Remove ignored index (special tokens) and convert to labels
    true_labels = [[label_names[l] for l in label if l != -100] for label
                   in labels]
    true_predictions = [
        [label_names[p] for (p, l) in zip(prediction, label) if l != -100]
        for prediction, label in zip(predictions, labels)
    ]
    all_metrics = metric.compute(predictions=true_predictions,
                   references=true_labels)
    return {
        "precision": all_metrics["overall_precision"],
        "recall": all_metrics["overall_recall"],
        "f1": all_metrics["overall_f1"],
        "accuracy": all_metrics["overall_accuracy"],
    }
```

Output

Initialize the Training Objects

Here, we will initialize the training objects and, more specifically, define the training hyperparameters using TrainingArguments. It is important to note that choosing the values for key hyperparameters can significantly impact model performance. Hyperparameters like the learning rate, number of epochs, and batch size must be carefully selected to strike a balance between effective learning and avoiding overfitting. Below are the key hyperparameters we explicitly set, along with their corresponding values:

- *learning_rate=2e-5*: A relatively small learning rate is used to ensure the model updates weights gradually during training, helping avoid overshooting minima and ensuring stable convergence.
- *num_train_epochs=5*: The model will train for five epochs, allowing multiple passes over the training data to ensure sufficient learning without overfitting.
- *weight_decay=0.01*: Weight decay is applied to prevent overfitting by adding a penalty to large weights, encouraging the model to prefer smaller, more generalized weights.
- *evaluation_strategy="epoch"*: The model will be evaluated at the end of each epoch to track performance on the validation set and monitor improvements.
- *save_strategy="epoch"*: A checkpoint of the model is saved at the end of each epoch, ensuring that progress can be stored and resumed, if necessary.
- *load_best_model_at_end=True*: This ensures that the best-performing model on the evaluation set is retained, rather than the model at the final epoch.
- *per_device_train_batch_size=8* (default): The batch size per device during training is set to eight by default, balancing memory usage and training speed.

Listing 6-18. Preparing the Model for Training

```
from transformers import DataCollatorForTokenClassification
from transformers import TrainingArguments

# Tokenize dataset
tokenized_datasets = healthcare_ner_dataset.map(
    tokenize_and_align_labels,
    batched=True,
    remove_columns=healthcare_ner_dataset["train"].column_names,
)

# init data collator
data_collator = DataCollatorForTokenClassification(tokenizer=tokenizer)

# init the model

id2label = {i: label for i, label in enumerate(label_names)}
label2id = {v: k for k, v in id2label.items()}

from transformers import AutoModelForTokenClassification

model = AutoModelForTokenClassification.from_pretrained(
    model_checkpoint,
    id2label=id2label,
    label2id=label2id,
)

# init the training arguments

models_base_folder = "/home/datascience/training_local_dir/models/healthcare_ner"
args = TrainingArguments(
    f"{models_base_folder}-{model_checkpoint}",
    evaluation_strategy="epoch",
    save_strategy="epoch",
    learning_rate=2e-5,
    num_train_epochs=5,
    weight_decay=0.01,
```

CHAPTER 6   MODEL FINE-TUNING

```
        load_best_model_at_end=True,
        push_to_hub=False,
)
```

Output

Starting the Training

Fine-tune the pretrained model into our Healthcare NER model.

Listing 6-19. Launching the Training

```
from transformers import Trainer

trainer = Trainer(
    model=model,
    args=args,
    train_dataset=tokenized_datasets["train"],
    eval_dataset=tokenized_datasets["validation"],
    data_collator=data_collator,
    compute_metrics=compute_metrics,
    tokenizer=tokenizer,
)
trainer.train()
```

Output

[2024-08-26 15:52:27,611] [INFO] [real_accelerator.py:158:get_accelerator] Setting ds_accelerator to cuda (auto detect)
 [4220/4220 06:03, Epoch 5/5]

Epoch	Training Loss	Validation Loss	Precision	Recall	F1	Accuracy
1	0.394700	0.221041	0.661832	0.724099	0.691567	0.918677
2	0.179100	0.201834	0.715773	0.747903	0.731486	0.932372
3	0.098300	0.212497	0.743618	0.752891	0.748226	0.936063
4	0.056400	0.252355	0.728846	0.783042	0.754973	0.936119
5	0.029800	0.275240	0.736583	0.781002	0.758143	0.937359

[9]:
TrainOutput(global_step=4220, training_loss=0.14149999550733522,
metrics={'train_runtime': 363.2984, 'train_samples_per_second': 92.899,
'train_steps_per_second': 11.616, 'total_flos': 557445304412724.0,
'train_loss': 0.14149999550733522, 'epoch': 5.0})

Below, a temporary fix for I/O error when trainer writes directly to the bucket mount.

Listing 6-20. Moving the Fine-Tuned Model to the Mounted Folder

```python
import shutil

# set the model local and mount directories
local_dir = f"/home/datascience/training_local_dir/models/
healthcare_ner-{model_checkpoint}"
mount_dir = f"/home/datascience/buckets/models-ckpt-bkt/
models/healthcare_ner-{model_checkpoint}"

# Move the model local folder to models-ckpt-bkt mount
shutil.move(local_dir, mount_dir)

print(f"Model folder moved to {mount_dir}")
```

Output

```
Model folder moved to /home/datascience/buckets/models-ckpt-bkt/models/
healthcare_ner-Dr-BERT/DrBERT-4GB
```

Analyzing Training and Evaluation Losses

After training a model, it is crucial to understand how well the model is learning over time.

Visual Analysis

By plotting losses over epochs, we can visually inspect the learning process and make informed decisions about which model checkpoint (i.e., which epoch) provides the best balance between learning and generalization.

CHAPTER 6   MODEL FINE-TUNING

In the plot, we can observe the following trends:

– The "training loss" decreases steadily across epochs, which suggests that the model is effectively learning and improving its performance on the training data.

– The "evaluation loss" initially decreases but then starts to stabilize and slightly increase in the later epochs. This is a common sign of "overfitting," where the model becomes too specialized in the training data and performs less well on the evaluation data.

The best checkpoint is typically the epoch where the evaluation loss is at its lowest, indicating the best generalization to unseen data. In this plot

– The evaluation loss is lowest around "epoch 2."

– However, starting from "epoch 3," the evaluation loss begins to increase slightly, suggesting that further training may not improve model generalization and might even harm it by overfitting.

Based on this analysis, "epoch 2" is likely the best checkpoint. It has the lowest evaluation loss, which means the model was best at generalizing to new data at this point. Continuing the training beyond this point appears to result in diminishing returns and increased risk of overfitting.

Listing 6-21. Plotting losses over epochs

```
import matplotlib.pyplot as plt

# Initialize lists to store the losses and epochs
train_loss = []
eval_loss = []
epochs = []

# Extract the losses and epochs from the log history
for log in trainer.state.log_history:
    if 'loss' in log and 'epoch' in log:  # Training loss
        train_loss.append(log['loss'])
```

```python
        epochs.append(log['epoch'])
    if 'eval_loss' in log and 'epoch' in log:  # Evaluation loss
        eval_loss.append(log['eval_loss'])

# Ensure the epochs list is aligned with eval_loss if needed
eval_epochs = epochs[:len(eval_loss)] if len(eval_loss) == len(epochs) else
list(range(1, len(eval_loss) + 1))

plt.figure(figsize=(10, 6))

# Plot training loss
plt.plot(epochs, train_loss, label='Training Loss', marker='o',
linestyle='-')

# Plot evaluation loss
plt.plot(eval_epochs, eval_loss, label='Evaluation Loss', marker='o',
linestyle='-')

# Add labels and title
plt.xlabel('Epochs')
plt.ylabel('Loss')
plt.title('Training and Evaluation Loss Over Epochs')
plt.legend()
plt.grid(True)

# Show the plot
plt.show()
```

Chapter 6 Model Fine-Tuning

Output (Figure 6-10)

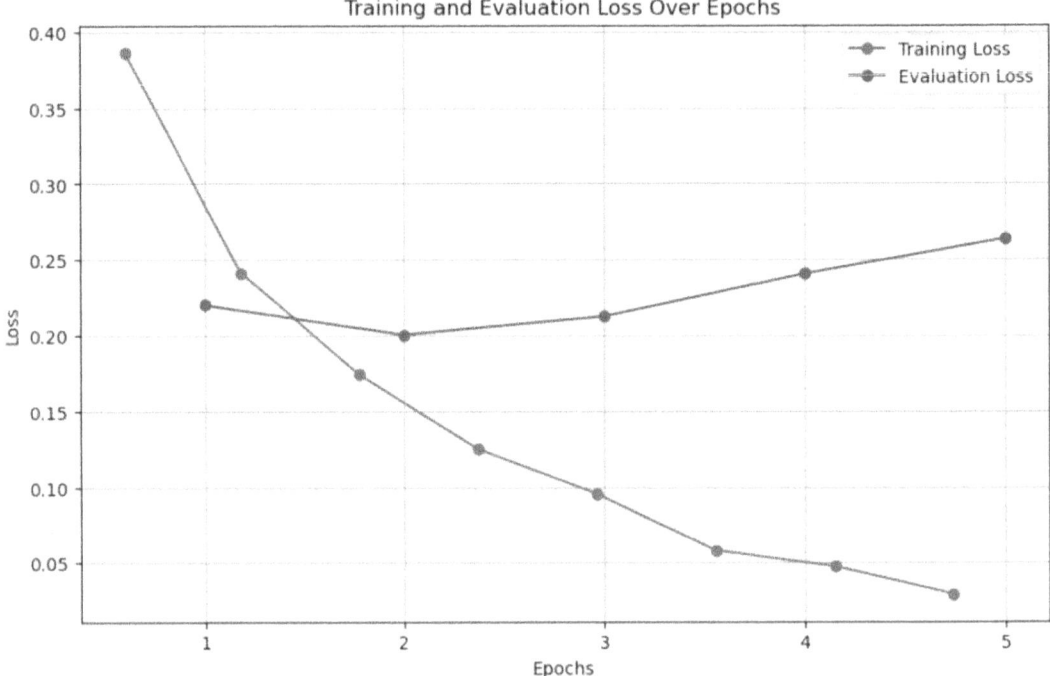

Figure 6-10. *Training loss over epoch plot*

Automated Checkpoint Selection

While effective, this manual approach has some limitations:

- Subjectivity: Visual inspection is subjective and may not be consistent across different viewers or datasets.

- Time-Consuming: Manually inspecting plots is feasible for a few experiments, but it becomes cumbersome when training multiple models or fine-tuning hyperparameters.

To streamline this process and make it more objective, we can leverage the "trainer.state.best_model_checkpoint" feature provided by the Hugging Face Trainer API. This feature automatically tracks the checkpoint with the best performance on the validation set during training, based on a specified metric (e.g., evaluation loss, F1 score, accuracy).

Listing 6-22. Tracking the best checkpoint

```
import re

# Get the best model checkpoint from the trainer state object
best_model_checkpoint = trainer.state.best_model_checkpoint

#temp fix
best_model_checkpoint = best_model_checkpoint.replace('training_local_dir', 
'buckets/models-ckpt-bkt')

# Extract the step number from the best model checkpoint name
# Assuming the format is like 'checkpoint-1688' at the end of the path
step_match = re.search(r'checkpoint-(\d+)', best_model_checkpoint)
best_checkpoint_step = int(step_match.group(1)) if step_match else None

# Initialize variables to store metrics
eval_f1, eval_recall, eval_accuracy = None, None, None

# Search for the relevant evaluation metrics in log_history
for log in trainer.state.log_history:
    if 'eval_f1' in log and log.get('step') == best_checkpoint_step:
        eval_f1 = log['eval_f1']
        eval_recall = log.get('eval_recall', None)
        eval_accuracy = log.get('eval_accuracy', None)
        break  # Once the metrics are found, exit the loop

# Print the metrics for the best model
if eval_f1 is not None:
    print(f"The best for checkpoint model {trainer.model.config.name_or_
    path} is:")
    print(f"Checkpoint: {best_model_checkpoint} (Step: {best_checkpoint_
    step}) :")
    print(f"F1 Score: {eval_f1:.4f}")
    print(f"Recall: {eval_recall:.4f}")
    print(f"Accuracy: {eval_accuracy:.4f}")
```

CHAPTER 6   MODEL FINE-TUNING

Output

```
The best for checkpoint model Dr-BERT/DrBERT-4GB is:
Checkpoint: /home/datascience/buckets/models-ckpt-bkt/models/healthcare_
ner-Dr-BERT/DrBERT-4GB/checkpoint-1688 (Step: 1688) :
F1 Score: 0.7315
Recall: 0.7479
Accuracy: 0.9324
```

Let's now test the best checkpoint for this Healthcare NER model.

Listing 6-23. Testing the best checkpoint

```
from transformers import pipeline

# test the healthcare NER model
token_classifier = pipeline("token-classification", model=best_model_
checkpoint, aggregation_strategy="first")
token_classifier("Le medecin donne des antibiotiques pour les infections
bactériennes.")
```

Output

```
[{'entity_group': 'MedicationVaccine',
  'score': 0.85936165,
  'word': 'antibiotiques',
  'start': 20,
  'end': 34}]
```

The newly fine-tuned Healthcare NER model successfully identified "antibiotiques" as a "MedicationVaccine" entity. This suggests that the model is effectively identifying medical entities within medical text.

Repeat the fine-tuning process for each of the top five MLM models:

1. Dr-BERT/DrBERT-4GB

2. abazoge/DrBERT-4096

3. Dr-BERT/DrBERT-7GB

4. almanach/camembert-bio-base

5. Dr-BERT/DrBERT-4GB-CP-PubMedBERT

To do this, restart this notebook and change the model_checkpoint at the cell: Set pretrained models for fine-tuning.

Caution When repeating the fine-tuning for the remaining MLM models, you can comment out the pip install commands for transformers==4.44.2, accelerate==0.33.0, and seqeval==1.2.2, as they are already installed. Commenting them out will save time.

Tip It is very important to deactivate your GPU-based OCI Data Science Notebook Session (e.g., cs-nlp-nbs-gpu) when you finish the fine-tuning process to stop billing and save on costs.

Healthcare NER Model Evaluation

This section will guide you through the high-level workflow for evaluating the fine-tuned Healthcare NER models (illustrated in Figure 6-11). Following the evaluation, the best-performing Healthcare NER model will be identified. As a final step, the selected model will be tested on our tiny handmade test set to validate its effectiveness.

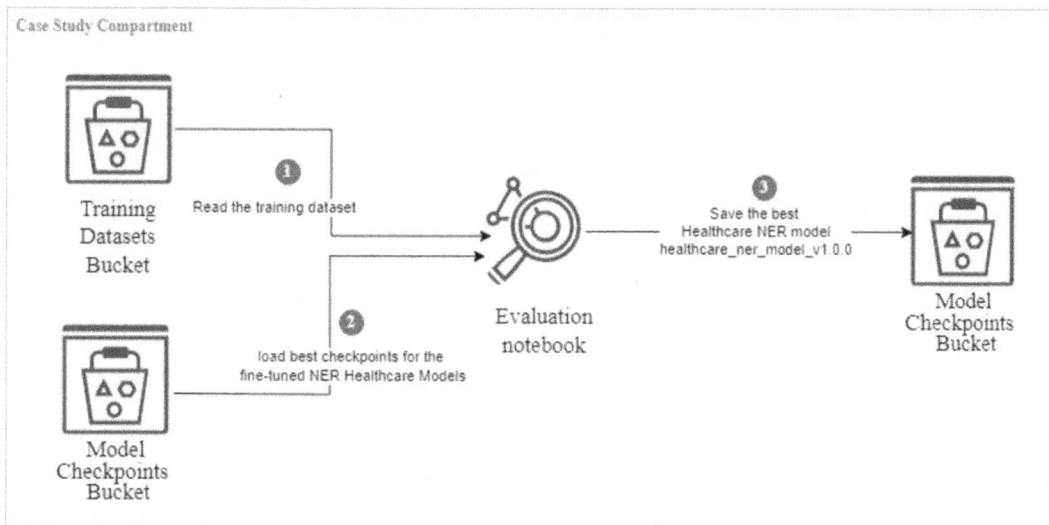

Figure 6-11. Evaluation high-level steps and data flow

CHAPTER 6   MODEL FINE-TUNING

Evaluation Notebook

This notebook guides you through the process of evaluating the fine-tuned Healthcare NER models.

For this evaluation step, we will need to open the GPU-based OCI Data Science Notebook Session.

1. Open the GPU-based notebook session:

    i. Go to **Analytics & AI** ➤ Machine Learning ➤ **Data Science** ➤ **Projects**.

    ii. Open the our OCI DS project, i.e., cs-nlp-prj.

    iii. If it is deactivated, activate the GPU-based OCI Data Science Notebook Session, i.e., cs-nlp-nbs-gpu.

    iv. Open the GPU-based OCI Data Science Notebook Session, i.e., cs-nlp-nbs-gpu.

2. From the JupyterLab file browser, open the notebook *evaluate_healthcare_ner_models.ipynb* under the folder (Figure 6-12; explained in Listings 6-24 to 6-34):

    /repos/john-doe-typica-ai/nlp-on-oci.git/chapt-6

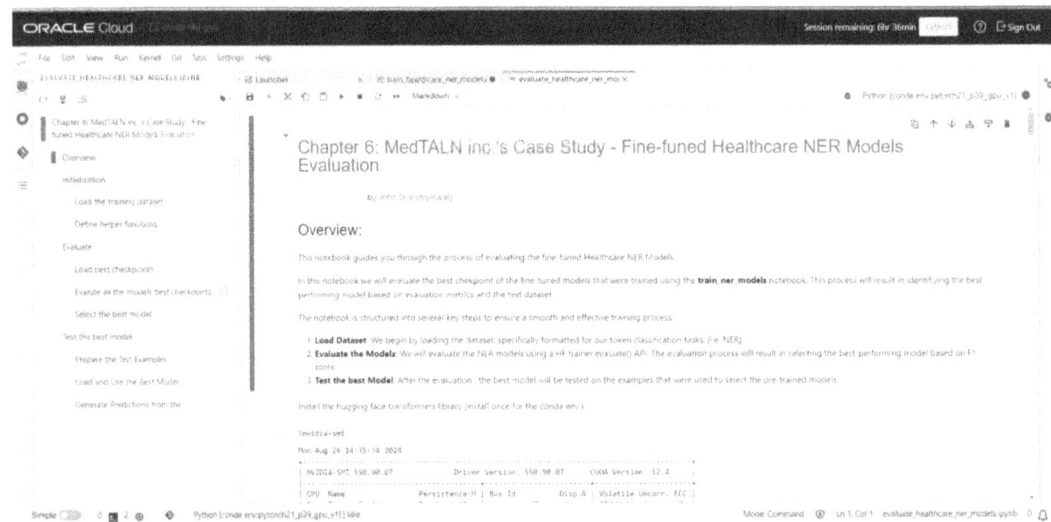

Figure 6-12. Healthcare NER model evaluation notebook

CHAPTER 6 MODEL FINE-TUNING

In this notebook, we will evaluate the best checkpoint of the fine-tuned models that were trained using the "train_ner_models" notebook. This process will result in identifying the best-performing model based on evaluation metrics and the test dataset.

The notebook is structured into several key steps to ensure a smooth and effective training process:

1. Load Dataset: We begin by loading the dataset, specifically formatted for our token classification tasks (i.e., NER).

2. Evaluate the Models: We will evaluate the NER models using a HF trainer.evaluate() API. The evaluation process will result in selecting the best-performing model based on F1 score.

3. Test the Best Model: After the evaluation , the best model will be tested on the examples that were used to select the pretrained models.

Initialization

Load the Training Dataset

Listing 6-24. Loading Dataset

```
from datasets import load_from_disk

healthcare_ner_dataset = load_from_disk('/home/datascience/buckets/training-datasets-bkt/healthcare_ner_dataset_v1.2.0')
healthcare_ner_dataset
```

　　Output

```
DatasetDict({
    train: Dataset({
        features: ['id', 'tokens', 'ner_tags'],
        num_rows: 6750
    })
    validation: Dataset({
        features: ['id', 'tokens', 'ner_tags'],
        num_rows: 1687
    })
```

CHAPTER 6   MODEL FINE-TUNING

```
    test: Dataset({
        features: ['id', 'tokens', 'ner_tags'],
        num_rows: 563
    })
})
```

Define Helper Functions

Listing 6-25. Defining Helper Functions

```
def align_labels_with_tokens(labels, word_ids):
    new_labels = []
    current_word = None
    for word_id in word_ids:
        if word_id != current_word:
            # Start of a new word!
            current_word = word_id
            label = -100 if word_id is None else labels[word_id]
            new_labels.append(label)
        elif word_id is None:
            # Special token
            new_labels.append(-100)
        else:
            # Same word as previous token
            label = labels[word_id]
            # If the label is B-XXX we change it to I-XXX
            if label % 2 == 1:
                label += 1
            new_labels.append(label)

    return new_labels

def tokenize_and_align_labels(examples):
    tokenized_inputs = tokenizer(
        examples["tokens"], truncation=True, is_split_into_words=True
    )
```

```
    all_labels = examples["ner_tags"]
    new_labels = []
    for i, labels in enumerate(all_labels):
        word_ids = tokenized_inputs.word_ids(i)
        new_labels.append(align_labels_with_tokens(labels, word_ids))

    tokenized_inputs["labels"] = new_labels
    return tokenized_inputs

def compute_metrics(eval_preds):
    logits, labels = eval_preds
    predictions = np.argmax(logits, axis=-1)

    # Remove ignored index (special tokens) and convert to labels
    true_labels = [[label_names[l] for l in label if l != -100] for label
                in labels]
    true_predictions = [
        [label_names[p] for (p, l) in zip(prediction, label) if l != -100]
        for prediction, label in zip(predictions, labels)
    ]
    all_metrics = metric.compute(predictions=true_predictions,
                references=true_labels)
    return {
        "precision": all_metrics["overall_precision"],
        "recall": all_metrics["overall_recall"],
        "f1": all_metrics["overall_f1"],
        "accuracy": all_metrics["overall_accuracy"],
    }
```

Output

Evaluate

Evaluate test split using the HF API trainer.evaluate().

Load Best Checkpoints

Listing 6-26. Retrieving the Best Checkpoints

```
import os
from transformers.trainer_callback import TrainerState

def get_best_checkpoint(root_dir):
    best_checkpoints = {}

    # Walk through all directories and subdirectories
    for root, dirs, files in os.walk(root_dir):
        # Filter directories that start with "healthcare_ner"
        relevant_dirs = [d for d in dirs if d.startswith("healthcare_ner")]

        for dir_name in relevant_dirs:
            model_dir = os.path.join(root, dir_name)
            for sub_root, sub_dirs, sub_files in os.walk(model_dir):
                ckpt_dirs = [d for d in sub_dirs if
                            d.startswith('checkpoint')]
                if ckpt_dirs:
                    ckpt_dirs = sorted(ckpt_dirs, key=lambda x: int(x.
                                    split('-')[1]))
                    last_ckpt = ckpt_dirs[-1]

                    state = TrainerState.load_from_json(f"{sub_root}/{last_
                            ckpt}/trainer_state.json")
                    best_model_checkpoint = state.best_model_checkpoint
                    #temp fix
                    best_model_checkpoint = best_model_checkpoint.
                                            replace('training_local_dir',
                                            'buckets/models-ckpt-bkt')

                    best_checkpoints[sub_root] = best_model_checkpoint

    return best_checkpoints

root_dir = "/home/datascience/buckets/models-ckpt-bkt/models"
```

```
best_models_checkpoints = get_best_checkpoint(root_dir)

for model_dir, best_ckpt in best_models_checkpoints.items():
    print(f"Best checkpoint for model in {model_dir}: {best_ckpt}")
```

Output

Best checkpoint for model in /home/datascience/buckets/models-ckpt-bkt/models/healthcare_ner-Dr-BERT/DrBERT-4GB: /home/datascience/buckets/models-ckpt-bkt/models/healthcare_ner-Dr-BERT/DrBERT-4GB/checkpoint-1688
Best checkpoint for model in /home/datascience/buckets/models-ckpt-bkt/models/healthcare_ner-Dr-BERT/DrBERT-4GB-CP-PubMedBERT: /home/datascience/buckets/models-ckpt-bkt/models/healthcare_ner-Dr-BERT/DrBERT-4GB-CP-PubMedBERT/checkpoint-2532
Best checkpoint for model in /home/datascience/buckets/models-ckpt-bkt/models/healthcare_ner-Dr-BERT/DrBERT-7GB: /home/datascience/buckets/models-ckpt-bkt/models/healthcare_ner-Dr-BERT/DrBERT-7GB/checkpoint-1688
Best checkpoint for model in /home/datascience/buckets/models-ckpt-bkt/models/healthcare_ner-abazoge/DrBERT-4096: /home/datascience/buckets/models-ckpt-bkt/models/healthcare_ner-abazoge/DrBERT-4096/checkpoint-1688
Best checkpoint for model in /home/datascience/buckets/models-ckpt-bkt/models/healthcare_ner-almanach/camembert-bio-base: /home/datascience/buckets/models-ckpt-bkt/models/healthcare_ner-almanach/camembert-bio-base/checkpoint-3376

Evaluate All the Models' Best Checkpoints

Loop on the fine-tuned model best checkpoints and evaluate.

Listing 6-27. Evaluating the Best Checkpoints

```
from transformers import AutoModelForTokenClassification, AutoTokenizer, Trainer, TrainingArguments
from transformers import DataCollatorForTokenClassification
import evaluate
from datasets import load_metric
from torch.utils.data import DataLoader
import numpy as np
```

```python
metric = evaluate.load("seqeval")

eval_dir = "/home/datascience/buckets/models-ckpt-bkt/models/healthcare_
ner/evaluation"

# Initialize TrainingArguments (to configure the evaluation)
evaluation_args = TrainingArguments(
    output_dir=f"{eval_dir}/eval_results",
    per_device_eval_batch_size=8,
    logging_dir="{eval_dir}/eval_logs",
    do_train=False,  # We are only evaluating
    do_eval=True,
)

label_names = healthcare_ner_dataset["test"].features["ner_tags"].
feature.names

# Evaluate each model and collect the results
model_results = {}

for model_dir, model_checkpoint in best_models_checkpoints.items():

    if model_checkpoint is not None:

        print(f"Evaluating model checkpoint : {model_checkpoint}")

        # Load the model and tokenizer
        model = AutoModelForTokenClassification.from_pretrained(model_
             checkpoint)
        tokenizer = AutoTokenizer.from_pretrained(model_checkpoint)
        data_collator = DataCollatorForTokenClassification(tokenizer=
                    tokenizer)

        test_tokenized_datasets = healthcare_ner_dataset["test"].map(
            tokenize_and_align_labels,
            batched=True,
            remove_columns=healthcare_ner_dataset["test"].column_names,
        )
```

CHAPTER 6 MODEL FINE-TUNING

```
eval_dataloader = DataLoader(
    test_tokenized_datasets, collate_fn=data_collator, batch_size=8
)

# Initialize the Trainer
trainer = Trainer(
    model=model,
    args=evaluation_args,
    eval_dataset=test_tokenized_datasets,#eval_dataset=test_dataset,
    data_collator=data_collator,   # Include the data collator here
    compute_metrics=compute_metrics,
    tokenizer=tokenizer,
    )

# Evaluate the model
results = trainer.evaluate()
model_results[model_checkpoint] = results
```

Output

Evaluating model checkpoint : /home/datascience/buckets/models-ckpt-bkt/
models/healthcare_ner-Dr-BERT/DrBERT-4GB/checkpoint-1688
Detected kernel version 5.4.17, which is below the recommended minimum of
5.5.0; this can cause the process to hang. It is recommended to upgrade the
kernel to the minimum version or higher.
[2024-08-26 16:16:55,437] [INFO] [real_accelerator.py:158:get_accelerator]
Setting ds_accelerator to cuda (auto detect)
 [71/71 00:00]
Evaluating model checkpoint : /home/datascience/buckets/models-ckpt-bkt/
models/healthcare_ner-Dr-BERT/DrBERT-4GB-CP-PubMedBERT/checkpoint-2532
Detected kernel version 5.4.17, which is below the recommended minimum of
5.5.0; this can cause the process to hang. It is recommended to upgrade the
kernel to the minimum version or higher.
 [71/71 00:00]
Evaluating model checkpoint : /home/datascience/buckets/models-ckpt-bkt/
models/healthcare_ner-Dr-BERT/DrBERT-7GB/checkpoint-1688

CHAPTER 6   MODEL FINE-TUNING

Detected kernel version 5.4.17, which is below the recommended minimum of 5.5.0; this can cause the process to hang. It is recommended to upgrade the kernel to the minimum version or higher.
 [71/71 00:00]
Evaluating model checkpoint : /home/datascience/buckets/models-ckpt-bkt/models/healthcare_ner-abazoge/DrBERT-4096/checkpoint-1688
Map: 100%
563/563 [00:00<00:00, 10483.43 examples/s]
Detected kernel version 5.4.17, which is below the recommended minimum of 5.5.0; this can cause the process to hang. It is recommended to upgrade the kernel to the minimum version or higher.
Input ids are automatically padded to be a multiple of 'config.attention_window': 512
 [71/71 00:10]
Evaluating model checkpoint : /home/datascience/buckets/models-ckpt-bkt/models/healthcare_ner-almanach/camembert-bio-base/checkpoint-3376
Detected kernel version 5.4.17, which is below the recommended minimum of 5.5.0; this can cause the process to hang. It is recommended to upgrade the kernel to the minimum version or higher.
 [71/71 00:00]

Listing 6-28. Creating a DataFrame with Evaluation Metrics

```
import pandas as pd
import matplotlib.pyplot as plt
from math import pi

# The data you provided
data = model_results

# Extract models and their metrics
# Extract models and their metrics
models = list(data.keys())
metrics_data = [list(metrics.values()) for metrics in data.values()]
metrics_names = list(data[models[0]].keys())  # Extract metric names
```

```
# Filter out samples and steps per second metrics
metrics_names = [metric for metric in metrics_names if metric not in
['eval_model_preparation_time','eval_runtime','eval_samples_per_second',
'eval_steps_per_second']]
df = pd.DataFrame(metrics_data, index=models, columns=list(data[models[0]].
keys()))[metrics_names]
df
```

Output

	eval_loss	eval_precision	eval_recall	eval_f1	eval_accuracy
/home/datascience/buckets/models-ckpt-bkt/models/healthcare_ner-Dr-BERT/DrBERT-4GB/checkpoint-1688	0.237051	0.732680	0.721364	0.726978	0.927640
/home/datascience/buckets/models-ckpt-bkt/models/healthcare_ner-Dr-BERT/DrBERT-4GB-CP-PubMedBERT/checkpoint-2532	0.211613	0.789223	0.763413	0.776104	0.941356
/home/datascience/buckets/models-ckpt-bkt/models/healthcare_ner-Dr-BERT/DrBERT-7GB/checkpoint-1688	0.240774	0.731629	0.729299	0.730463	0.924297
/home/datascience/buckets/models-ckpt-bkt/models/healthcare_ner-abazoge/DrBERT-4096/checkpoint-1688	0.236985	0.728912	0.721019	0.724944	0.924297
/home/datascience/buckets/models-ckpt-bkt/models/healthcare_ner-almanach/camembert-bio-base/checkpoint-3376	0.283445	0.782905	0.775584	0.779228	0.927894

Visualizing Results: You can feed in the results list above into the plot_radar() function to visualize different aspects of their performance and choose the model that is the best fit, depending on the metric(s) that are relevant to your use case:

Listing 6-29. Plotting Evaluation Metrics

```
# Radar plot
def radar_plot(data, model_names, metrics):
    categories = metrics
    N = len(categories)

    # Calculate angle of each axis
    angles = [n / float(N) * 2 * pi for n in range(N)]
    angles += angles[:1]

    fig, ax = plt.subplots(figsize=(8, 8), subplot_kw=dict(polar=True))

    # Loop through each model's results
    for i, model_name in enumerate(model_names):
        values = data.iloc[i].tolist()
        values += values[:1]   # to close the plot
        ax.plot(angles, values, linewidth=2, linestyle='solid',
        label=model_name)
        ax.fill(angles, values, alpha=0.25)

    # Add labels to axes
    ax.set_xticks(angles[:-1])
    ax.set_xticklabels(categories, fontsize=12)

    # Draw one axe per variable and add labels
    ax.set_rlabel_position(0)

    # Show legend
    plt.legend(loc='upper right', bbox_to_anchor=(0.1, 0.1))
    plt.show()

# Filter out samples and steps per second metrics
metrics_names = [metric for metric in metrics_names if metric not in
['eval_loss','eval_model_preparation_time','eval_runtime','eval_samples_
per_second', 'eval_steps_per_second']]
```

CHAPTER 6   MODEL FINE-TUNING

```
#metrics_names = [metric for metric in metrics_names if metric not in ['eval_
model_preparation_time','eval_runtime','eval_samples_per_second', 'eval_steps_
per_second']]
df = pd.DataFrame(metrics_data, index=models, columns=list(data[models[0]].
keys()))[metrics_names]

# Plot all metrics on the radar chart
radar_plot(data=df, model_names=models, metrics=metrics_names)
```

Output (Figure 6-13)

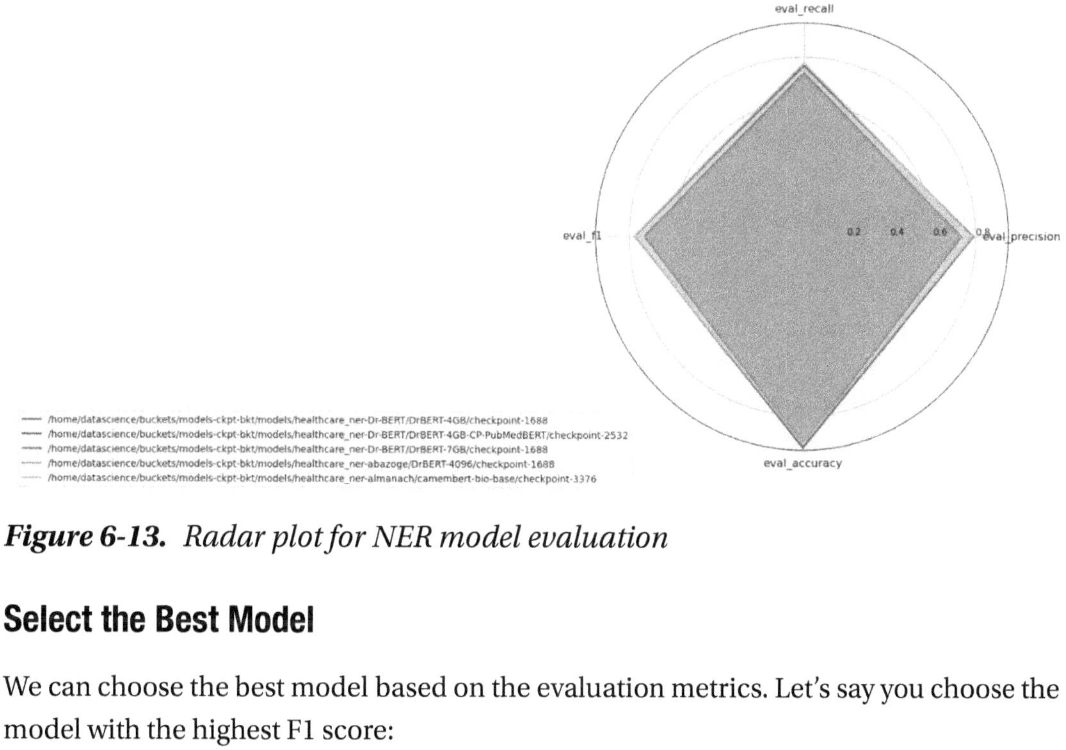

Figure 6-13. *Radar plot for NER model evaluation*

Select the Best Model

We can choose the best model based on the evaluation metrics. Let's say you choose the model with the highest F1 score:

Now that we have a DataFrame df with the evaluation metrics, we can select the best model based on one of these metrics, such as F1 score.

Listing 6-30. Initializing the Model and Tokenizer from the Best Performing Checkpoint

```
from transformers import AutoTokenizer, AutoModelForTokenClassification

best_model_row = df.loc[df['eval_f1'].idxmax()]
best_model_path = best_model_row.name   # This gets the index, which is the
model path
```

```
from transformers import AutoTokenizer, AutoModelForTokenClassification,
pipeline

# Load the best model and tokenizer
tokenizer = AutoTokenizer.from_pretrained(best_model_path)
model = AutoModelForTokenClassification.from_pretrained(best_model_path)

print(f"Model and Tokenizer initialized from the best performing model
checkpoint:\n{best_model_path}")
```

Output

```
Model and Tokenizer initialized from the best performing model checkpoint:
/home/datascience/buckets/models-ckpt-bkt/models/healthcare_ner-almanach/
camembert-bio-base/checkpoint-3376
```

Save the Best Model

Save the trained healthcare NER model (version 1.0.0) and its tokenizer to the models-ckpt-bkt bucket under the name: /home/datascience/buckets/models-ckpt-bkt/models/healthcare_ner_model_v1.0.0.

Using the save_pretrained method, both the model's weights and configuration files, as well as the tokenizer, are saved to this specified directory. This ensures that the model and tokenizer can be easily reloaded or shared later.

Listing 6-31. Saving Healthcare NER Model

```
# Define the directory where you want to save the model

# Modify the _name_or_path attribute
model.config._name_or_path = "healthcare_ner_model_v1.0.0"

model_save_dir = f'/home/datascience/buckets/models-ckpt-bkt/models/{model.
config._name_or_path}'

# Save the model
model.save_pretrained(model_save_dir)

# Save the tokenizer
tokenizer.save_pretrained(model_save_dir)

print(f"Model and tokenizer saved to {model_save_dir}")
```

Output

```
Model and tokenizer saved to /home/datascience/buckets/models-ckpt-bkt/
models/healthcare_ner_model_v1.0.0
```

Test the Best Model

Given the evaluation metrics in the DataFrame, the model with the best performance across precision, recall, F1, and accuracy is healthcare_ner-almanach/camembert-bio-base.

Prepare the Test Examples

We will reuse the test sentences that we already used in the select_pretrained_model.

Listing 6-32. Preparing Test Examples

```
test_examples = [
    "Le medecin donne des antibiotiques en cas d'infections des voies
    respiratoires."
    ,"Le médecin recommande des corticoïdes pour réduire l'inflammation
    dans les poumons."
    ,"Pour soulager les symptômes d'allergie, le médecin prescrit des
    antihistaminiques."
    ,"Pour gérer le diabète, le médecin prescrit une insulinothérapie."
    ,"Après une blessure musculaire, le patient doit suivre une
    physiothérapie."
    ,"En cas d'infection bactérienne, le médecin recommande une
    antibiothérapie."
]
test_examples
```

Output

```
["Le medecin donne des antibiotiques en cas d'infections des voies
respiratoires.", "Le médecin recommande des corticoïdes pour réduire
l'inflammation dans les poumons.", "Pour soulager les symptômes d'allergie,
```

le médecin prescrit des antihistaminiques.", 'Pour gérer le diabète, le médecin prescrit une insulinothérapie.', 'Après une blessure musculaire, le patient doit suivre une physiothérapie.', "En cas d'infection bactérienne, le médecin recommande une antibiothérapie."]

Load and Use the Best Model

We can now generate predictions using this dynamically selected model.

Listing 6-33. Loading Healthcare NER Model

```
from transformers import AutoTokenizer, AutoModelForTokenClassification,
pipeline

# Load the best model and tokenizer from the model_save_dir
tokenizer = AutoTokenizer.from_pretrained(model_save_dir)
model = AutoModelForTokenClassification.from_pretrained(model_save_dir)

# Define the pipeline, for named entity recognition (NER) pipeline
# Set the device parameter to use GPU (0 for the first GPU, or -1 for CPU)
ner_pipeline = pipeline('token-classification',
                        model=model,
                        tokenizer=tokenizer,
                        aggregation_strategy="first",
                        device=0  # Use the first GPU
                       )
print(f"pipeline initialized with model {model.config._name_or_path} from  {model_save_dir}")
```

   Output

```
pipeline initialized with model /home/datascience/buckets/models-ckpt-bkt/models/healthcare_ner_model_v1.0.0 from  /home/datascience/buckets/models-ckpt-bkt/models/healthcare_ner_model_v1.0.0
```

CHAPTER 6 MODEL FINE-TUNING

Generate Predictions

Generate predictions for each test example using the best-performing model.

Running this code will print the predictions from the best-performing model in a human-readable format, allowing you to see how it labels each token in your test sentences.

Listing 6-34. Performing Inference with the Healthcare NER Model

```
def format_predictions(predictions):
    formatted_output = []
    for entity in predictions:
        formatted_output.append(f"Entity: {entity['word']}, Label:
        {entity['entity_group']}, Score: {entity['score']:.2f}")
    return "\n".join(formatted_output)

for example in test_examples:
    print(f"Input: {example}")
    predictions = ner_pipeline(example)
    formatted_output = format_predictions(predictions)
    print(formatted_output)
    print("\n")
```

Output

Input: Le medecin donne des antibiotiques en cas d'infections des voies respiratoires.
Entity: antibiotiques, Label: MedicationVaccine, Score: 0.74

Input: Le médecin recommande des corticoïdes pour réduire l'inflammation dans les poumons.
Entity: corticoïdes, Label: MedicationVaccine, Score: 0.75

Input: Pour soulager les symptômes d'allergie, le médecin prescrit des antihistaminiques.
Entity: antihistaminiques., Label: MedicationVaccine, Score: 0.75

Input: Pour gérer le diabète, le médecin prescrit une insulinothérapie.
Entity: insulinothérapie., Label: MedicalProcedure, Score: 0.83

CHAPTER 6   MODEL FINE-TUNING

Input: Après une blessure musculaire, le patient doit suivre une physiothérapie.
Entity: physiothérapie., Label: MedicalProcedure, Score: 0.83

Input: En cas d'infection bactérienne, le médecin recommande une antibiothérapie.
Entity: antibiothérapie., Label: MedicalProcedure, Score: 0.83

Tip It is very important to deactivate your GPU-based OCI Data Science Notebook session (e.g., cs-nlp-nbs-gpu) when you finish the evaluation process to stop billing and save on costs.

Our evaluation process has identified the highest-performing model based on F1 score: */healthcare_ner-almanach/camembert-bio-base/checkpoint-3376*. This model is built upon **CamemBERT-bio-base** (Rian & Eric, 2024), a pretrained model specifically developed for French biomedical applications. CamemBERT-bio-base is an enhanced version of CamemBERT-base, having undergone additional pretraining on a diverse corpus of biomedical texts to better understand scientific and medical language. As shown in Figure 6-14 and described in the Hugging Face Model Card,[1] Camem BERT-bio-base was trained on an extensive range of French biomedical literature, including scientific publications, clinical cases, and drug leaflets.

---

[1] The model can be found at https://huggingface.co/almanach/camembert-bio-base

CHAPTER 6   MODEL FINE-TUNING

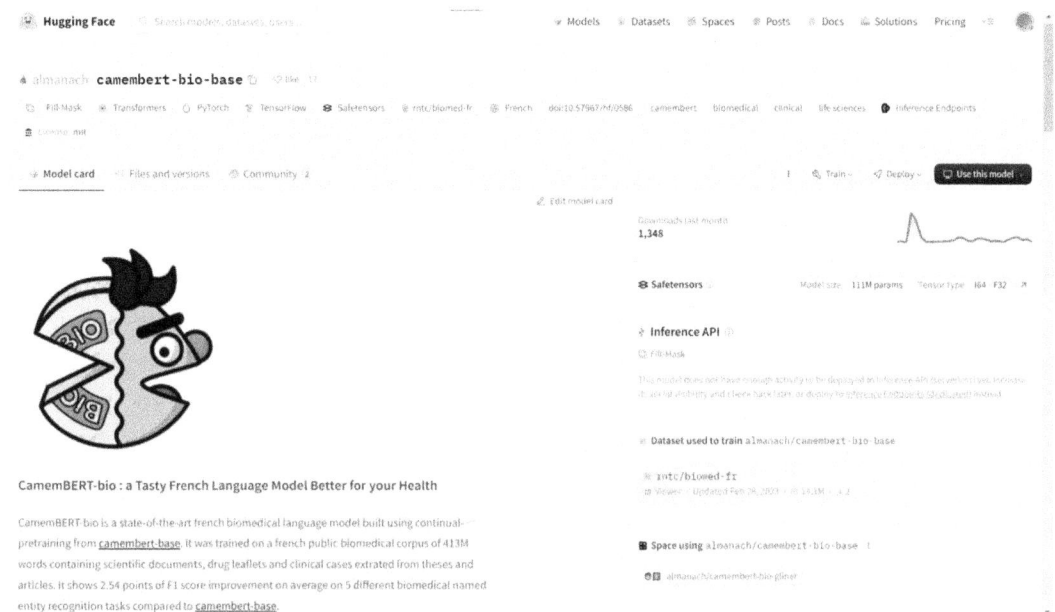

Figure 6-14.   The pretrained MLM model almanach/camembert-bio-base on which our best fine-tuned NER model is based

Our systematic approach to developing a Named Entity Recognition (NER) model involved selecting the top five Masked Language Models (MLMs), fine-tuning each of these models, and evaluating their performance. We conducted this process without manual intervention or prior knowledge of specific pretrained models. This methodology led us to an unexpected discovery: our best-performing NER model was fine-tuned on CamemBERT-bio-base, a recent and robust pretrained model that we were not previously aware of. This added to our confidence that our best fine-tuned Healthcare NER Model will achieve optimal performance in accurately and reliably identifying medical terms and entities in French language healthcare texts.

Summary

This chapter focused on the critical training and evaluation stages in the Natural Language Processing (NLP) life cycle, essential for developing a robust model tailored to specific needs. In our MedTALN Inc. case study, we aimed to build a Healthcare Named Entity Recognition (NER) model for the French language.

CHAPTER 6   MODEL FINE-TUNING

The process began with the efficient selection of the top five pretrained language models for the healthcare domain, supporting the French language, from the Hugging Face Hub. These selected models were then fine-tuned using the OCI Data Science GPU-based Notebook Session.

After fine-tuning, we moved to the evaluation phase, a crucial step in assessing the model's performance on unseen data. This involved using our training dataset's test split to benchmark the fine-tuned Healthcare NER models based on the F1 measure.

We didn't cover hyperparameter optimization in this chapter, but our fine-tuned model performed well enough to meet the objectives of this case study. For those seeking to improve model performance further, the Hugging Face Transformers Trainer class provides a built-in API for hyperparameter search. This allows users to experiment with different configurations, such as learning rates and batch sizes, with the API handling the search for optimal hyperparameters, making the process of hyperparameters optimization more efficient and straightforward.

By adopting a transfer learning approach, leveraging open source pretrained models, and taking advantage of OCI Data Science's GPU-based Notebook Sessions, which can be deactivated after training is complete, we successfully built a robust and cost-effective Healthcare NER model. The next chapter will focus on steps involved in deploying our fine-tuned Healthcare NER model.

References

Ashish Vaswani, N. S. (2017). Attention Is All You Need. *Advances in Neural Information Processing Systems 30 (NeurIPS 2017)*. https://arxiv.org/abs/1706.03762

Bengio, Y., Ducharme, R., Vincent, P., & Jauvin, C. (2003). A Neural Probabilistic Language Model. *Journal of Machine Learning Research*

Jacob Devlin, M.-W. C. (2018). BERT: Pre-training of Deep Bidirectional Transformers for Language Understanding. https://arxiv.org/abs/1810.04805

Jeffrey Pennington, R. S. (2014). {G}lo{V}e: Global Vectors for Word Representation. *Proceedings of the 2014 Conference on Empirical Methods in Natural Language Processing ({EMNLP})*. https://aclanthology.org/D14-1162

Radford, A. N. (2018). Improving Language Understanding by Generative Pre-Training

Rian, T., & Eric, d. l. (2024). CamemBERT-bio: Leveraging Continual Pre-training for Cost-Effective Models on {F}rench Biomedical Data. *Proceedings of the 2024 Joint International Conference on Computational Linguistics, Language Resources and Evaluation (LREC-COLING 2024)*

Tomas Mikolov, K. C. (2013). Efficient Estimation of Word Representations in Vector Space. https://arxiv.org/abs/1301.3781

PART III

Case Study Deployment and Wrap-Up

In **Part 3**, we transition from implementation to deployment, guiding you through the final stages of operationalizing MedTALN Inc.'s Healthcare NER model on Oracle Cloud Infrastructure (OCI).

Chapter 7, "Model Deployment and Monitoring," covers the practical steps necessary to deploy the model on OCI. You'll learn how to deploy our Healthcare NER model using OCI Data Science and its library Oracle Accelerated Data Science (ADS) and invoke it using OCI's Model deployment inference endpoint, completing the technical implementation of our case study.

Chapter 8, "MLOps and Conclusion," wraps up our journey by summarizing the key concepts and takeaways from previous chapters and providing insights into potential future enhancements. This chapter reflects on the lessons learned throughout the case study and discusses the broader implications of building NLP solutions (such as cost and carbon emissions).

Part 3 completes our journey from theoretical foundations to a fully operational NLP solution on OCI. It ties together all concepts and practices discussed throughout the book, serving as the culmination of our learning journey. This final section provides you with a comprehensive picture of developing, deploying, and operationalizing an NLP solution on OCI.

CHAPTER 7

Model Deployment and Monitoring

In this chapter, we transition from model development to real-world application as we explore the crucial steps of deploying and invoking our Healthcare NER model on Oracle Cloud Infrastructure (OCI). This marks the final stage of our technical implementation, bringing our case study to completion.

We'll guide you through the practical process of deploying the model using OCI Data Science and leveraging the powerful Oracle Accelerated Data Science (ADS) library. You'll learn how to set up the necessary environment, configure the deployment, and ensure your model is ready for production use.

Furthermore, we'll delve into the intricacies of model invocation, demonstrating how to utilize OCI's model deployment inference endpoint. This will enable you to seamlessly integrate the deployed model into client applications, allowing for real-time Named Entity Recognition in medical texts.

By the end of this chapter, you'll have a comprehensive understanding of

- The deployment process on OCI Data Science
- Utilizing the ADS library for Hugging Face pipeline deployment
- Invoking OCI's model deployment inference endpoint

Let's begin the final technical phase of our journey, turning our Healthcare NER model into a fully operational API-based service bringing NLP capabilities to healthcare applications.

CHAPTER 7 MODEL DEPLOYMENT AND MONITORING

Model Inference Preliminaries

In this MedTALN Inc.' journey to infuse NLP into its healthcare analytics, the NLP consultant and the IT team have successfully navigated through the crucial stages of dataset preparation and model fine-tuning for the Named Entity Recognition (NER) model. Now, they stand at the threshold of bringing the healthcare NER model to life in a real-world setting.

In MedTALN Inc.'s journey to infuse NLP into its healthcare analytics, the NLP consultant and IT team have successfully navigated through the crucial stages of dataset preparation and model fine-tuning for the Named Entity Recognition (NER) model. Now, they stand at the threshold of bringing the healthcare NER model to life in a real-world setting. This section provides preliminaries for model inference, a critical phase that bridges the gap between a well-trained model and its practical application in healthcare environments.

Model inference preliminaries focus on the essential considerations and preparations required to deploy and operate the NER model effectively in production.

Understanding Inference vs. Training

In the context of our healthcare Named Entity Recognition (NER) model, understanding the distinction between inference and training is crucial for effective deployment and operation.

Training, which we've completed in the previous chapter, involved exposing our model to large volumes of annotated healthcare data to learn patterns and relationships. This process was computationally intensive and time-consuming, requiring significant resources to fine-tune our model for accurate entity recognition in medical texts.

Inference, on the other hand, is the operational phase where our trained NER model applies its learned knowledge to new, unseen healthcare data. This process is typically faster and less resource-intensive than training, but it comes with its own set of challenges that require careful consideration.

One of the primary concerns is resource allocation and optimization. While training the model demands significant computational resources, typically utilizing GPUs or TPUs to process large datasets and complex model's neural networks architectures, the inference phase necessitates a shift toward lower-latency environments. This often means deploying the model on CPUs, which may require optimization techniques such as model compression—quantization or pruning—to ensure that the model remains efficient without sacrificing performance.

Another critical aspect is monitoring model drift. NLP models are susceptible to performance degradation over time, particularly as domain-specific terminology changes. To mitigate this risk, it is essential to implement continuous monitoring systems that track key performance metrics, such as F1 score, precision, and recall, on a validation dataset. Setting up alerts for significant drops in these metrics can help identify instances of model drift, allowing for timely intervention through retraining or fine-tuning.

Accuracy and reliability are paramount in the healthcare context, where incorrect entity recognition can have significant implications. Implementing confidence thresholds for entity predictions can help ensure that only high-confidence outputs are accepted.

As the usage of our Healthcare NER model grows, scalability and throughput become increasingly important. Management of the model deployment's number of instances or their compute shapes can provide greater flexibility and better cost control while handling the increased workload. This approach allows for dynamic scaling of resources based on demand, ensuring optimal performance during peak times and cost-efficiency during periods of lower activity.

Finally, interpretability and debugging play a crucial role in maintaining and improving the model's performance. Implementing attention visualization techniques can help elucidate which parts of the input text contribute most to the model's predictions. Additionally, developing tools for error analysis can identify patterns in misclassifications, guiding targeted improvements to the model.

Finally, interpretability and debugging are crucial for maintaining our BERT-based Healthcare NER model's performance. Given BERT's complexity, understanding its decision-making process is essential. We employ error analysis using a confusion matrix to identify misclassification patterns, such as consistently mislabeling certain medical entities. This method reveals the model's weaknesses, like struggles with rare diseases or ambiguous abbreviations. By focusing on these insights, we can guide targeted improvements by fine-tuning specific data subsets or adjusting preprocessing steps to address common errors.

By addressing these NLP-specific challenges during the transition from training to inference, we can ensure that our healthcare NER model not only performs well initially but also continues to deliver accurate and reliable results over time.

CHAPTER 7 MODEL DEPLOYMENT AND MONITORING

Cost-Saving Strategies for the Inference Phase

In alignment with MedTALN Inc.'s strategic requirement of building a performant yet cost-effective NLP solution, our NLP consultant, John Doe, has provided comprehensive cost-saving strategies throughout the model development life cycle. Having addressed cost optimization during the data preparation and model training phases, we now focus on the critical inference stage.

As MedTALN Inc. prepares to deploy its Healthcare NER model, optimizing costs during inference becomes critical. This phase is particularly important as it represents the ongoing operational expenses of the NLP initiative. By implementing effective strategies at this stage, we can ensure that the model performs well and remains financially viable in the long term.

John Doe emphasizes that while maintaining high performance is crucial, there are several approaches we can take to significantly reduce the infrastructure costs associated with model inference as depicted in in Figure 7-1. These strategies are designed to balance the need for quick and accurate Named Entity Recognition in healthcare texts with the imperative of managing operational expenses.

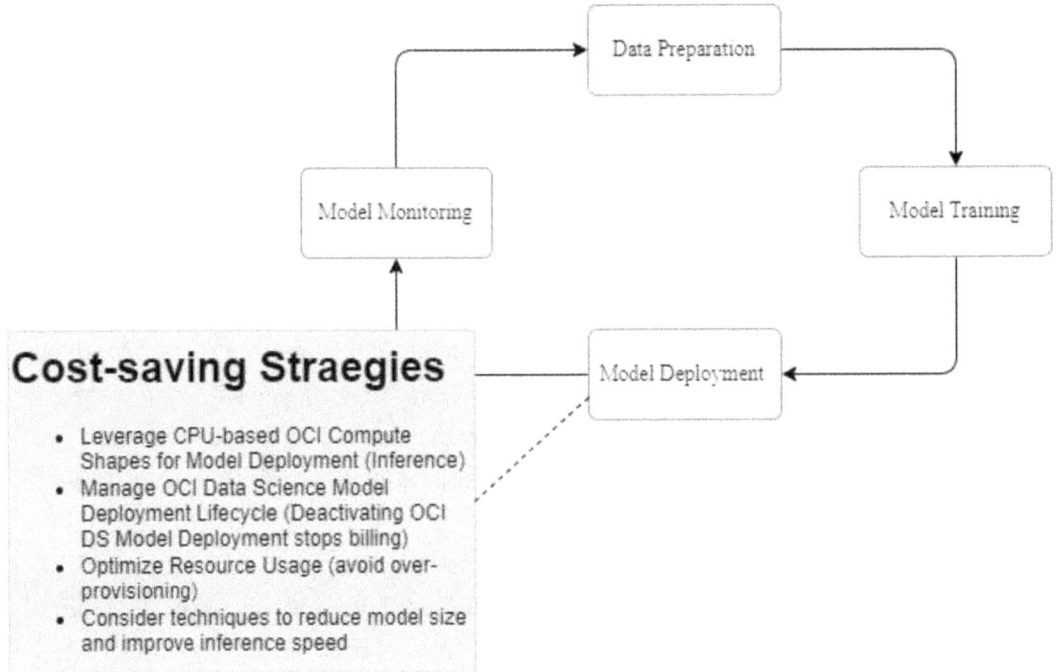

Figure 7-1. Cost-saving strategies for inference

CHAPTER 7   MODEL DEPLOYMENT AND MONITORING

Let's explore the key cost-saving strategies for the inference phase of our Healthcare NER model:

- Leveraging CPU-Based OCI Compute Shapes for Model Deployment: For many NLP tasks, modern CPUs can provide sufficient performance at a lower cost. Our Healthcare NER model can achieve acceptable performance on CPU instances, which are often more cost-effective than GPUs. Additionally, OCI allows us to choose the appropriate CPU compute shape based on our model's requirements and expected workload.

- Managing OCI Data Science Model Deployment Life Cycle: Deactivate the OCI Data Science Model Deployments when not in use to pause billing. OCI allows you to stop model deployments to avoid incurring charges when they're not in use.

- Optimizing Resource Usage: Regularly review and adjust the size of your compute instances. Ensure you're using the most cost-effective instance types that meet your performance requirements.

- Monitoring and Analyzing Usage Patterns and Costs: Continuously monitor key performance metrics such as latency, throughput, and resource utilization. Use this data to identify bottlenecks and optimization opportunities by adjusting resources based on actual usage to avoid overprovisioning. Also, regularly analyze spending patterns and look for opportunities to reduce costs.

- Leveraging Batch Inference: For non-real-time applications, batch processing can be more cost-effective. By aggregating requests and processing them in batches, you can optimize resource utilization and reduce overall compute costs.

While not implemented in our current case study, another potential cost-saving strategy worth considering is model optimization for inference through techniques such as quantization, which could potentially reduce model size and computational needs without significantly impacting accuracy.

> **Tip** Using flexible shapes for you model deployment allows you to customize the number of OCPUs and the amount of memory when launching or resizing the model deployment virtual machine (VM).

By implementing these strategies, MedTALN Inc. can significantly reduce the infrastructure costs associated with inference for its Healthcare NER model.

Preparing the Environment

Before initiating the deployment of our Healthcare MER model, we need to carry out some preparatory steps for our data science. This will involve creating a custom log for the model deployment to integrate with the logging service. Additionally, it is necessary to republish our custom conda environment that was used during the model training phase.

Setting Up Policies

After deploying our model, data scientists will need to monitor its performance. To enable this, we need to add a new policy to our data science policies, as illustrated in Figure 7-2.

Listing 7-1. New data science policy

```
allow group data-scientists-users-grp to read metrics in compartment
case-study-cmpt
```

CHAPTER 7   MODEL DEPLOYMENT AND MONITORING

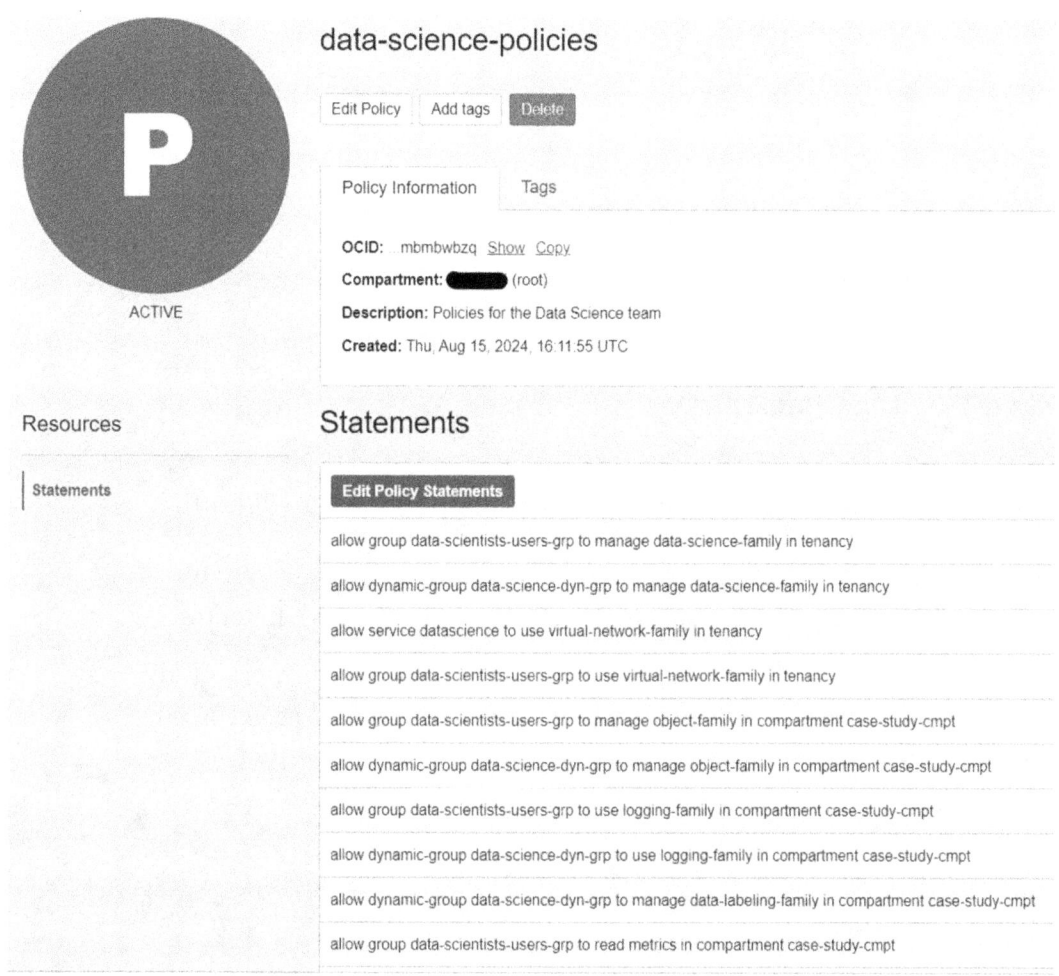

Figure 7-2. data-science-policies

Setting Up Logging

The OCI administrator is responsible for creating the logs for the model deployment.
To creation the logs, perform the following:

1. Open the navigation menu, and click **Observability & Management**. Under **Logging**, click **Log Groups** (Figure 7-3).

CHAPTER 7　MODEL DEPLOYMENT AND MONITORING

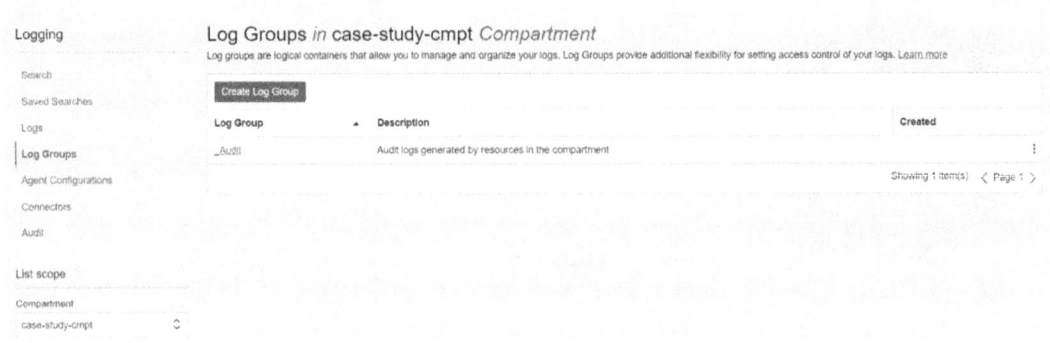

Figure 7-3. Log Groups list page

2. Select the compartment for our case study: *case-study-cmpt*.

3. Click ***Create Log Group***.

4. The ***Create Log Group*** panel is displayed (Figure 7-4). Enter the following:

    a. Compartment: The compartment field is prefilled based on our compartment, i.e., *case-study-cmpt*.

    b. Name: cs-log-group.

    c. Description: Log Group for Case Study.

    d. Click ***Create***.

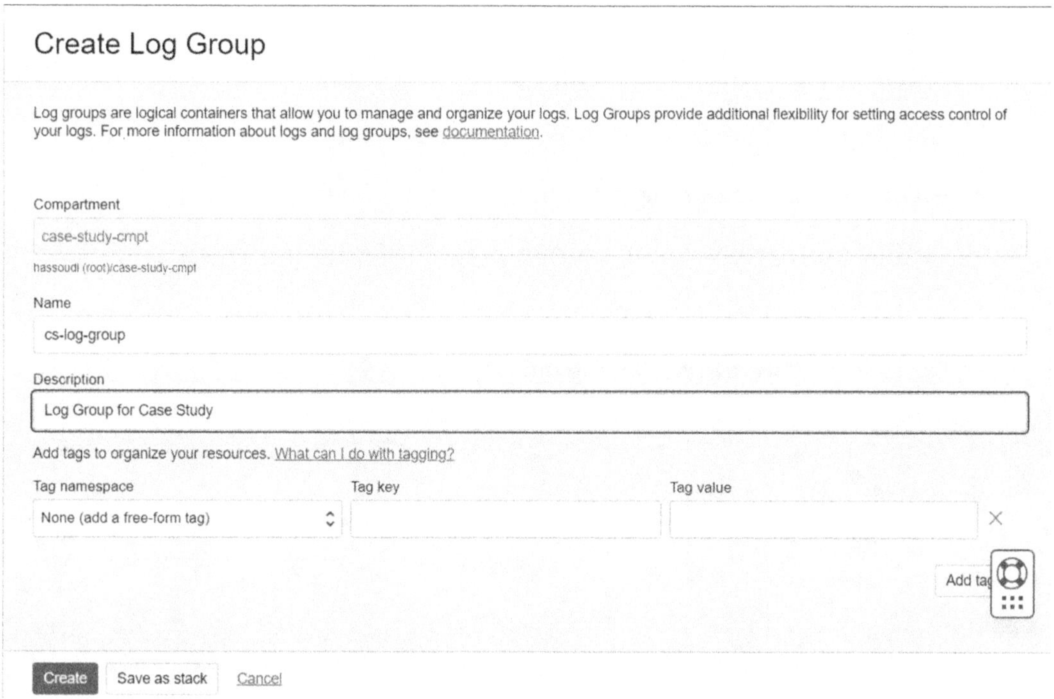

CHAPTER 7 MODEL DEPLOYMENT AND MONITORING

Figure 7-4. Log group creation

5. On the ***Log Group*** detail page, select the option ***Logs*** (Figure 7-5).

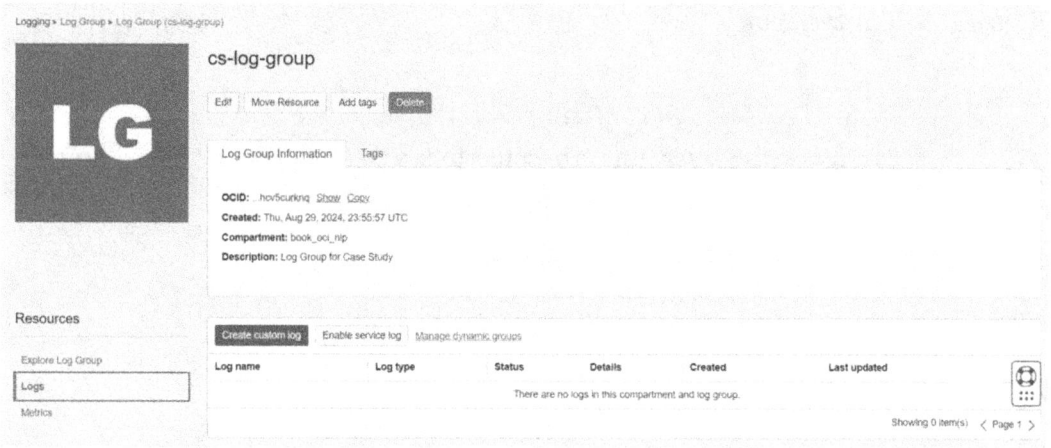

*Figure 7-5. **Log Group** detail page*

CHAPTER 7   MODEL DEPLOYMENT AND MONITORING

Click **Logs**.

6. Click the button **Create custom log**.

7. The **Log Creation** dialog is displayed (Figure 7-6).

8. In the **Create custom log** step, enter

   a. Custom log name : Cs-custom-log.

   b. Log group: cs-log-group (which is already prefilled).

   c. Click the **Create custom log** button.

Figure 7-6. *Log creation*

   d. On the **Create agent configuration** step, select the option **Add configuration later** (this is because jobs and model deployments aren't integrated into the Logging service agent configuration).

   e. Click the button **Create agent config** (Figure 7-7).

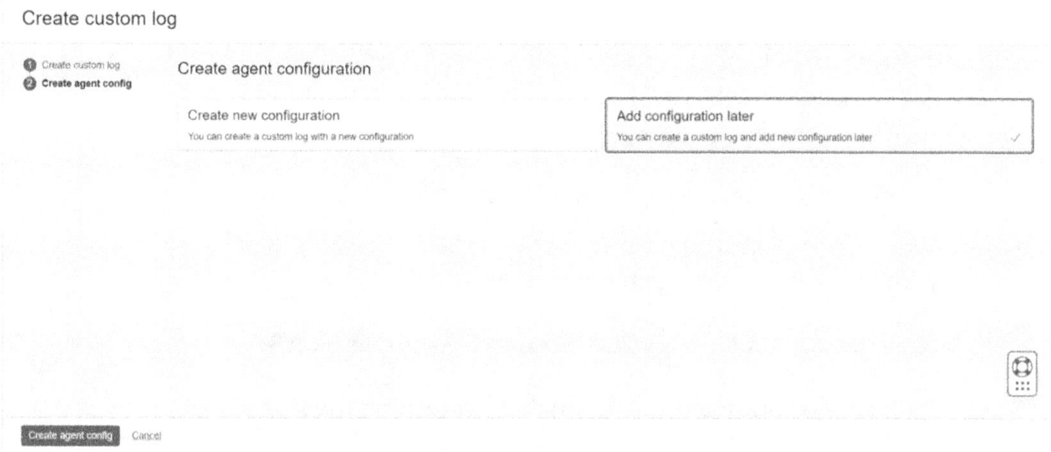

CHAPTER 7   MODEL DEPLOYMENT AND MONITORING

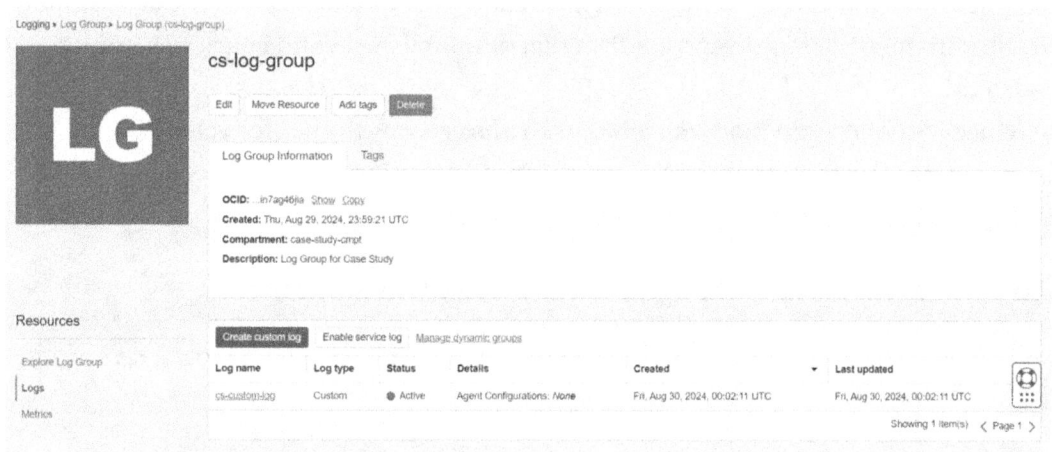

Figure 7-7. Log creation (agent configuration step)

Figure 7-8. Log Group detail page

Publish Custom Conda Env.

During the training, the NLP consultant upgraded the Transformers library from the original version 4.37.2 shipped with the **pytorch21_p39_gpu_v1** conda env. to the latest Transformers version 4.44.2.[1] As a result, we need to refresh our published custom

---

[1] At the time of writing, the latest version of the Transformers library is 4.44.2.

333

conda environment to include the new Transformers version. We'll do this by publishing the conda environment from the training notebook session, using the --force flag to overwrite the environment we previously published in Chapter 4.

To publish our modified custom conda env., the steps are as follows:

1. Configure *odsc conda* to use an Object Storage bucket *conda-envs-bkt* using this command (replace yz2wwgkgt8eh with your Object storage namespace):

   odsc conda init -b conda-envs-bkt -n yz2wwgkgt8eh

2. Publish the conda env. to our bucket by running the following command line:

   odsc conda publish -s pytorch21_p39_gpu_v1 --force

Once the publishing process is done, we can go to our Object Storage bucket conda-envs-bkt in the OCI console and confirm that our published conda pack was properly updated.

Once it's done, from the **Environment Explorer**, copy the source value (Listing 7-2) from the published conda env. "PyTorch 2.1 for GPU on Python 3.9" (replace yz2wwgkgt8eh with your Object storage namespace).

Listing 7-2. Conda env. "PyTorch 2.1 for GPU on Python 3.9"

```
oci://conda-envs-bkt@yz2wwgkgt8eh/conda_environments/gpu/PyTorch 2.1 for GPU on Python 3.9/1.0/pytorch21_p39_gpu_v1
```

Deployment Process

In Chapter 6, we successfully fine-tuned our Healthcare NER model using OCI Data Science Notebooks, leveraging both CPU- and GPU-based notebook sessions. We utilized the Hugging Face ecosystem, including the Hugging Face Hub, models, and datasets, along with key libraries such as Transformers and datasets. The resulting model was saved in the "Model Checkpoint Bucket" under the path *<buckets mount>/models-ckpt-bkt/models/healthcare_ner_model_v1.0.0*.

Now, we'll leverage the OCI Data Science Deployment (Oracle, 2024) features and capabilities to efficiently transition this fine-tuned model from a trained solution to a fully operational web service ready for integration into MedTALN's healthcare

CHAPTER 7   MODEL DEPLOYMENT AND MONITORING

applications. The OCI Data Science Service provides a managed resource for deploying our machine learning model as an HTTP endpoint in OCI, making it easily accessible for real-time predictions.

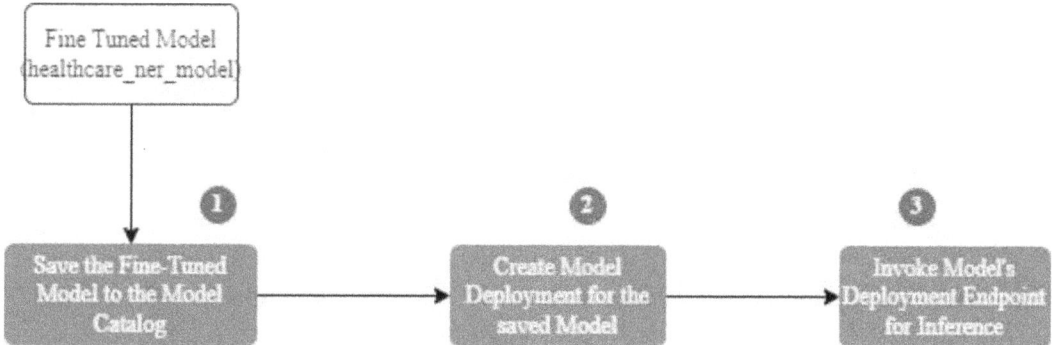

Figure 7-9. *High-level deployment process*

As illustrated by Figure 7-9, our deployment process will cover the following key steps:

1. Save our best fine-tuned Healthcare NER model (i.e., healthcare_ner_model_v1.0.0) to the OCI Data Science Model Catalog.

2. Create the OCI Data Science Model Deployment (with specified compute shape).

3. Invoke the OCI Data Science Model Deployment endpoint to get model predictions.

Through the upcoming sections, we'll demonstrate how to properly and easily handle the deployment of Transformer-based models built with Hugging Face libraries within the OCI environment.

Oracle Data Science Model Catalog

The Oracle Data Science Model Catalog serves as a centralized repository within the OCI Data Science Service, allowing for efficient storage, management, and versioning of machine learning models. It acts as a central hub for all models developed within an organization, enabling easy access, sharing, and deployment of these models.

One of its key features is the immutable storage of model artifacts, ensuring the integrity and reliability of stored models. This immutability is complemented by provenance tracking, which allows users to monitor models throughout their life cycle, enhancing reproducibility and transparency.

A model artifact within the catalog includes not only the model itself but also essential metadata, input and output schemas, and scripts for loading the model and making predictions. This comprehensive structure facilitates easy sharing among team members and ensures that models can be effectively reproduced and deployed across different environments.

The Model Catalog can be accessed directly in a notebook session using Oracle's Accelerated Data Science (ADS) or through the OCI console. It supports extensive documentation capabilities, allowing users to detail the model's use case, algorithm, custom metadata, provenance, and input/output schemas. This is particularly useful for documenting the specific entities recognized by the NER model and the healthcare standards it adheres to.

Furthermore, the Model Catalog supports taxonomy metadata, enabling the specification of critical information such as use case type, framework, algorithm, and hyperparameters. This feature is valuable for maintaining clear records of model versions and their specific configurations.

It's important to note that model artifacts have size limitations, with a maximum of 100 MB when accessed from the console and up to 400 GB for larger models. This should be sufficient for our NER models which has a size of around 400 MB.

By leveraging the Model Catalog in the deployment process, organizations can ensure a more organized, efficient, and reliable workflow from model development to production deployment.

Oracle Data Science Model Deployment

Let's discuss the model deployment flow and the related architecture in OCI Data Science. We'll examine the various components of the model deployment.

After the model training and evaluation process is complete, the best model is saved to the Model Catalog. Model deployments in OCI Data Science are managed resources that deploy machine learning models as HTTP endpoints.

Figure 7-10 depicts the OCI Data Science Model Deployment Architecture for MedTALN Inc.'s Healthcare NER model, illustrating the deployment process which consists of several key stages.

CHAPTER 7   MODEL DEPLOYMENT AND MONITORING

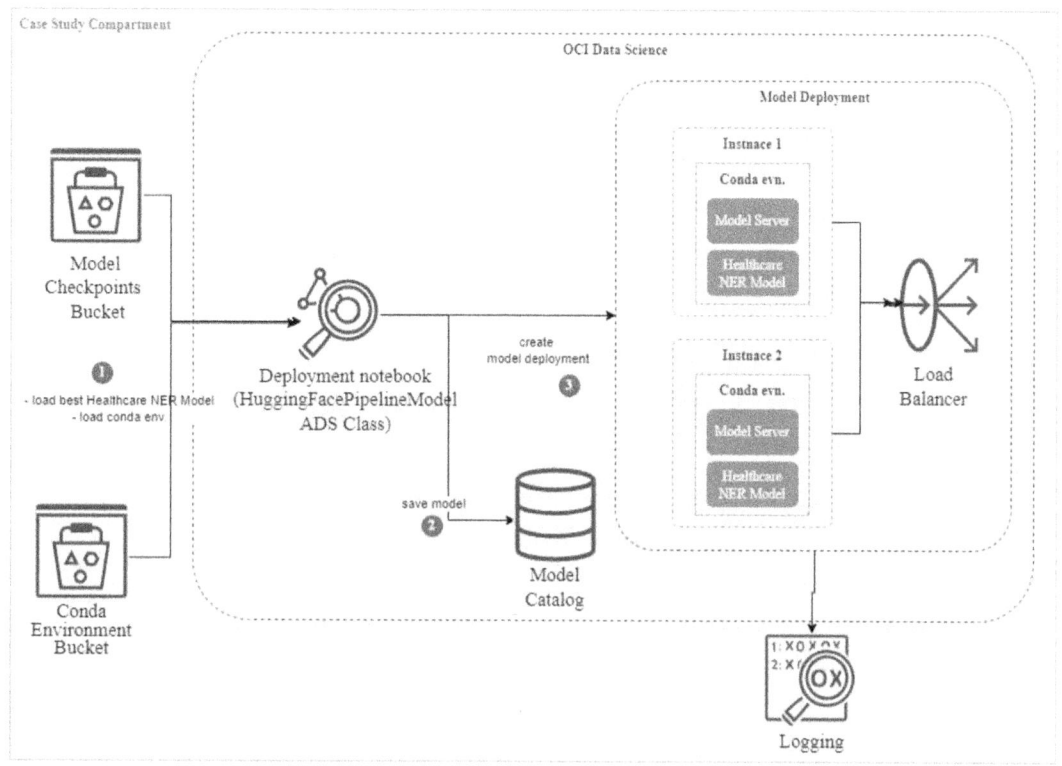

Figure 7-10. *OCI Data Science Model Deployment Architecture*

The key components of the model deployment architecture are

- Load Balancer: Provides an automated way to distribute traffic from one entry point to multiple model servers running in a pool of virtual machines

- VM Instances Pool: A pool of VM instances hosting the model server, the conda environment, and the model itself

- Model Artifact: The actual model file to load and its predict code for inferencing

- Conda Environment: Encapsulates all the third-party Python dependencies, like Hugging Face Transformers, that our model requires

- Logs: Emit logs from the inference code to OCI logging, helpful for monitoring and debugging

The high-level key steps to creating and invoking our model deployment using OCI Data Science are

1. Loading the Model and Environment: Initialize the HuggingFacePipelineModel class (Hugging Face) from the Oracle Accelerated Data Science (ADS) library with our Healthcare NER model and its artifacts. Call the prepare method with our conda environment.

2. Saving the Model: Use the save method to store the model in the OCI Data Science Model Catalog, making it accessible for deployment.

3. Model Deployment: Create the model deployment by calling the deploy method. This deploys the model artifact from the Model Catalog with the necessary conda environment onto multiple OCI instances, sets up each instance, and connects them to a load balancer for scalability.

Once deployment is complete and in an active state, it can be invoked to generate predictions on new data via HTTP requests to the endpoints. The model deployment returns an HTTP response with the predictions.

To invoke the model deployment, you can pass healthcare text samples to the predict endpoint, and the model will return predictions for the extracted medical entities. You can use the sample code from the model deployment details to invoke the model endpoint using OCI CLI or, alternatively, use the OCI Python SDK or Java SDK.

> **Caution** When invoking the model, be aware of the following limitations: the payload size is limited to 10 MB, the invocation timeout is 60 seconds, and the payload must be encoded in base64 format. Keeping these constraints in mind will help avoid errors when interacting with the deployed model.

Oracle ADS HuggingFacePipelineModel

Before diving into the step-by-step notebook for deploying our Healthcare NER model on OCI, we need to understand the crucial role of the Oracle ADS HuggingFacePipelineModel (Oracle, 2022) from Oracle's Accelerated Data Science

CHAPTER 7   MODEL DEPLOYMENT AND MONITORING

(ADS) library. This class is the cornerstone of our automated deployment process and is specifically designed to simplify the deployment of Hugging Face models, such as our fine-tuned Healthcare NER model, on Oracle Cloud Infrastructure (OCI).

Using Oracle ADS HuggingFacePipelineModel offers several key benefits:

- Automated Artifact Generation: The .prepare() method automatically creates necessary model artifacts for deployment without manual configuration.

- Customization Flexibility: While artifacts are auto-generated, you can still customize the score.py file if needed.

- Easy Debugging: The .verify() method allows you to test and debug your model without actual deployment, saving time and resources.

- Streamlined Deployment: The .save() method simplifies model artifact deployment to the catalog, while .deploy() easily creates a REST endpoint for the model.

In the upcoming sections, we'll walk through the process of using HuggingFacePipelineModel to deploy our Healthcare NER model. We'll explore key methods such as "prepare()" for creating deployment artifacts, "verify()" for predeployment testing, "save()" for storing the model in the OCI Model Catalog, and "deploy()" for creating a REST endpoint to serve our model.

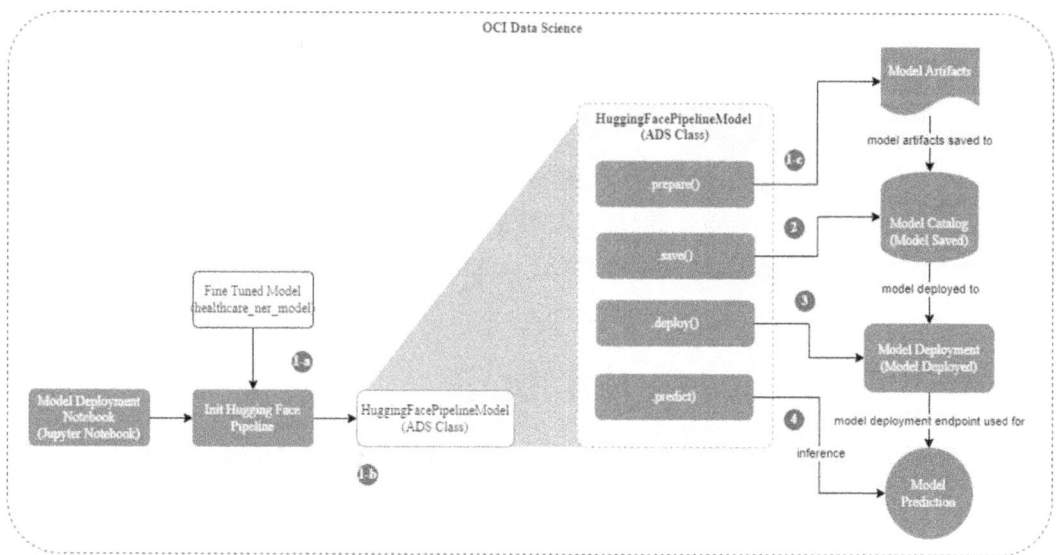

Figure 7-11. High-level deployment process

339

CHAPTER 7   MODEL DEPLOYMENT AND MONITORING

As illustrated by Figure 7-11, our deployment process will cover the following key steps:

1. Initializing the ADS Class HuggingFacePipelineModel:

   a. Load the Hugging Face pipeline with our best fine-tuned Healthcare NER model (i.e., healthcare_ner_model_v1.0.0).

   b. Initialize the HuggingFacePipelineModel Class with our model.

   c. Generate model's artifacts.

2. Saving the model's artifact to the OCI Data Science Model Catalog.

3. Creating a the OCI Data Science Model Deployment (with specified compute shape).

Tip   Our case study shows how the OCI Data Science Service supports the Hugging Face ecosystem, effectively accommodating state-of-the-art NLP technologies. This integration enables us to leverage advanced models like BERT for our Healthcare NER application, enabling us to train, deploy, and manage NLP models efficiently.

By leveraging the HuggingFacePipelineModel, we're poised to efficiently transform MedTALN's Healthcare NER model from a trained solution into a robust, production-ready tool. This deployment will enhance MedTALN's healthcare applications with advanced Named Entity Recognition capabilities, improving data processing and insight extraction from medical texts.

As we proceed, we'll provide a detailed notebook demonstrating each step of the deployment process, ensuring that MedTALN's team can replicate and adapt this process for future model iterations or additional NLP tasks in their healthcare ecosystem.

Deployment Process Notebook

This notebook guides you through the process of deploying our Healthcare NER model. The development of this notebook was based on the following resources:

- ADS v2.10.0 documentation for the class HuggingFacePipelineModel
- Train, Register, and Deploy Hugging Face Pipeline Example Notebook from Notebook Explorer (train-register-deploy-huggingface-pipeline.ipynb)

The deployment process is composed of the following key steps (detailed in Listings 7-3 to 7-15):

1. Initializing the ADS Class HuggingFacePipelineModel:

    a. Load the Hugging Face pipeline with our best fine-tuned Healthcare NER model (i.e., healthcare_ner_model_v1.0.0).

    b. Initialize the HuggingFacePipelineModel class with our model.

    c. Generate the model's artifacts.

2. Saving the model's artifacts to the OCI Data Science Model Catalog

3. Creating the OCI Data Science Model Deployment:

    a. Provision the necessary OCI resources for the model deployment and create model prediction endpoint.

    b. Invoke the model's endpoint to test its inference capability.

Initializing the ADS Class "HuggingFacePipelineModel"

We start our model deployment process by initializing the ADS Class HuggingFacePipelineModel.

Authenticate

Authentication to the OCI Data Science Service is required. Here, we default to resource principals.

Listing 7-3. Authentication

```
import ads
ads.set_auth(auth="resource_principal")
```

Output

Initialize Hugging Face Pipeline

Initialize "transformers.pipeline" with our best fine-tuned Healthcare NER model (i.e., "healthcare_ner_model_v1.0.0").

Listing 7-4. Initializing Hugging Face Pipeline

```
from transformers import pipeline
import warnings

model_checkpoint = "/home/datascience/buckets/models-ckpt-bkt/models/healthcare_ner_model_v1.0.0"
print(f"model checkpoint: {model_checkpoint}")

data = "Le medecin donne des antibiotiques en cas d'infections des voies respiratoires."
pipeline = pipeline(
    "token-classification", model=model_checkpoint, aggregation_strategy="first"
)
preds = pipeline(data)
preds
```

Output

```
model checkpoint: /home/datascience/buckets/models-ckpt-bkt/models/healthcare_ner_model_v1.0.0
[6]:
[{'entity_group': 'MedicationVaccine',
  'score': 0.7428154,
  'word': 'antibiotiques',
  'start': 21,
  'end': 34}]
```

Prepare Model Artifact

Instantiate a HuggingFacePipelineModel() object with Hugging Face pipelines. All the pipelines related files are saved under the artifact_dir (replace yz2wwgkgt8eh with your Object storage namespace).

Listing 7-5. Preparing Model Artifact

```
from ads.model import HuggingFacePipelineModel
from ads.model.model_metadata import UseCaseType
import tempfile

# Create a temporary directory for the model artifacts
artifacts_temp_dir = tempfile.mkdtemp()

print(f"Model path {pipeline.model.config._name_or_path} and the Model artifacts temp dir {artifacts_temp_dir}")

# Initialize the model
huggingface_pipeline_model = HuggingFacePipelineModel(pipeline, artifact_dir=artifacts_temp_dir)

#Prepare the model
conda_env_source = "oci://conda-envs-bkt@yz2wwgkgt8eh/conda_environments/gpu/PyTorch 2.1 for GPU on Python 3.9/1.0/pytorch21_p39_gpu_v1"

huggingface_pipeline_model.prepare(
  inference_conda_env=conda_env_source,
  inference_python_version="3.9",
  training_conda_env=conda_env_source,
  use_case_type=UseCaseType.OTHER,
  force_overwrite=True,
)
```

  Output

```
Model path /home/datascience/buckets/models-ckpt-bkt/models/healthcare_ner_model_v1.0.0 and the Model artifacts temp dir /tmp/tmp9ph1hgv0
[2024-08-30 01:12:44,108] [INFO] [real_accelerator.py:158:get_accelerator] Setting ds_accelerator to cuda (auto detect) ?, ?it/s]
algorithm: TokenClassificationPipeline
artifact_dir:
  /tmp/tmp9ph1hgv0:
  - - .model-ignore
    - special_tokens_map.json
```

        - tokenizer.json
        - config.json
        - runtime.yaml
        - score.py
        - model.safetensors
        - sentencepiece.bpe.model
        - tokenizer_config.json
framework: transformers
model_deployment_id: null
model_id: null
```

## Manually Correct score.py

The generated score.py contains some bad code that needs to be fixed. Our consultant, John, fixed the issues and provided a corrected version, score_fixed.py, which contains the necessary code adjustments to make it work.

*Listing 7-6.* Fixing the file score.py

```
import shutil

# Specify the path to the source file (score_fixed.py) and the target location (/tmp/score.py)
source_file = './score_fixed.py'  # Update with the actual path to score_fixed.py
target_file = f'{artifacts_temp_dir}/score.py'

# Copy the source file to the target location, effectively replacing it
shutil.copyfile(source_file, target_file)
```

   Output

```
'/tmp/tmp9ph1hgv0/score.py'
```

Run Introspection

Run an introspection test to perform a sanity check on the model's artifacts, including tests on the score.py and runtime.yaml files, with the goal of capturing common errors and issues in the model artifacts. When the model is saved, introspection tests are included in the model metadata.
```

Listing 7-7. Running Introspection Test

huggingface_pipeline_model.introspect()

Output

['.model-ignore', 'special_tokens_map.json', 'tokenizer.json', 'config.json', 'runtime.yaml', 'score.py', 'model.safetensors', 'sentencepiece.bpe.model', 'tokenizer_config.json']
[6]:

	Test key	Test name	Result	Message
0	runtime_env_path	Check that field MODEL_DEPLOYMENT.INFERENCE_ENV_PATH is set	Passed	
1	runtime_env_python	Check that field MODEL_DEPLOYMENT.INFERENCE_PYTHON_VERSION is set to a value of 3.6 or higher	Passed	
2	runtime_path_exist	Check that the file path in MODEL_DEPLOYMENT.INFERENCE_ENV_PATH is correct.	Passed	
3	runtime_version	Check that field MODEL_ARTIFACT_VERSION is set to 3.0	Passed	
4	runtime_yaml	Check that the file "runtime.yaml" exists and is in the top level directory of the artifact directory	Passed	
5	score_load_model	Check that load_model() is defined	Passed	
6	score_predict	Check that predict() is defined	Passed	
7	score_predict_arg	Check that all other arguments in predict() are optional and have default values	Passed	
8	score_predict_data	Check that the only required argument for predict() is named "data"	Passed	

(*continued*)

CHAPTER 7   MODEL DEPLOYMENT AND MONITORING

Test key		Test name	Result	Message
9	score_py	Check that the file "score.py" exists and is in the top level directory of the artifact directory	Passed	
10	score_syntax	Check for Python syntax errors	Passed	

Call Model Summary

The .summary_status() method returns a Pandas DataFrame that guides you through the entire workflow. It shows which methods are available to call and which ones aren't. Plus, it outlines what each method does. If extra actions are required, it also shows those actions.

Listing 7-8. Running Model Summary

huggingface_pipeline_model.summary_status()

Output

Step	Status	Details
initiate	Done	Initiated the model
prepare()	Done	Generated runtime.yaml
		Generated score.py
		Serialized model
		Populated metadata(Custom, Taxonomy and Provenance)
verify()	Available	Local tested .predict from score.py
save()	Available	Conducted Introspect Test
		Uploaded artifact to model catalog
deploy()	UNKNOWN	Deployed the model
predict()	Not Available	Called deployment predict endpoint

CHAPTER 7   MODEL DEPLOYMENT AND MONITORING

Verify the Generated Model Artifacts

Verify the generated model artifacts before deploying the model to model catalog.

Listing 7-9. Verifying the Model Artifacts

```
print(data)
huggingface_pipeline_model.verify(data)
```

   Output

```
Le medecin donne des antibiotiques en cas d'infections des voies
respiratoires.
Model is successfully loaded.
[12]:
{'prediction': [{'entity_group': 'MedicationVaccine',
   'score': 0.742815375328064,
   'word': 'antibiotiques',
   'start': 21,
   'end': 34}]}
```

Save the Model to the Model Catalog

At this step, we create Model Version Set and save our model to the Model Catalog.

Create a Model Version Set

The Model Version Set, which acts as a container by assigning sequential version numbers to models, makes it easier to track their evolution and relationships.

Listing 7-10. Creating Version Set for the Model

```
from ads.model import ModelVersionSet

# Create a model version set
mvs = ModelVersionSet(
    name = "healthcare-ner-model-ver-set",
    description = "A model version set for the Healthcare NER Model")
mvs.create()
```

CHAPTER 7 MODEL DEPLOYMENT AND MONITORING

Output

```
kind: modelVersionSet
spec:
  compartmentId: ocid1.compartment.oc1..
aaaaaaaaceavj5r6agl5e2mysyxg6twnvwh6cw7s2pi6nobiv6nynjcwmhxa
  definedTags:
    Oracle-Tags:
      CreatedBy: ocid1.datasciencenotebooksession.oc1.ca-toronto-1.
      amaaaaaa3hvgr2qaresmmqmzbu3i3mo6npudc3nf6ltgu43in4su7ogons4a
      CreatedOn: '2024-08-30T01:39:55.445Z'
  description: A model version set for the Healthcare NER Model
  id: ocid1.datasciencemodelversionset.oc1.ca-toronto-1.
  amaaaaaa3hvgr2qa4gck2mvzy6r5nbaumfmdbv7rsilu6x5dnggvg5kxrhxa
  name: healthcare-ner-model-ver-set
  projectId: ocid1.datascienceproject.oc1.ca-toronto-1.
  amaaaaaa3hvgr2qaqd5gstwgcmxycii3q7zi4jemjwhb7wienmlapx2ni6ja
type: modelVersionSet
```

Save the Model

Save the model to the Model Catalog.

Listing 7-11. Saving the Model

```
# Register the model
model_id = huggingface_pipeline_model.save(display_name="Healthcare NER Model", model_version_set=mvs, version_label="Version 1")
model_id
```

CHAPTER 7   MODEL DEPLOYMENT AND MONITORING

Output

```
Model is successfully loaded.
['.model-ignore', 'special_tokens_map.json', 'tokenizer.json', 'config.
json', 'runtime.yaml', 'score.py', 'test_json_output.json', 'model.
safetensors', 'sentencepiece.bpe.model', 'tokenizer_config.json']
[27]:
'ocid1.datasciencemodel.oc1.ca-toronto-1.
amaaaaaa3hvgr2qaxmjd5sf3kv3vri4rsgdo3hzw66bq6e4in65taft2y3yq'
```

At this step, go to our Data Science Project, and click **Models** to verify if our model was properly created, as shown in Figure 7-12. You will see that model versioning is activated through the Model Version Set, which acts as a container by assigning sequential version numbers to models, making it easier to track their evolution and relationships.

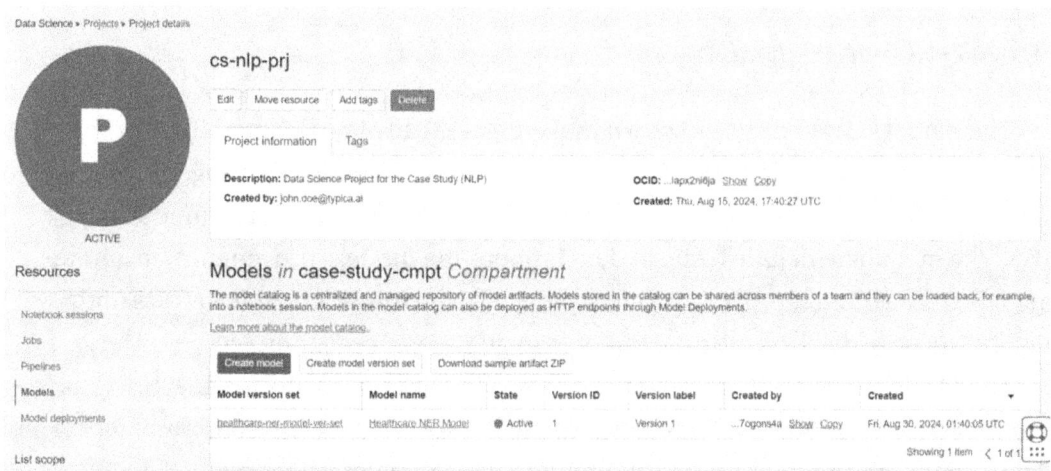

Figure 7-12. *Model created*

Click the model name to open the model's detail page.

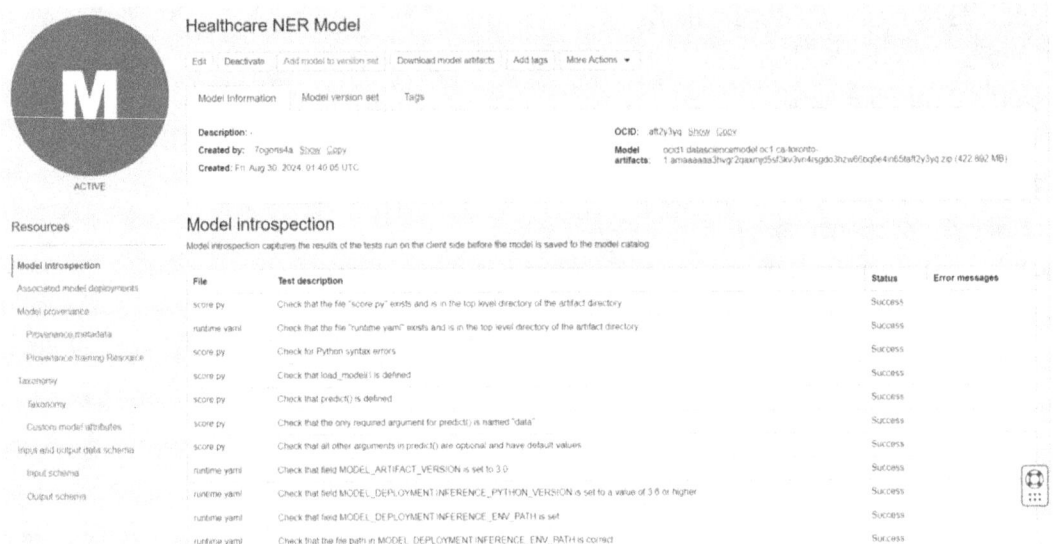

Figure 7-13. Model detail page

The model's detail page, as shown in Figure 7-13, displays the metadata for the Healthcare NER model in the OCI Data Science Model Catalog. This page showcases the comprehensive information automatically captured when saving a model using ADS. The metadata includes crucial details about the model's provenance, taxonomy, and custom attributes, providing a clear picture of the model's characteristics and history.The page displays the model metadata in different sections.

The **Model introspection** section, as shown in Figure 7-13, documents the tests performed on the model artifacts, particularly the score.py and runtime.yaml files. These tests help identify potential issues before the model deployment.

The **Provenance metadata** section, as shown in Figure 7-14, documents the Git information including the repository URL, branch, and commit hash. This information enhances the reproducibility and auditability of the model development process.

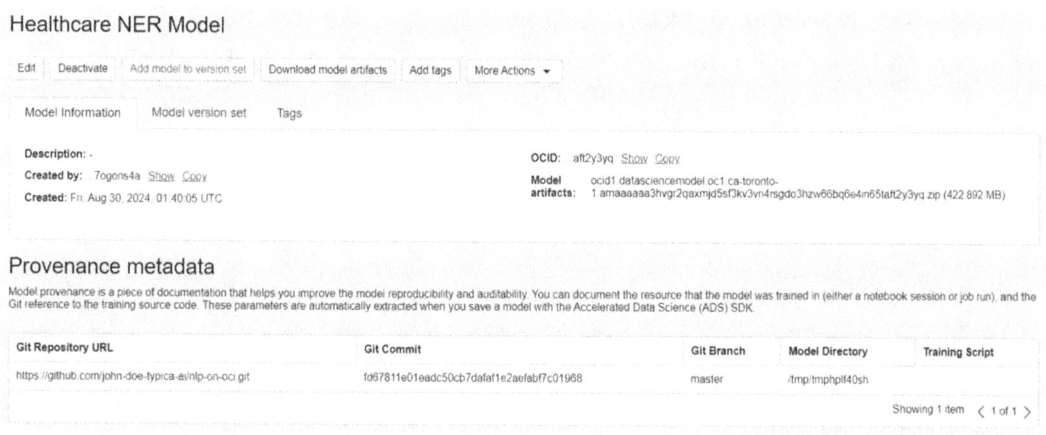

CHAPTER 7   MODEL DEPLOYMENT AND MONITORING

*Figure 7-14. Model **Provenance metadata***

The provenance resource section, as shown in Figure 7-15, documents the model's origin, including details about the training environment. Specifically, it indicates that our Healthcare NER model was trained in the notebook session named *cs-nlp-nbs-cpu*. This section also records who trained the model and when, providing a clear trail of the model's creation and updates. Together, these provenance details offer a comprehensive history of the model's development, crucial for version control and collaborative work in the MLOps pipeline.

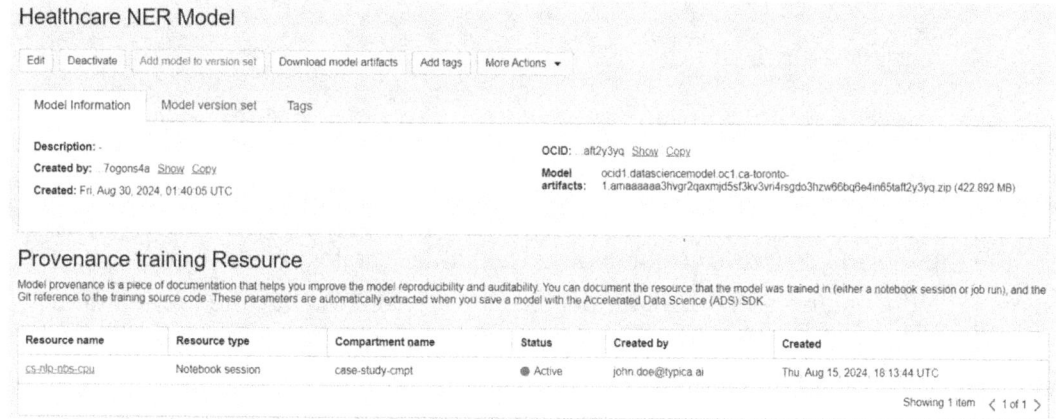

Figure 7-15. Model provenance resource

CHAPTER 7   MODEL DEPLOYMENT AND MONITORING

The taxonomy section, as shown in Figure 7-16, shows key attributes for our Healthcare NER model, this includes the framework used, which is Hugging Face Transformers, its version (4.44.2), and the model task (TokenClassificationPipeline). Additionally, it displays the model's prediction labels along with some of the model internal parameters (e.g., few hyperparameters).These model specifications are particularly valuable when troubleshooting the model's predictions or optimizing its training hyperparameters.

Figure 7-16. Model taxonomy

The **Custom model attributes** section of the Model Catalog, as illustrated in Figure 7-17, documents additional, model-specific information beyond the standard taxonomy fields, such as the conda environment used for training our Healthcare NER model.

Having this detailed information readily available in the taxonomy section allows for quick reference and facilitates efficient model management and fine-tuning processes. This level of detail is crucial for maintaining and improving the performance of the Healthcare NER model throughout its life cycle.

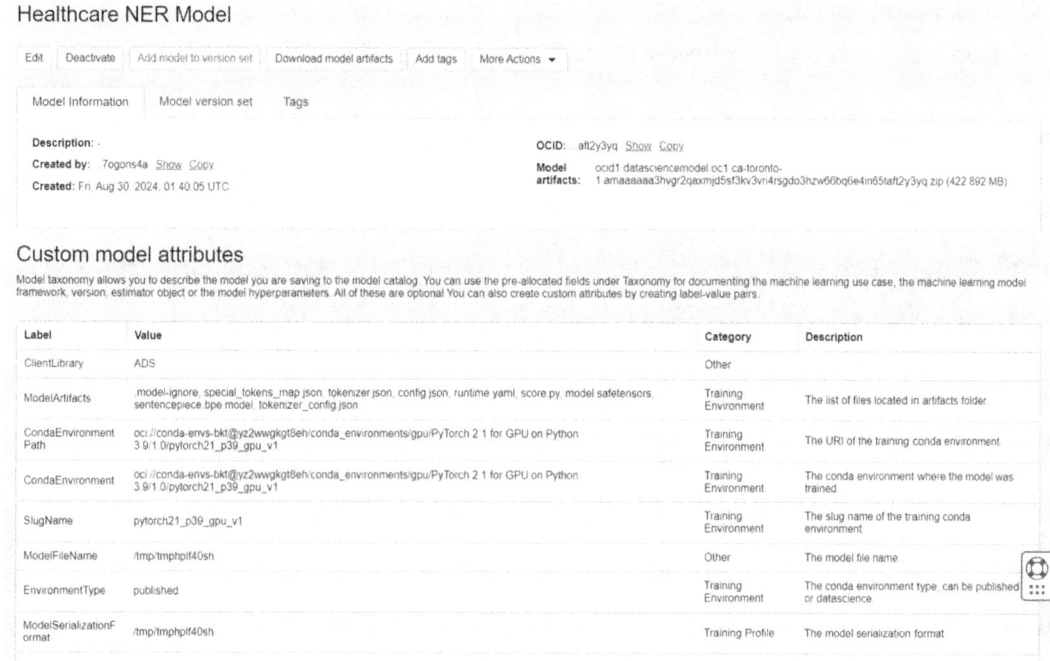

*Figure 7-17. Model taxonomy: **Custom model attributes***

Deploy and Invoke

Now that our model is saved to the Model Catalog, we can start the creation of the model deployment which will provision the necessary OCI resources to deploy our Healthcare NER models as an HTTP endpoint (REST-based API).

Deploy and Generate Endpoint

Listing 7-12. Deploying the Model

```
huggingface_pipeline_model.deploy(
    display_name="Healthcare NER Mode Deployment",
    description="Healthcare NER Mode Deployment",
    deployment_log_group_id="ocid1.loggroup.oc1.ca-toronto-1.
    amaaaaaa3hvgr2qafnnj4nyxd3bxxxmfx35bbmsmftedosa3fxin7ag46jia",
    deployment_access_log_id="ocid1.log.oc1.ca-toronto-1.
    amaaaaaa3hvgr2qah2mx7uxr5nl7rtqtaygsui3liy3uqyor5wp6zs7xsdua",
```

CHAPTER 7   MODEL DEPLOYMENT AND MONITORING

```
    deployment_predict_log_id="ocid1.log.oc1.ca-toronto-1.
    amaaaaaa3hvgr2qah2mx7uxr5nl7rtqtaygsui3liy3uqyor5wp6zs7xsdua"
)
print(f"Endpoint: {huggingface_pipeline_model.model_deployment.url}")
```

Output

```
Model Deployment OCID: ocid1.datasciencemodeldeployment.oc1.ca-toronto-1.
amaaaaaa3hvgr2qauukxaeb3pel6d575orfdxngmkhy4fsmbz3bezlvqvalq
Endpoint: https://modeldeployment.ca-toronto-1.oci.customer-
oci.com/ocid1.datasciencemodeldeployment.oc1.ca-toronto-1.
amaaaaaa3hvgr2qauukxaeb3pel6d575orfdxngmkhy4fsmbz3bezlvqvalq
```

Listing 7-13. Retrieving Model Deployment Details

```
print(huggingface_pipeline_model.model_deployment.display_name)
print(huggingface_pipeline_model.model_deployment.description)
print(huggingface_pipeline_model.model_deployment.time_created)
```

Output

```
Healthcare NER Mode Deployment
Healthcare NER Mode Deployment
2024-08-30 01:49:54.888000+00:00
```

At this step, go to our Data Science Project, and click **Model deployments** to verify if our model deployment was properly created, as shown in Figure 7-18. You will notice that the default compute shape is VM.Standard.E4.Flex and the initial instance count is 1.

CHAPTER 7 MODEL DEPLOYMENT AND MONITORING

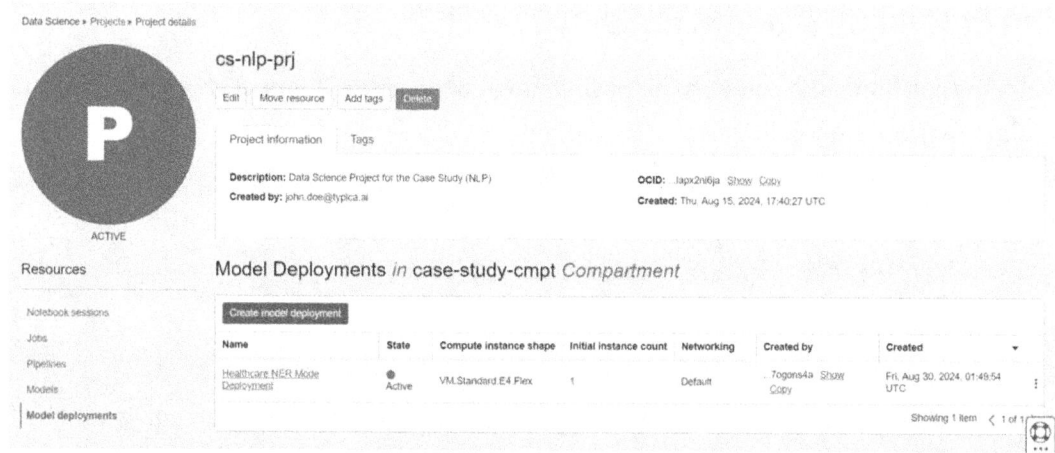

Figure 7-18. *Model deployment created*

As shown in Figure 7-19, a work request is generated to track the progress of deploying our Healthcare NER model. The work request, accessible from the ***Model deployments*** detail page, details the deployment process steps, such as provisioning compute instances, configuring the load balancer, and deploying the model artifacts. The work request's log and error messages (if any) allow us to identify and troubleshoot any issues arising during the process quickly.

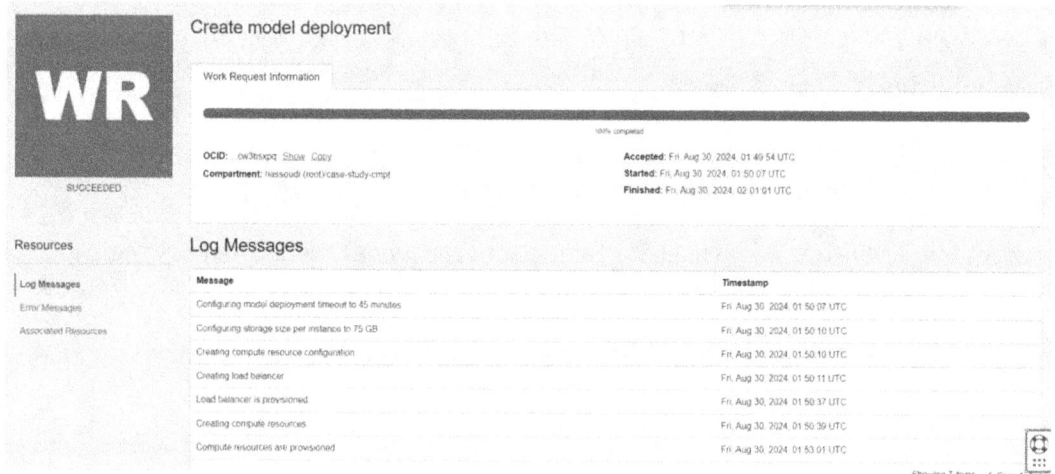

Figure 7-19. *Model deployment work request*

Run Prediction Against Endpoint

Let's now continue in our notebook by invoking our model ADS pipeline class's predict method to generate predictions.

Listing 7-14. Generating predictions

```
preds = huggingface_pipeline_model.predict(data)

#print predictions
for pred in preds['prediction']:
    print(pred['word'],pred['entity_group'], pred['score'], pred['start'],
    pred['end'])
```

Output

```
antibiotiques MedicationVaccine 0.742815375328064 21 34
```

Now, we invoke our model and generate predictions by calling the model's HTTP endpoint (REST-based API).

Listing 7-15. Invoking the Model Endpoint for Predictions

```
import requests
import oci
from oci.signer import Signer

# Get notebook session's resource principal
signer = oci.auth.signers.get_resource_principals_signer()

prediction_endpoint = f'{huggingface_pipeline_model.model_deployment.url}/
                      predict'

print(f"Invoking Model Endpoint: {prediction_endpoint}")
print(f"Medical sample text : {data}")

body = {"inputs":data} # payload
headers = {} # headers

preds = requests.post(prediction_endpoint, json=body, auth=signer,
headers=headers).json()

print(f"Extracted medical entities : {preds}")
```

CHAPTER 7 MODEL DEPLOYMENT AND MONITORING

Output

```
Invoking Model Endpoint: https://modeldeployment.ca-toronto-1.oci.
customer-oci.com/ocid1.datasciencemodeldeployment.oc1.ca-toronto-1.
amaaaaaa3hvgr2qauukxaeb3pel6d575orfdxngmkhy4fsmbz3bezlvqvalq/predict
Medical sample text : Le medecin donne des antibiotiques en cas
d'infections des voies respiratoires.
Extracted medical entities : {'prediction': [{'entity_group':
'MedicationVaccine', 'score': 0.742815375328064, 'word': 'antibiotiques',
'start': 21, 'end': 34}]}
```

Tip To avoid being billed for model deployment resources, such as instances and the load balancer, you can deactivate your model deployment. Deactivating a model deployment shuts down the associated instances, stops metering and billing for those instances and the load balancer, and makes the deployment's HTTP endpoint unavailable. However, the model deployment's metadata is preserved. If needed, you can reactivate the deactivated model deployment, restoring the same HTTP endpoint and enabling requests to be made to it again.

The model's prediction endpoint can be found in the console. Navigate to the **Model deployments** detail page, as shown in Figure 7-20, and click the model deployment endpoint link (Invoking your model).

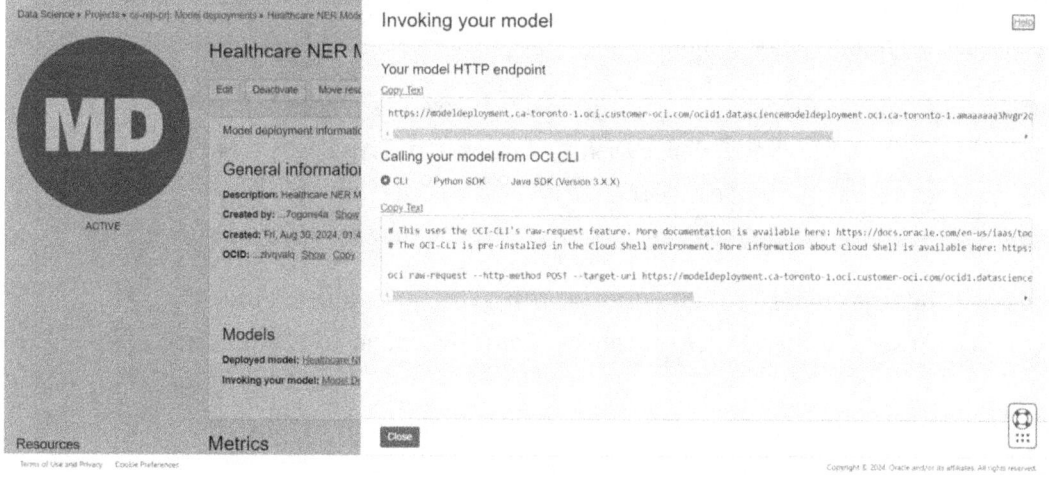

Figure 7-20. Model deployments detail page

CHAPTER 7   MODEL DEPLOYMENT AND MONITORING

Monitoring and Maintenance

After deploying our Healthcare NER model, we can start monitoring it through logs and metrics. To showcase briefly those monitoring capabilities, let's say we used our testing notebook, as show in Figure 7-21, to send inference calls to our NLP model.

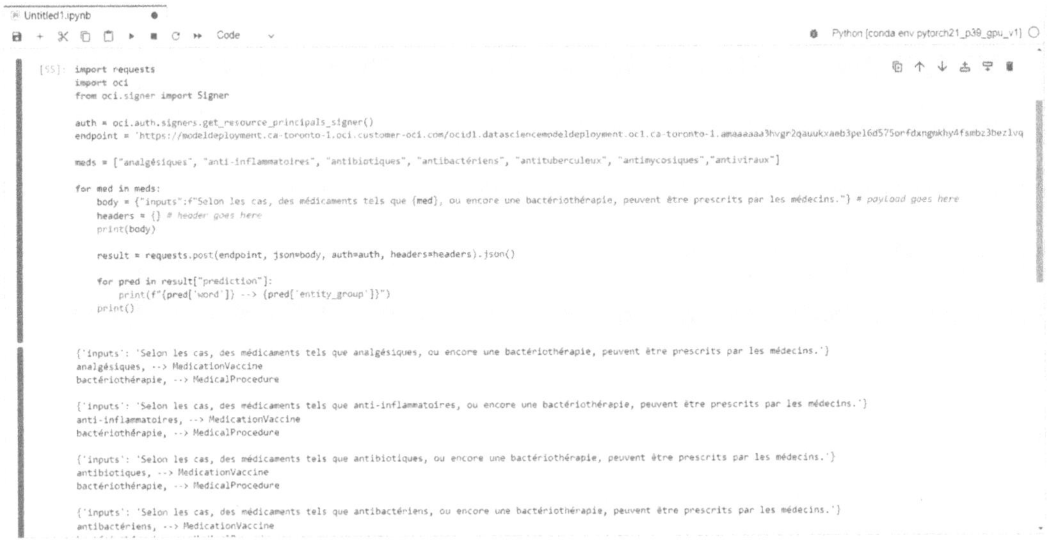

Figure 7-21. Testing notebook

Listing 7-16. Testing the Model Endpoint

```
import requests
import oci
from oci.signer import Signer

auth = oci.auth.signers.get_resource_principals_signer()
endpoint = 'https://modeldeployment.ca-toronto-1.oci.customer-oci.com/ocid1.datasciencemodeldeployment.oc1.ca-toronto-1.amaaaaaa3hvgr2qauukxaeb3pel6d575orfdxngmkhy4fsmbz3bezlvqvalq/predict'

meds = ["analgésiques", "anti-inflammatoires", "antibiotiques", "antibactériens", "antituberculeux", "antimycosiques","antiviraux"]
```

```
for med in meds:
    body = {"inputs":f"Selon les cas, des médicaments tels que {med}, ou
            encore une bactériothérapie, peuvent être prescrits par les
            médecins."} # payload goes here
    headers = {} # header goes here
    print(body)

    result = requests.post(endpoint, json=body, auth=auth,
            headers=headers).json()

    for pred in result["prediction"]:
        print(f"{pred['word']} --> {pred['entity_group']}")
    print()
```

Output

{'inputs': 'Selon les cas, des médicaments tels que analgésiques, ou encore une bactériothérapie, peuvent être prescrits par les médecins.'}
analgésiques, --> MedicationVaccine
bactériothérapie, --> MedicalProcedure

{'inputs': 'Selon les cas, des médicaments tels que anti-inflammatoires, ou encore une bactériothérapie, peuvent être prescrits par les médecins.'}
anti-inflammatoires, --> MedicationVaccine
bactériothérapie, --> MedicalProcedure

...

Logs

As shown in Figure 7-22, OCI Logging service is used to capture important information.

CHAPTER 7   MODEL DEPLOYMENT AND MONITORING

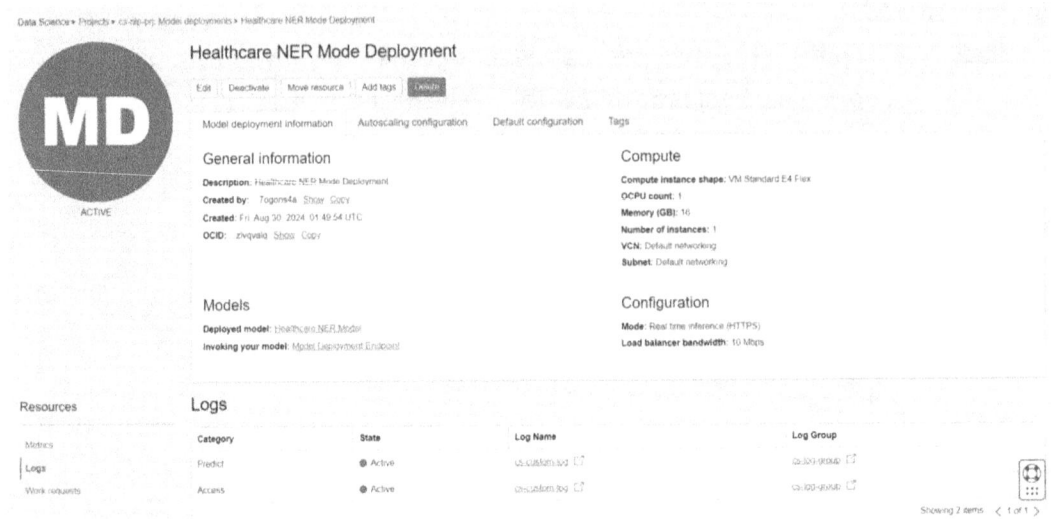

Figure 7-22. Model deployment logs

The access log details information about requests sent to the model endpoint, while the predict log captures the score.py logging calls (Figure 7-23).

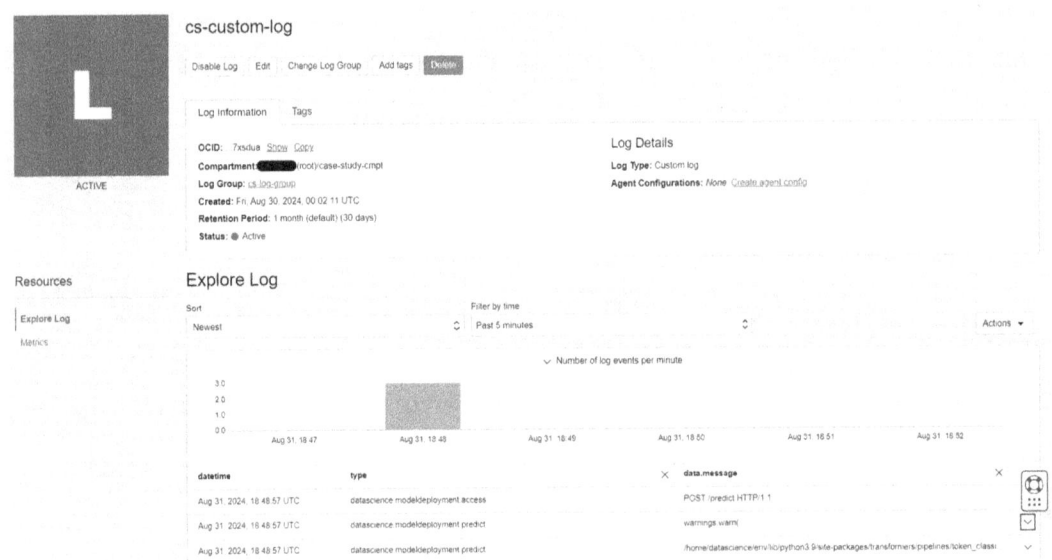

Figure 7-23. Predict log

CHAPTER 7   MODEL DEPLOYMENT AND MONITORING

Metrics

We can also monitor the health, capacity, and performance of model deployments with the built-in metrics using OCI Monitoring. Model deployments has metrics for CPU utilization, memory utilization, and network utilization, which includes parameters such as request count, latency, and bandwidth. As show in Figure 7-24, from the console, we can open the metric space (under the model deployment resources) to view all the built-in metrics.

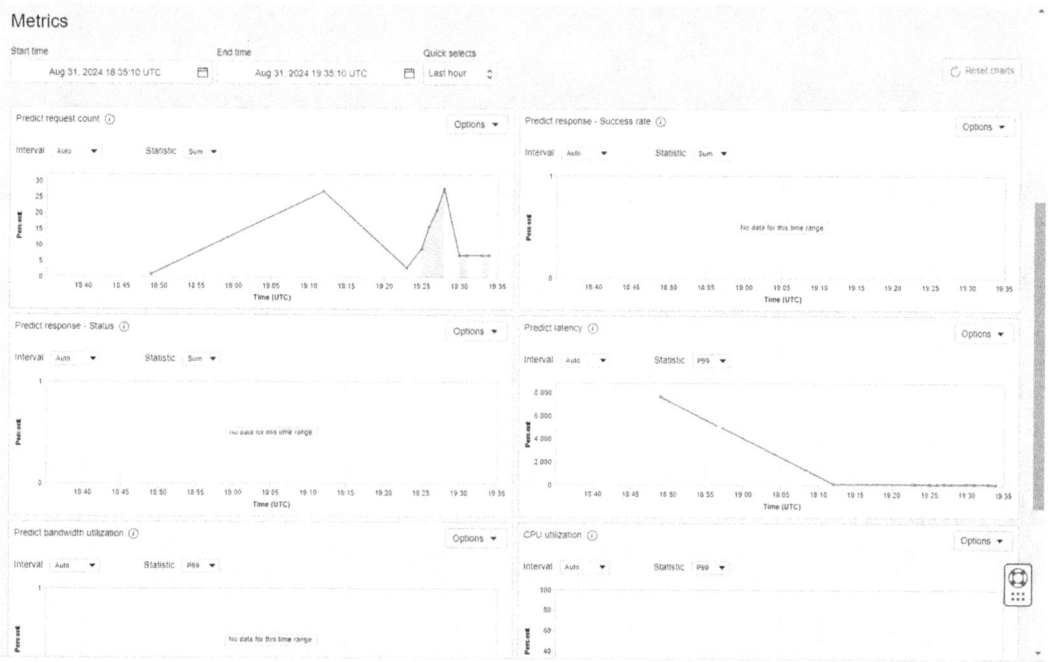

Figure 7-24. Model deployment metrics

We can also monitor our model from OCI Monitoring feature.

To do that, navigate to the menu option **Observability & Management ➤ Monitoring**.

You can then use **Service Metrics**, as shown in Figure 7-25, to monitor metrics for the service **oci_datascience_modeldeploy.**

Additionally, you can explore each metric in more detail by opening the metrics query in the **Metrics Explorer** or create an alarm based on the metric when it crosses a specified threshold.

CHAPTER 7   MODEL DEPLOYMENT AND MONITORING

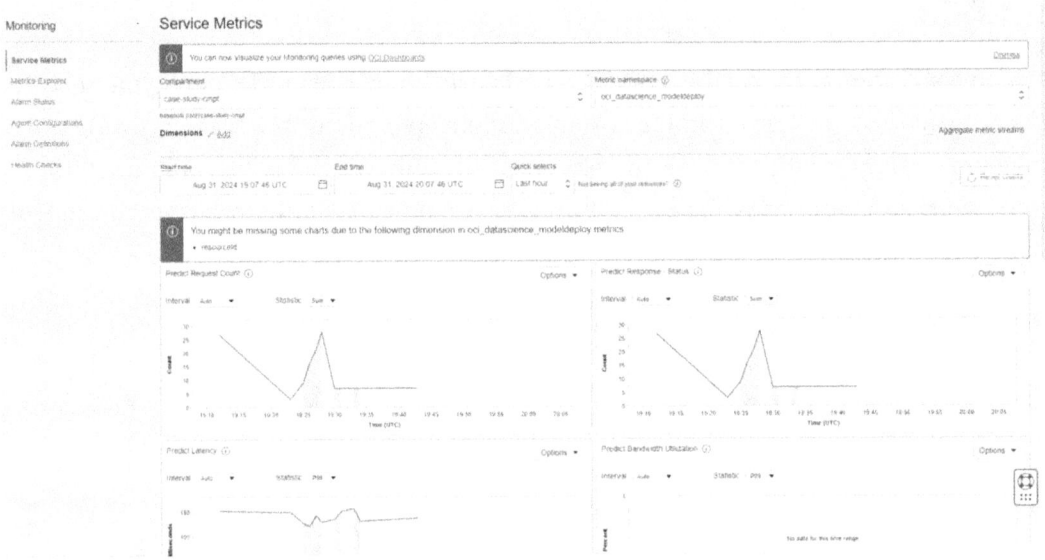

Figure 7-25. OCI Monitoring

By effectively monitoring our Healthcare NER model on OCI, data scientists or OCI administrators can ensure optimal performance and compliance.

Summary

This chapter marks the end of our journey in building our Healthcare NER model on Oracle Cloud Infrastructure (OCI). This chapter focused on the practical aspects of deploying and operationalizing our model using OCI Data Science and its Oracle Accelerated Data Science (ADS) library.

Key points covered in this chapter include

- Detailed steps for deploying the Healthcare NER model on OCI Data Science

- Utilization of the ADS library to streamline the deployment process

- Instructions on how to invoke the deployed model using OCI's model deployment inference endpoint

While there are advanced postdeployment topics relevant to our case study, we've kept the focus on the fundamentals. For those interested in exploring further, strategies such as rolling updates, blue-green deployments, and A/B testing in OCI can help minimize downtime and enhance reliability. These techniques, although not covered in depth in this book, are valuable for ensuring smooth and reliable model updates in production environments.

Batch inference is another powerful technique, especially for processing large datasets. While this book emphasizes real-time inference, the OCI Data Science Jobs service offers robust tools for batch processing. This can be particularly beneficial for tasks like analyzing large sets of medical records in batch mode, which is highly relevant to our healthcare-focused case study. These advanced areas provide opportunities for future exploration.

References

Hugging Face. (n.d.). *Pipelines*. Retrieved from Transformers: https://huggingface.co/docs/transformers/en/main_classes/pipelines

Oracle. (2022). *HuggingFacePipelineModel*. Retrieved from Oracle Accelerated Data Science (ADS): https://accelerated-data-science.readthedocs.io/en/v2.10.0/user_guide/model_registration/frameworks/huggingfacemodel.html

Oracle. (2024, 04 17). *Model Deployments*. Retrieved from Oracle Cloud Infrastructure Documentation: https://docs.oracle.com/en-us/iaas/data-science/using/model-dep-about.htm

CHAPTER 8

MLOps and Conclusion

In this chapter, we conclude the case study by first exploring how MLOps can be implemented using Data Science Pipelines, focusing on its significance in enhancing the efficiency and effectiveness of NLP workflows.

We will then revisit the development of our Healthcare Named Entity Recognition (NER) model, reflecting on each key step—from dataset preparation to fine-tuning pretrained models and, finally, the deployment and monitoring of the fine-tuned model.

By synthesizing these experiences, we will highlight the overarching themes and lessons learned throughout the journey. Lastly, we will provide key takeaways and discuss the broader implications of building NLP solutions, including important considerations like carbon emissions in Transformer-based models.

MLOps with OCI Data Science

Before concluding our case study, it's essential to explore MLOps, as it plays a crucial role in streamlining and automating the life cycle of NLP models. By integrating MLOps using Data Science Pipelines, we can enhance the efficiency and reliability of deploying, monitoring, and maintaining NLP models, ensuring smoother workflows and better scalability. This makes it a vital topic to address before wrapping up our case study.

OCI Data Science Pipelines

Our NLP consultant, John Doe, will focus on implementing MLOps as the final step in MedTALN Inc.'s NLP initiative. MLOps integrates DevOps principles into the machine learning life cycle to automate workflows, improve team collaboration, and ensure reproducibility. This streamlines the entire process from development to production and ensures that models perform well over time, not just during deployment.

The power of Data Science Pipelines lies in their ability to automate the main steps in the training and deployment processes. By leveraging the MLOps principles and the notebooks provided, you can streamline the process of taking NLP models from the experimentation phase to production, ensuring that they remain reliable and effective in the long run.

Note Automating the NLP workflow using MLOps principles allows for a repeatable process, facilitating rapid experimentation, model retraining, and efficient deployment.

This section will guide you through implementing MLOps using Data Science Pipelines. We will provide a straightforward yet practical example to equip you with the knowledge to build more advanced pipelines for real-world projects.

Pipeline Example

Throughout this book, we've provided Jupyter notebooks with Python code to automate key tasks in the NLP life cycle. These tasks range from building datasets using prelabeled Hugging Face datasets, to fine-tuning pretrained models, and deploying models on Oracle Cloud Infrastructure (OCI). By integrating these notebooks with OCI's Data Science Pipelines, data scientists can adopt a structured approach to training and deploying machine learning models, allowing for faster experimentation and iteration.

For the sake of simplicity, we will implement a straightforward yet practical pipeline to demonstrate techniques that can be used to build more sophisticated pipelines. We've chosen to create a simple, two-step pipeline to showcase how OCI Data Science Pipelines can effectively automate processes, ensuring the explanation remains clear and conducive to effective learning.

This pipeline builds on the pretrained model selection notebook from Chapter 6, where we focused on fine-tuning our Healthcare NER model. In that chapter, we manually explored various pretrained models to find the best fit for our task. Now, we take that process a step further by automating model selection through a two-step pipeline.

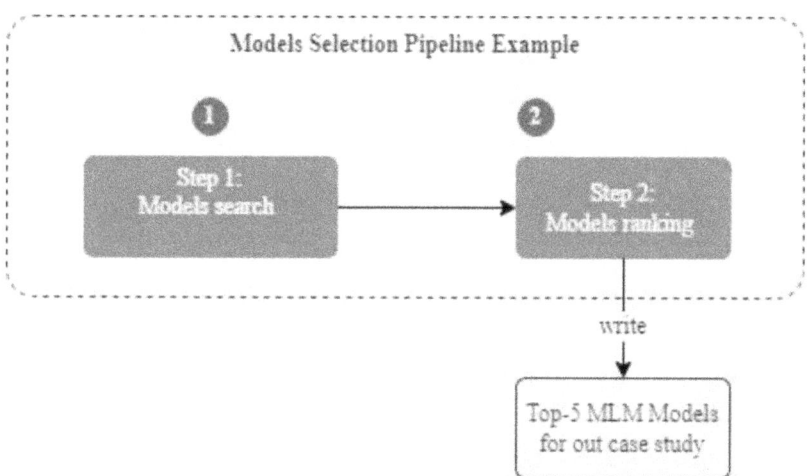

Figure 8-1. *Pipeline example for our case study*

As show in Figure 8-1, the first step of the pipeline will identify a list of candidate Masked Language Models (MLM) from the Hugging Face Hub. We'll programmatically search for models that support the French language and filter the results to include only monolingual models with at least one healthcare-related tag. In the second step, the pipeline will evaluate and rank these models based on their ability to predict medical entities using the "fill-mask" pipeline.

This simple yet powerful pipeline serves as a foundation to show how OCI Data Science Pipelines can automate these tasks, offering a clear and practical learning experience.

Pipeline Creation Step-by-Step

Figure 8-2 illustrates how MLOps and Data Science Pipelines can be used to automate and streamline the pretrained model selection task.

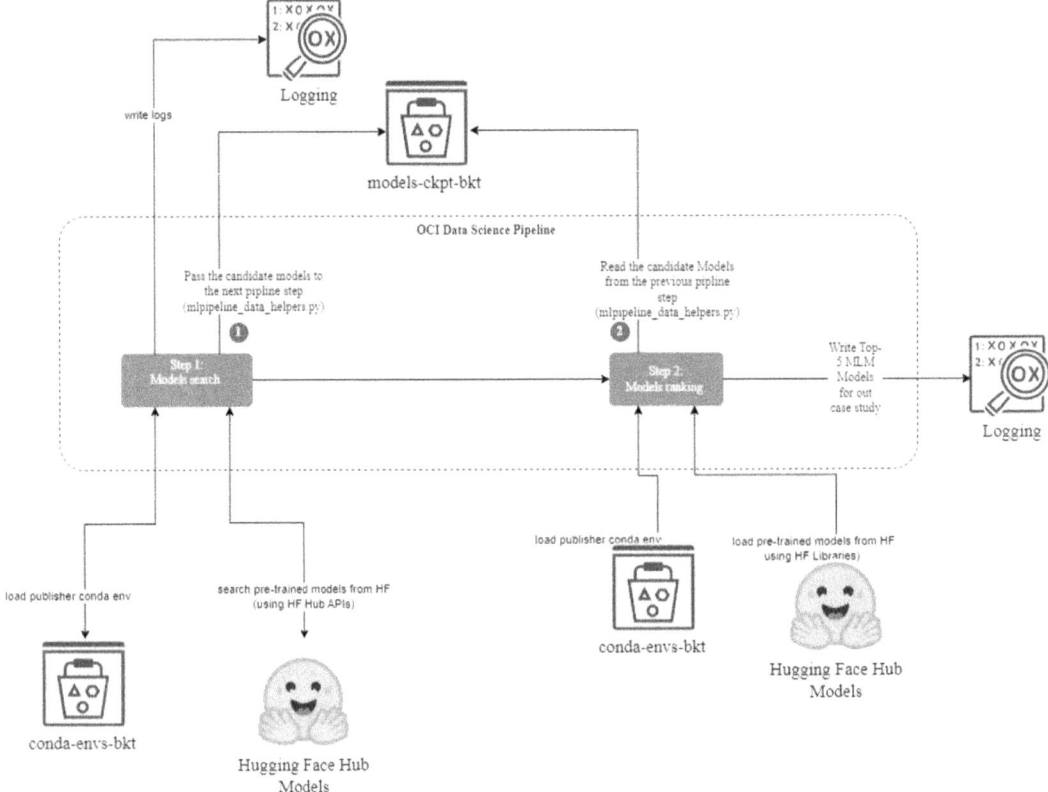

Figure 8-2. *Pipeline flow for the case study*

This two-step pipeline shows how data can be shared and processed sequentially in OCI Data Science Pipelines. It provides a simple template for creating more complex pipelines, serving as a foundation for building efficient and reproducible NLP workflows.

Pipeline Creation Prerequisites

Before we begin implementing our Model Selection Pipeline using a Data Science pipeline, we need to complete the following preparatory tasks:

- Set Up Tenancy: The OCI admin should add a new rule for OCI Data Science Pipelines to the Data Science Dynamic Group.

- Create Pipeline Step Artifacts: The NLP consultant will write and code the Python scripts and then package each step as a ZIP file.

Create Pipeline Step Artifacts

To implement our simple pipeline with data sharing between steps, we will repurpose and reorganize code from the JupyterLab notebook "select_mlm_models.ipynb" located in the "chapt-6" folder. This code will be divided into two separate Python files, each representing a distinct step in the pipeline:

1. Step 1: Model Search and Identification: In this step, we will extract and organize code from the Jupyter notebook into a Python script named *pipeline_step1_mlm_search.py*. This script will handle the task of searching for pretrained models that support the French language, using the Hugging Face Transformers library. The results will be filtered to retain only monolingual models with healthcare-related tags. The list of candidate models will then be passed to the next step for evaluation. We will use helper functions from the *mlpipeline_data_helpers.py*[1] module to facilitate data transfer between steps. The script, along with *mlpipeline_data_helpers.py*, will be packaged into a ZIP file for execution as part of the pipeline.

2. Step 2: Model Evaluation and Ranking: In the second step, the Python script *pipeline_step2_mlm_ranking.py* will be created to evaluate and rank the models identified in Step 1. This script will use the "fill-mask" pipeline from Hugging Face to assess each model's ability to predict masked medical terms. Each model's performance will be scored based on its accuracy, and the top five models will be logged in the pipeline for further analysis. To ensure smooth data handoff from Step 1, this script will also use helper functions from *mlpipeline_data_helpers.py*. Like the first step, this script and the helper file will be packaged into a ZIP file, making it ready for execution within the OCI Data Science Pipeline.

---

[1] The helper functions used to transfer data between steps, *mlpipeline_data_helpers.py*, is adapted from the code available at Oracle's OCI Data Science AI samples repository (https://github.com/oracle-samples/oci-data-science-ai-samples/blob/main/pipelines/samples/simple/mlpipeline_data_helpers.py).

CHAPTER 8 MLOPS AND CONCLUSION

Below the ZIP file creation for each step (Listing 8-1 and Listing 8-2):

For the pipeline Step 1, first, create a folder named pipeline_step1_mlm_search. Then, copy the following Python files into this folder:

- mlpipeline_data_helpers.py
- pipline_step1_mlm_search.py

Listing 8-1. Creating Pipeline Artifacts ZIP

```
(/home/datascience/conda/pytorch21_p39_gpu_v1) bash-4.2$ cd
./pipline_step1_mlm_search
(/home/datascience/conda/pytorch21_p39_gpu_v1) bash-4.2$ zip -r
../pipline_step1_mlm_search.zip *.py
```

Then, create the artifacts for the pipeline Step 2. To do that, create another folder named pipline_step2_mlm_ranking. Then, copy the following python files into this folder:

- mlpipeline_data_helpers.py
- adding: pipline_step2_mlm_ranking.py

Listing 8-2. Creating Pipeline Artifacts ZIP (for step 2)

```
(/home/datascience/conda/pytorch21_p39_gpu_v1) bash-4.2$ cd
../pipline_step2_mlm_ranking
(/home/datascience/conda/pytorch21_p39_gpu_v1) bash-4.2$ zip -r
../pipline_step2_mlm_ranking.zip *.py
```

Note Ready-to-use ZIP files are provided in the GitHub repository under /nlp-on-oci/chapt-8.

Data Science Dynamic Group Rule

Make sure you have added the following rule to the data science dynamic group rules (Figure 8-3; Listing 8-3).

CHAPTER 8 MLOPS AND CONCLUSION

Listing 8-3. Adding a Dynamic Group Rule

```
ALL {resource.type='datasciencepipelinerun'}
```

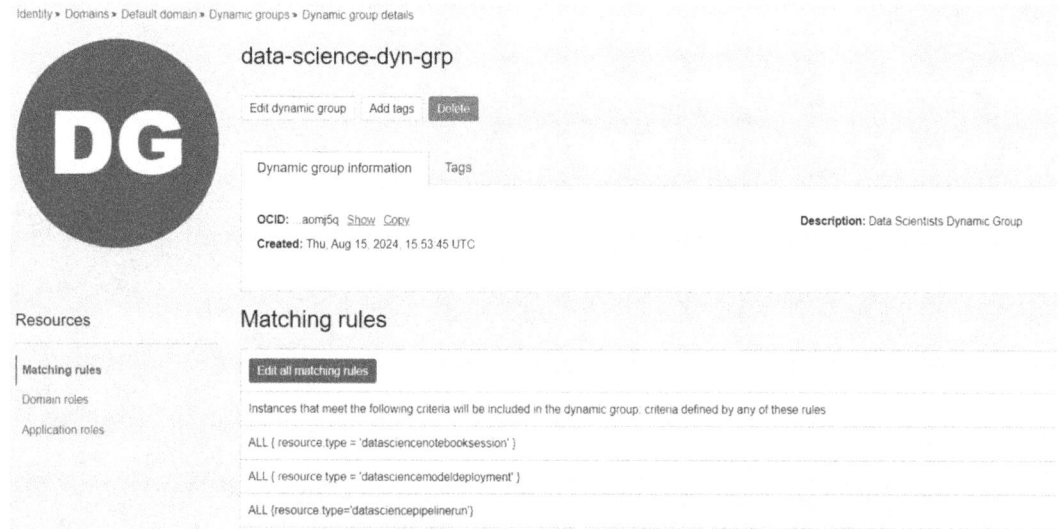

Figure 8-3. *Data scientist dynamic group rule for pipeline*

Create Pipeline

Below are the steps for creating our pipeline:

1. Open the navigation menu, and click ***Analytics & AI***. Under ***Machine Learning***, click ***Data Science***.

2. Select our case study compartment, i.e., case-study-cmpt.

3. From the ***Projects*** list page, click the name of our case study project, i.e., cs-nlp-prj.

4. From the ***Project details*** page, under ***Resources***, click ***Pipelines***.

5. Click ***Create pipeline*** (Figure 8-4).

CHAPTER 8 MLOPS AND CONCLUSION

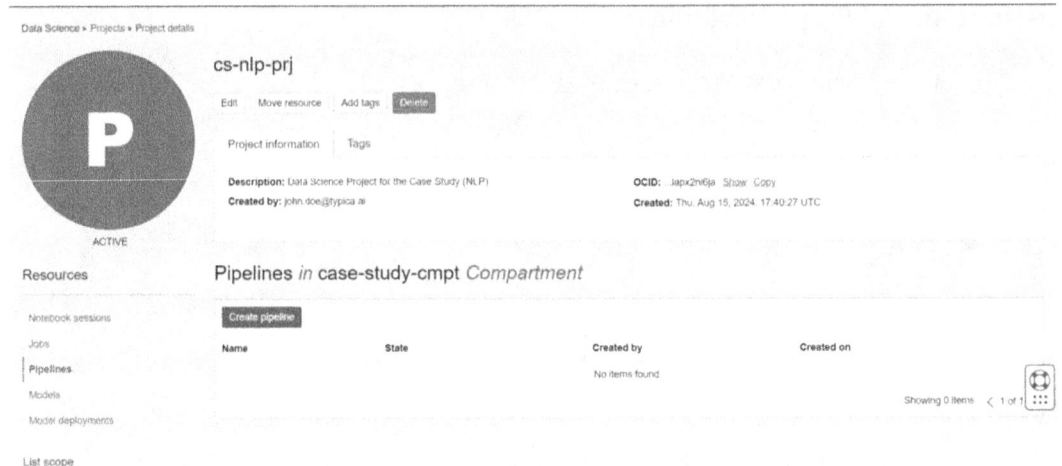

Figure 8-4. *DS project page with pipelines*

6. In the pipeline creation dialog, enter

    a. The compartment for the pipeline is already prefilled.

    b. Enter a name and description for the pipeline (limit of 255 characters): cs-nlp-pipeline.

    c. Description: Example of a pipeline for the case study.

Note The name and description are optional. If you don't provide a name, a name is automatically generated.

---

7. Click **Add pipeline step** to start defining the workflow for the pipeline (see Figure 8-5). In the **Add pipeline step** panel, do the following:

    a. Select the option **From script**.

    b. Enter a unique name for the step: pipline_step1_mlm_search.

    You can't repeat a step name in a pipeline.

    c. Enter a step description: Healthcare NER Model pipeline - pretrained MLM models search step.

    d. This is the first step; there is no other step that should run before this step.

CHAPTER 8 MLOPS AND CONCLUSION

  e. Drag a job step file into the box, or click *Select a file* to navigate to it for selection, i.e., pipline_step1_mlm_search.zip.

  f. In *Entry point*, select one file to be the entry run point of the step, i.e., pipline_step1_mlm_search.py.

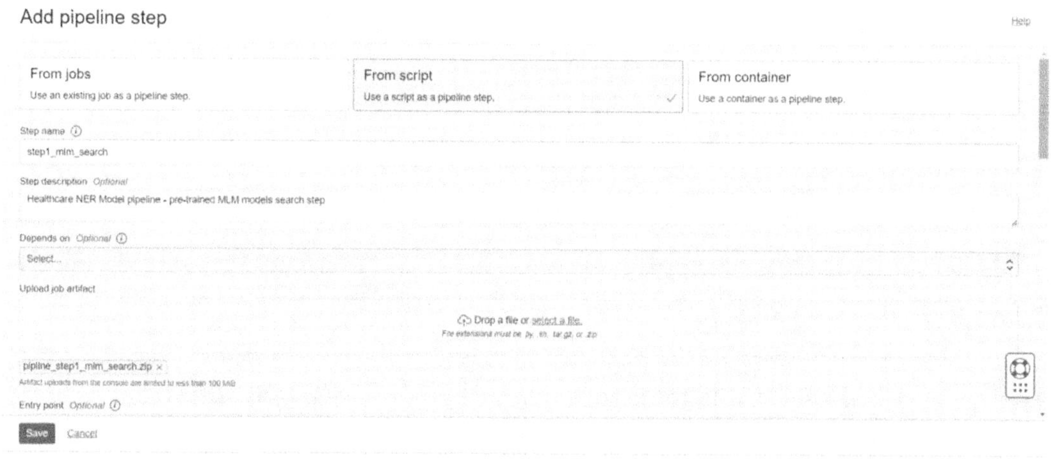

Figure 8-5. Add pipeline step

  8. Click *Save* to add the step, and return to the *Create pipeline* page.

  9. Use *+Add pipeline step* to add more steps to complete your workflow by repeating the preceding steps (refer to Figure 8-6):

  a. Select the option *From script*.

  b. Enter a unique name for the step: pipline_step2_mlm_ranking.

  You can't repeat a step name in a pipeline.

  c. Enter a step description: Healthcare NER Model pipeline - pretrained MLM models ranking step (Step 2).

  d. Depends on: pipline_step1_mlm_search.

  Select the step that should run before this step.

  e. Drag a job step file into the box, or click *Select a file* to navigate to it for selection, i.e., pipline_step2_mlm_ranking.zip.

  f. In *Entry point*, select one file to be the entry run point of the step, i.e., pipline_step2_mlm_ranking.py.

373

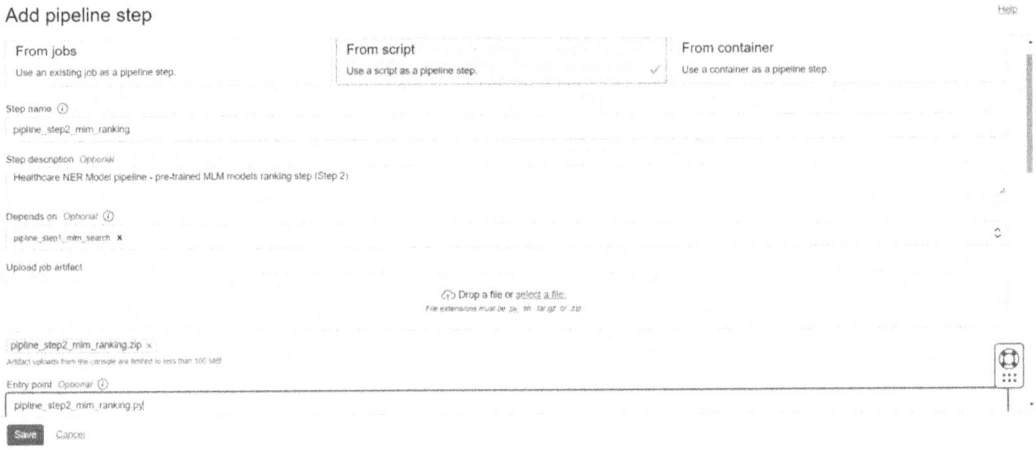

CHAPTER 8　MLOPS AND CONCLUSION

Figure 8-6. Pipeline Step 2

g. Click **Save** to add the step, and return to the **Create pipeline** page (Figure 8-7).

Figure 8-7. Create pipeline

10. Create a default pipeline configuration that's used when the pipeline is run:

a. Under the configuration section (refer to Figure 8-8), add the following custom environment variable keys and values (replace yz2wwgkgt8eh with your Object storage namespace):

CHAPTER 8   MLOPS AND CONCLUSION

DATA_LOCATION = oci:// models-ckpt-bkt@yz2wwgkgt8eh/
CONDA_ENV_TYPE = published
CONDA_ENV_OBJECT_NAME = conda_environments/gpu/
PyTorch 2.1 for GPU on Python 3.9/1.0/pytorch21_p39_gpu_v1
CONDA_ENV_NAMESPACE = yz2wwgkgt8eh
CONDA_ENV_BUCKET = conda-envs-bkt

b. Leave **Command line arguments** empty.

c. Leave the default **Maximum runtime** (in minutes).

The maximum number of minutes that the pipeline step is allowed to run. The service cancels the pipeline run if its runtime exceeds the specified value. The maximum runtime is 30 days (43,200 minutes). We recommend that you configure a maximum runtime on all pipeline runs to prevent runaway pipeline runs.

d. Leave default value for block storage

The amount of storage can be between 50 GB and 10, 240 GB (10 TB). You can change the value by 1 GB increments. The default value is 100 GB.

Figure 8-8. Pipeline configuration section

CHAPTER 8 MLOPS AND CONCLUSION

11. Click **Save** to add the step, and return to the **Create pipeline** page.

12. To configure the network type (Figure 8-9), select **Custom networking**—Select the **VCN** and **Subnet**:

    VCN: cs-vcn

    Subnet: private subnet-cs-vcn

13. For logging, click **Select**, and then ensure that **Enable logging** is selected.

    a. Select a log group from the list: cs-log-group.

    b. Log name: cs-custom-log.

    Select one of the following to store all stdout and stderr messages:

    c. Click **Select** to return to the job run creation page.

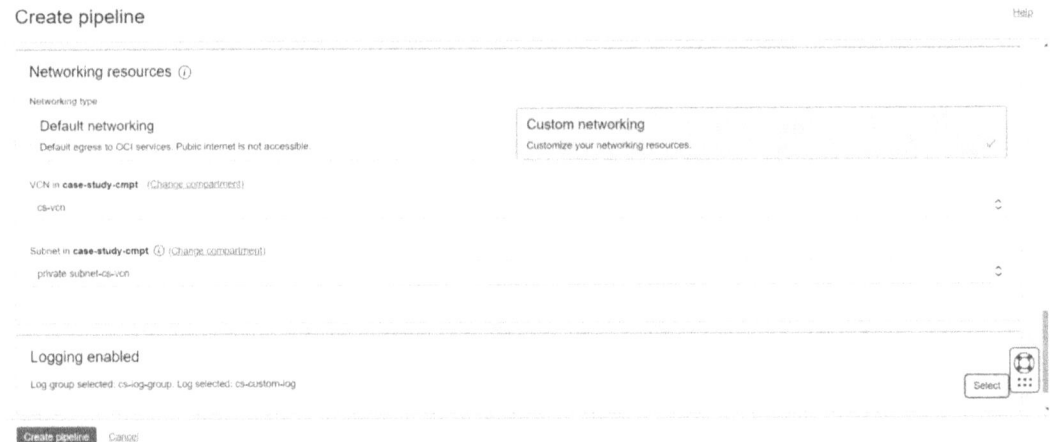

Figure 8-9. Pipeline networking

14. Click **Create pipeline**.

15. After the pipeline is in an active state (as shown in Figure 8-10), you can use pipeline runs to repeatedly run the pipeline.

CHAPTER 8   MLOPS AND CONCLUSION

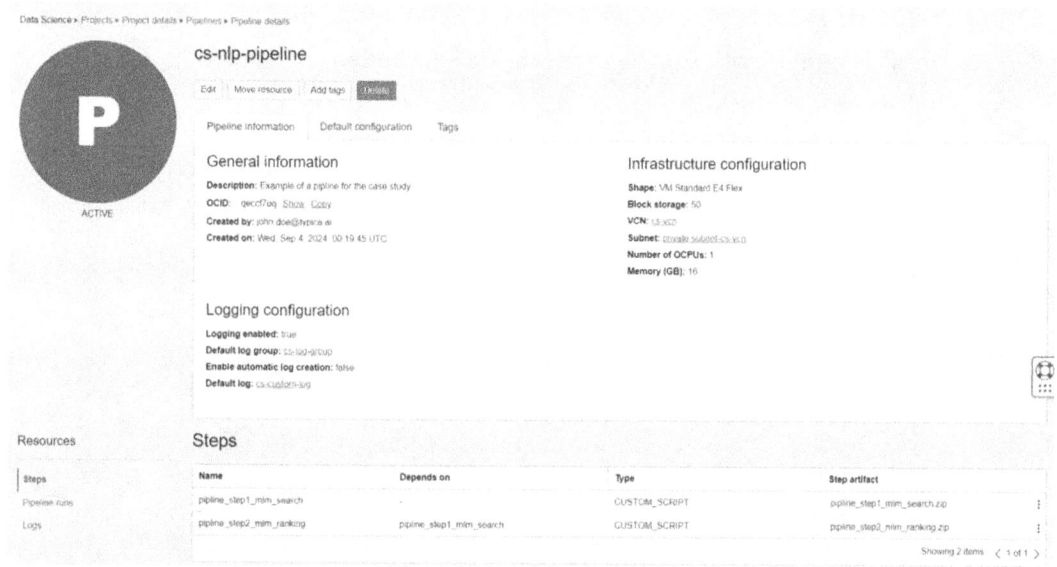

Figure 8-10. Pipeline details page

16. From the **Pipeline details** page, under **Resources**, click **Pipeline runs**.

    In the **Start a pipeline run** panel (Figure 8-11), enter the name of the run, leave everything else, and click the **Start** button.

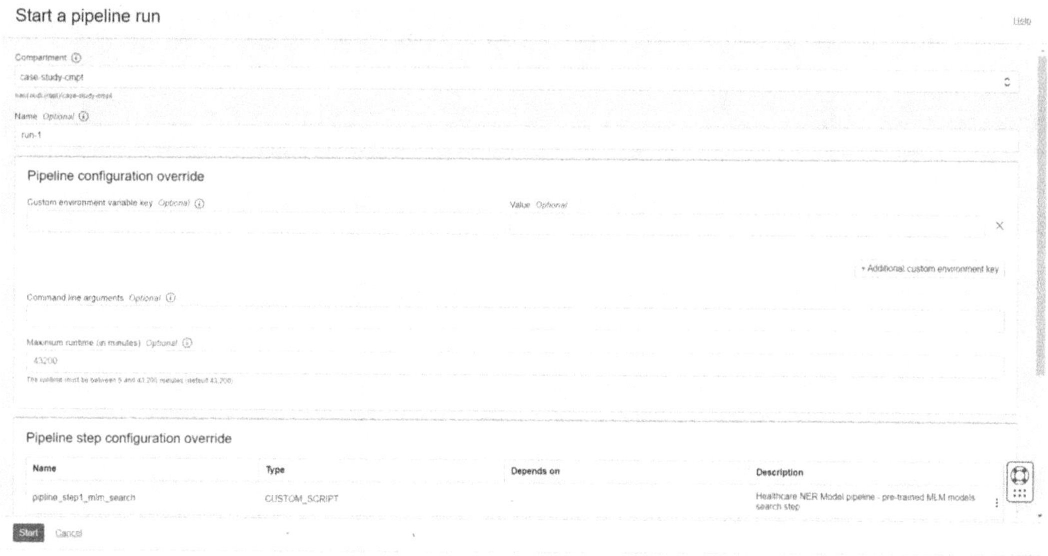

Figure 8-11. Start a pipeline run

The pipeline run is in the "Accepted" state until the run begins, and then, it changes to "In Progress." When the run finishes, it's either "Succeeded" or "Failed."

Our pipeline run, as show in Figure 8-12, has succeeded.

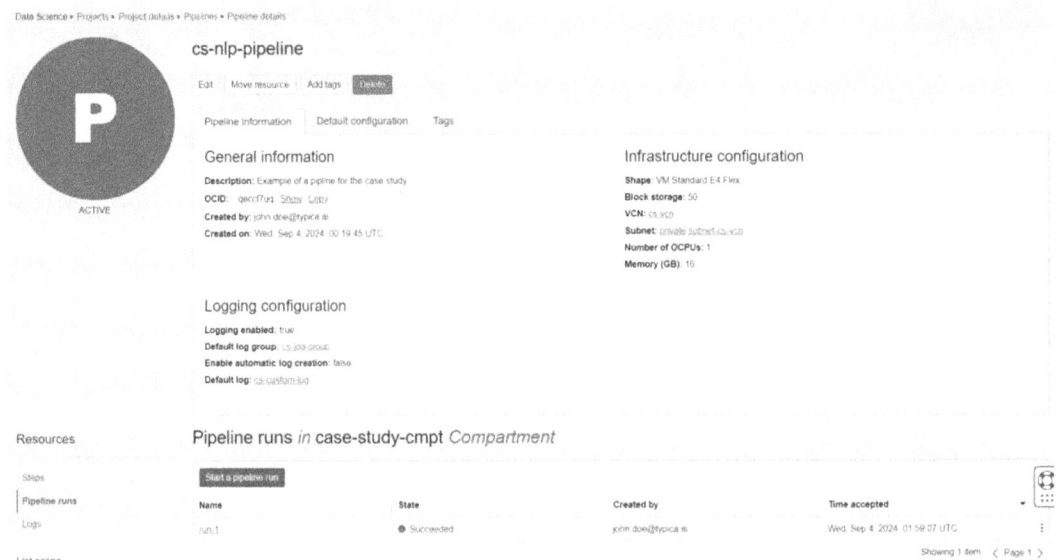

Figure 8-12. *Pipeline runs: first test (run 1)*

As illustrated in Figure 8-13, the **Pipeline runs** detail page offers detailed, fine-grained information about each step of the pipeline execution.

The status of each pipeline step is listed. Pipeline steps are in the "Waiting" state until they run, and then they change to "In Progress." When a step finishes, it's either "Succeeded" or "Failed."

CHAPTER 8 MLOPS AND CONCLUSION

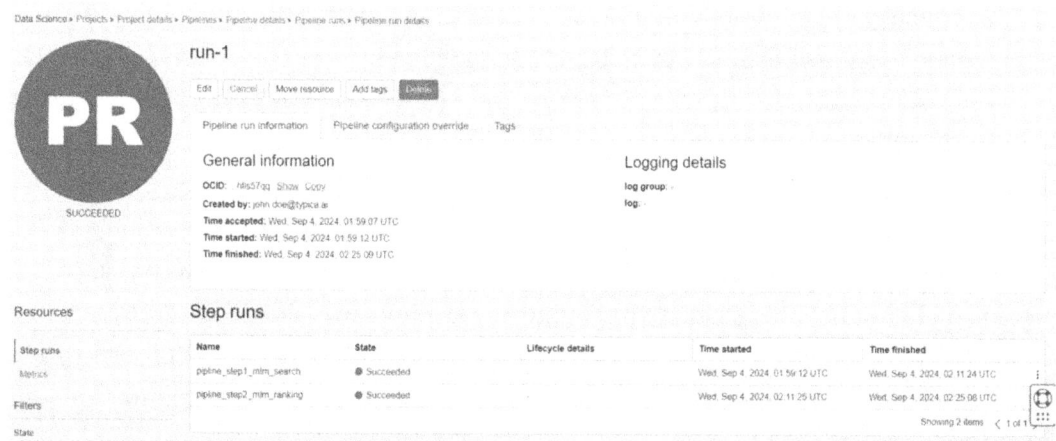

Figure 8-13. Pipeline runs detail page

As illustrated in Figure 8-14, by opening the cs-custom-log, we can see that the list of the top five MLM models is output to stdout using Python's print function. Additionally, the log shows that the data helper is used to pass these results to the next step, which would logically be the fine-tuning phase (though this step is not implemented in this particular pipeline example).

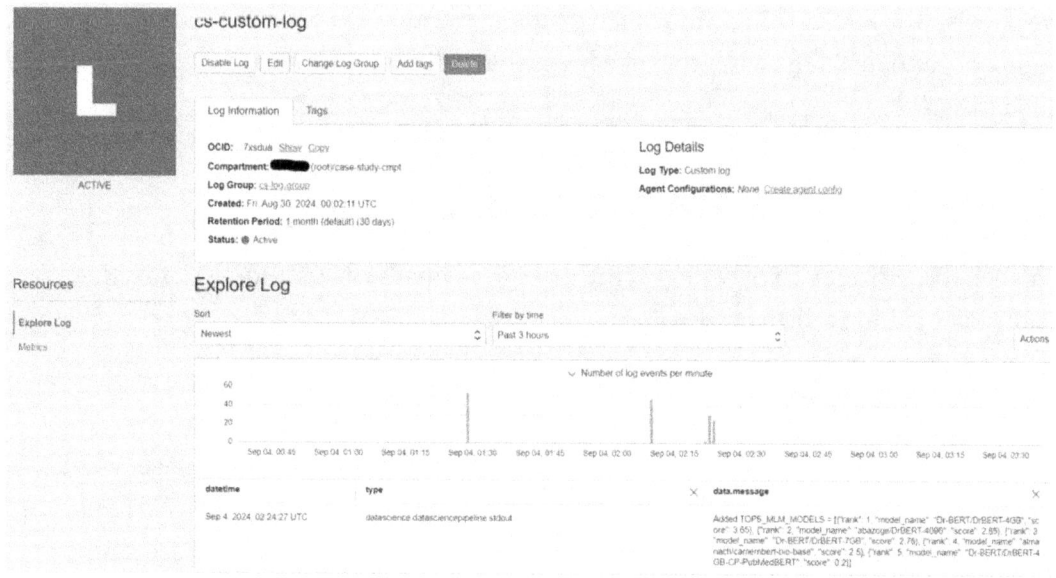

Figure 8-14. Pipeline run logs

This concludes the implementation of our example pipeline. We used a simple two-step pipeline to showcase how Oracle Data Science Pipelines can be utilized effectively. However, it's important to note that MLOps goes beyond this basic example. It's not just about getting models into production; it's about ensuring they remain performant and reliable over time.

By leveraging OCI Data Science Pipelines, you can automate and streamline much more complex workflows—from data preparation to model deployment, as well as ongoing maintenance and monitoring. While our example was simple, it serves as a foundation for building more sophisticated pipelines that can handle the full end-to-end machine learning life cycle, making it easier to experiment, deploy, and maintain models in a scalable and efficient manner.

Journey Through NLP: From Theory to Practice

As we conclude our exploration of NLP on Oracle Cloud Infrastructure, let's reflect on the path we've traveled. This journey, guided by the fictional yet expertly crafted narrative of John Doe and MedTALN Inc., has taken us from the foundational concepts of NLP to the intricacies of implementing advanced solutions on OCI. In this section, we'll revisit the key learnings from each chapter, examine the power of our case study approach, and distill the essential takeaways that will serve as pillars for your future NLP endeavors.

Healthcare NER Model Life Cycle Summary

This section provides an overview of the key stages involved in building our Named Entity Recognition (NER) model.

Data Preparation

In our journey to build the Healthcare NER model, preparing the right dataset was a critical step. This wasn't just about finding any dataset but about carefully curating, refining, and enhancing data that aligned perfectly with the goals of our case study. From defining the problem to finalizing the training data, each stage played a key role in ensuring the success of the model.

We began with a clear **problem definition**. The task was to train a Healthcare NER model that could identify medical entities like conditions, medications, and symptoms. To achieve this, we adopted a transfer learning approach, leveraging a pretrained model and adapting it to our specific task. Defining the problem in this way ensured that all subsequent decisions, particularly around dataset selection, aligned with the overall objective.

Next came the challenge of **dataset selection**. While there are many prelabeled datasets available, choosing the right one for our project required careful consideration. We needed a dataset that was not only labeled for Named Entity Recognition (NER) but also focused on the medical domain and contained a significant number of French examples. After evaluating several options, we selected the **TypicaAI/MedicalNER_Fr** dataset from Hugging Face, which met all our criteria. This choice gave us a solid foundation, allowing us to focus on adapting and refining the data rather than starting from scratch.

With the dataset selected, we moved on to **dataset collection**. We stored the chosen dataset in our Labeling Datasets Buckets, a step that ensured easy access for further processing. This organization made it simple for the next stages of preparation, particularly as we moved toward enriching and refining the data for our specific use case.

Dataset wrangling was where the real transformation began. This step involved cleaning and refining the dataset to make it more relevant to our case study. First, we removed irrelevant examples—entries that did not contain any medical entities of interest. It was important to ensure that the dataset was focused on the right kinds of data for training. We also took care to balance the labeled entities, ensuring that no particular type of entity was over- or underrepresented. Following the cleaning, we transformed the dataset into a format compatible with OCI Data Labeling. This step was crucial to ensure that the data was ready for the labeling phase.

The **dataset labeling** phase was a manual yet essential part of the process. Even though the dataset we selected was prelabeled, it needed further refinement. Our labeling team worked to enrich the dataset by manually identifying and labeling any missing medical entities. This step ensured that the dataset was not only complete but also tailored specifically to the needs of our Healthcare NER model. The labeling process was carefully reviewed to ensure consistency and accuracy, as these qualities directly impact the performance of the final model.

Finally, we arrived at **dataset creation**. Once labeling was complete, the final dataset was exported from OCI Data Labeling and stored in the Training Datasets Bucket.

Here's a more concise version of the takeaways, with related points grouped together:

Takeaways:

- Clear Problem Definition and Dataset Selection: A well-defined problem and clear selection criteria are essential to guide the dataset preparation process. Establishing task-, domain-, and language-specific requirements ensures that the dataset is aligned with the project's goals. In our case, selecting the TypicaAI/MedicalNER_Fr dataset was key to creating a relevant, cost-effective training dataset.

- Leveraging Community Resources and Specialized Data: Using community-curated datasets, such as those from Hugging Face, accelerates the process while minimizing costs. This is particularly valuable in niche domains or non-English languages where data is scarce, helping to overcome the challenges of finding specialized annotators.

- Data Enrichment and Labeling Quality: Even with prelabeled datasets, manual data enrichment is often necessary to meet project-specific needs. Ensuring consistency and quality in labeling is critical for building a robust training dataset that will drive high model performance.

Lessons Learned:

- Data Quality Directly Impacts Model Performance: Poor-quality or irrelevant data leads to suboptimal model performance. Cleaning the dataset to remove irrelevant examples and ensuring a balance of labeled entities is essential for successful training.

- The Need for Automation and Tools: Automating the dataset preparation steps—especially collection, cleaning, and transformation—using tools like OCI Data Labeling APIs and Python code reduces manual work and increases efficiency. Automation also allows for easier scaling in larger projects.

- Specialized Data for Domain-Specific Projects: In domain-specific projects like healthcare, finding the right data can be challenging, particularly for non-English languages. Specialized annotators may be necessary, and securing them can be time-consuming and costly. By using community-curated datasets, we can offset some of these challenges.

Model Training and Evaluation

The training and evaluation phases are the heart of our journey in developing a robust Healthcare Named Entity Recognition (NER) model tailored for the French language. These steps are crucial, as they ultimately determine how well our model performs in recognizing and understanding medical entities within French texts. Throughout this phase, we leveraged the powerful tools available within OCI Data Science, ensuring a methodical and efficient process from start to finish.

Our journey began with a clear understanding of the problem we were aiming to solve. In this case study, the challenge was to create an effective NER model that could handle the complexities of medical terminology in French. This clarity was essential because it guided us in selecting the right pretrained language models to build upon. We knew that starting with a solid foundation was critical, so we turned to the Hugging Face Hub, which offers a wide array of pretrained Masked Language Models (MLMs). Our goal was to find models that not only supported the French language but also had the potential to perform well in the healthcare context.

To make our selection process systematic, we utilized a dedicated notebook designed to guide us through evaluating potential models. This notebook served as a central hub where we conducted experiments, recorded results, and ultimately made informed decisions. We compiled a list of candidate models, considering factors like language support, relevance to the healthcare domain, and model architecture. These models were then evaluated and ranked based on their ability to accurately identify medical entities in French text. This evaluation was critical, as it allowed us to narrow down our choices to the model most likely to succeed in our specific task.

With our pretrained model selected, we moved on to preparing our training data. This step involved converting CoNLL files into a format optimized for training Transformer models, specifically tailored for our NER task. We created splits for the dataset, dividing it into training, validation, and test sets. This preparation was crucial

CHAPTER 8 MLOPS AND CONCLUSION

to ensure that our model could learn effectively and generalize well to new, unseen data. Once the dataset was ready, we saved it for future use, ensuring that our work was preserved and could be easily accessed during the training and evaluation processes.

Loading the training dataset into the environment was the next step, followed by initializing the various components necessary for training. This included setting up the model, optimizer, and learning rate scheduler, all of which play a vital role in ensuring the training process runs smoothly. With everything in place, we began the actual training. This phase involved iteratively updating the model's weights based on the training data, gradually improving its ability to recognize entities in French healthcare texts. Throughout this process, we closely monitored the training and evaluation losses. Keeping an eye on these metrics was essential for gauging the model's progress and identifying any potential issues, such as overfitting or underfitting, which could hinder the model's performance.

After the training phase, our attention turned to evaluation. We needed to ensure that our model wasn't just performing well on the training data but could also handle new, unseen examples with the same level of accuracy. To achieve this, we used an evaluation notebook to assess the model's performance on both validation and test datasets. This step provided us with key insights into the model's capabilities, allowing us to fine-tune hyperparameters and make any necessary adjustments.

Once we identified the best-performing model, we saved it for deployment. However, before moving forward, we conducted additional testing on a separate test set to confirm the model's reliability. This final step was crucial, as it ensured that the model was not only accurate but also robust enough to be deployed in real-world healthcare applications.

Looking back, the training and evaluation phases were integral to shaping the effectiveness of our Healthcare NER model. By carefully selecting a pretrained model, fine-tuning it with relevant data, and rigorously evaluating its performance, we laid a strong foundation for the model's success. The key takeaway from this phase is the importance of a thoughtful, methodical approach to training. Each decision, from model selection to hyperparameter tuning, plays a significant role in the model's final performance. As we move forward to deploy this model, we do so with confidence, knowing that we have built something precise, reliable, and ready to make a meaningful impact in the field of healthcare.

Here are the takeaways and lessons learned from the training and evaluation phase:

- Importance of Clear Problem Definition: Starting with a well-defined problem was essential in guiding the selection of appropriate pretrained models. Understanding the specific challenges of recognizing medical entities in French set the direction for the entire process.

- Systematic Model Selection: Using a structured approach to evaluate and rank pretrained models ensured that the chosen model was well-suited for our task. This systematic evaluation saved time and increased the likelihood of success.

- Preparation of High-Quality Data: Converting and splitting the dataset into training, validation, and test sets was crucial for effective model training. Proper data preparation is foundational to achieving good model performance.

- Continuous Monitoring of Training Progress: Regularly tracking training and evaluation losses helped in identifying and addressing potential issues early, such as overfitting or underfitting.

- Rigorous Evaluation: The thorough evaluation process, including testing on unseen data, ensured that the model was not only accurate but also generalizable and robust enough for real-world applications.

- Model Fine-Tuning and Hyperparameter Optimization: Fine-tuning the model and optimizing hyperparameters were key to improving the model's performance, particularly in handling the nuances of the French language in a healthcare context.

- Final Validation and Testing: Conducting final tests on a separate test set provided confidence in the model's reliability, confirming that it was ready for deployment.

Lessons Learned:

- Thoughtful Selection of Pretrained Models is Critical: Not all pretrained models are created equal, especially when dealing with specialized domains like healthcare. Choosing the right model requires careful consideration of domain relevance and language support.

- Data Quality Directly Impacts Model Success: The quality and preparation of the dataset are just as important as the model itself. Poor data can lead to poor model performance, regardless of the sophistication of the model.

- The Value of a Methodical Approach: Following a systematic, step-by-step process in training and evaluation leads to better outcomes. Skipping steps or rushing through the process can result in suboptimal models that don't perform well in real-world scenarios.

- Adaptability Is Key: Each project has its unique challenges, and being able to adapt the training and evaluation process to meet those challenges is vital. Flexibility in approach allows for the fine-tuning of methods to achieve the best possible results.

- Thorough Evaluation Cannot Be Overlooked: It's easy to focus on training, but rigorous evaluation is what ultimately determines the model's readiness for deployment. Skipping or underestimating this phase can lead to unexpected failures in production.

These takeaways and lessons underscore the importance of a careful, deliberate approach to training and evaluating NLP models, especially in specialized fields like healthcare.

Model Deployment and Monitoring

Transitioning from a well-trained model to a fully integrated solution within a cloud environment is a critical step in bringing the benefits of our work to life. For MedTALN Inc., the challenge was not just to deploy our finely tuned Healthcare NER model but to ensure it seamlessly fits within the Oracle Cloud Infrastructure (OCI) ecosystem. The focus here is to deploy the model in a way that maximizes both performance and utility, making it ready for real-world application.

Deploy

After rigorous training and evaluation, we arrived at a set of candidate models. The best-performing model was selected for deployment. Before we could deploy it, the first step was to save this model in the OCI Model Catalog. The Model Catalog acts as a central

CHAPTER 8   MLOPS AND CONCLUSION

repository where the model is safely stored. This not only ensures that the model is easily shareable among team members for collaboration but also allows us to reload the model into various working environments whenever needed.

With the model stored securely in the Model Catalog, the next step was to deploy it as an HTTP endpoint using OCI's Data Science service. This service handles the complex operations of deploying the model, transforming it into a responsive and scalable HTTP endpoint. OCI takes care of all the underlying infrastructure, including compute provisioning and load balancing, which ensures that the model is always ready to handle incoming requests efficiently. Although deploying models as serverless functions is an alternative within OCI, for our case study, we focused on the traditional HTTP endpoint deployment method.

Deployment involves provisioning several key OCI resources. Given MedTALN Inc.'s expertise with OCI, we had an advantage in managing and fine-tuning these resources to ensure that the solution could handle the expected real-world data volumes. The essential resources provisioned included

- Load Balancer: This component distributes incoming traffic across multiple model servers hosted on virtual machines, ensuring that no single server becomes a bottleneck.

- Virtual Machine Pool: Each virtual machine in the pool hosts an identical copy of the model server, along with a dedicated conda environment. For our project, we used the "Natural Language Processing for CPU on Python" conda environment, which includes key deep learning libraries like Hugging Face and PyTorch.

Once the model was deployed and the endpoint was set up, it was ready to handle incoming requests. Using this model involves sending text data to its "predict" endpoint, which then returns predictions. In our case study, these predictions are focused on identifying healthcare entities within French texts, and our primary goal was to ensure the model could perform this task with high accuracy.

For the deployment phase, we followed a series of concrete steps:

1. Model Storage: We deposited the trained NER model into the OCI Model Catalog.

2. Deployment via OCI Data Science Service: We deployed the model as an HTTP endpoint, making necessary adjustments to the load balancer and VM pool to meet our deployment requirements.

CHAPTER 8   MLOPS AND CONCLUSION

3. Model Invocation: We tested the deployed model by sending sample French texts to its predict endpoint and assessed its ability to accurately recognize named entities.

Deploying and invoking an NLP model is about more than just making it operational; it's about ensuring that the model consistently delivers reliable results in real-world applications. OCI's robust infrastructure provided the perfect platform for our Healthcare NER model, enabling real-time entity recognition for French texts. Even with MedTALN Inc.'s existing OCI expertise, the insights from our external NLP consultant were invaluable during this phase, helping us fine-tune and optimize the deployment for the best possible performance.

As we move forward, monitoring and maintaining the deployed model will be crucial to ensure it continues to operate at peak performance. This next phase will focus on setting up logging, tracking metrics, and implementing strategies to maintain the model's accuracy and efficiency over time.

Here are the takeaways and lessons learned from the model deployment phase:

- Importance of Choosing the Right Compute Shape: For many NLP tasks, modern CPUs can provide sufficient performance at a lower cost. Using flexible shapes for model deployment allows you to customize the number of OCPUs and the amount of memory when launching or resizing the model deployment virtual machine (VM).

- Optimizing Resource Usage: Regularly review and adjust the size of your compute instances. Ensure you're using the most cost-effective instance types that meet your performance requirements. Deactivate the OCI Data Science Model Deployments when not in use to pause billing.

Lessons Learned:

- Oracle ADS HuggingFacePipelineModel Class: Leveraging this class was key to accelerate the development of our automated deployment process. This class is specifically designed to simplify the deployment of Hugging Face models, such as our fine-tuned Healthcare NER model, on Oracle Cloud Infrastructure (OCI).

- Fine-Tuning Deployment Settings: Adjusting settings for load balancers and virtual machine pools maybe necessary to tailor the deployment to a specific needs.

CHAPTER 8   MLOPS AND CONCLUSION

Monitor

After deploying our Healthcare NER model, the next crucial step is to ensure it continues to perform at its best through effective monitoring and maintenance. This phase is all about keeping a close eye on how the model operates in real-world conditions, ensuring that it remains reliable and responsive, and making any necessary adjustments to maintain its accuracy and efficiency.

Monitoring is a cornerstone of the MLOps approach, which emphasizes a disciplined, standardized, and automated method for managing the entire machine learning life cycle. In this context, monitoring goes beyond just keeping the model running; it involves actively overseeing various job performance metrics, setting up alarms, and maintaining a comprehensive log repository. These practices help us understand how the system behaves under different conditions and allow us to react promptly to any changes that could affect performance.

For MedTALN Inc., monitoring infrastructure utilization is key. By observing metrics like CPU usage during peak NLP activities, the team can scale resources up or down as needed, ensuring the model can handle varying data loads without any hitches. Logging plays a crucial role in this process, providing detailed records of events that can be analyzed later to optimize processes and troubleshoot issues. For instance, by channeling logs from both training and inference activities into the OCI Logging service, debugging becomes much more straightforward, and tracking the model's performance over time is simplified.

Continuously monitoring an NLP model helps ensure that any deviations in model predictions—such as those caused by changes in the input data—are detected early. When these drifts occur, the model can be retrained and redeployed promptly, minimizing any negative impact on performance. This ongoing management and optimization are crucial for maintaining the relevance and accuracy of the NER model as it adapts to evolving data patterns and demands.

To showcase the monitoring capabilities in action, consider a scenario where we use our testing notebook to send inference calls to the deployed Healthcare NER model. By doing so, we can observe how the model processes these requests and make sure it's performing as expected. For instance, we could send various French medical texts to the model and analyze its predictions. Monitoring these interactions helps us verify that the model correctly identifies healthcare entities, such as medications or medical procedures, ensuring its predictions align with our expectations.

The OCI Logging service is indispensable here, capturing all the essential information about requests sent to the model endpoint. The access log, for example, records the details of each request, while the predict log captures the output from the model, such as the results of the "score.py" script. This detailed logging not only helps in verifying the model's performance but also aids in identifying any anomalies or issues that may need attention.

In addition to logging, we can monitor the model's health, capacity, and performance using OCI's built-in metrics. These metrics include CPU utilization, memory usage, and network throughput, which give us a clear picture of how the model is performing under different conditions. From the OCI console, we can access the metric space under the model deployment resources to view these metrics in real time. This allows us to track request counts, latency, and bandwidth, all of which are critical for ensuring the model's responsiveness and efficiency.

If any metric crosses a specified threshold, alarms can be set up to alert the team, prompting immediate action to prevent potential issues from escalating. This proactive monitoring ensures that the model remains stable and continues to meet the performance standards required for real-world healthcare applications.

By effectively monitoring our Healthcare NER model on OCI, we can ensure that it operates optimally and continues to deliver accurate and reliable predictions. This not only supports MedTALN Inc.'s objectives in maintaining data protection and compliance but also ensures that the model remains a valuable tool in the organization's broader healthcare analytics efforts.

Takeaways:

- Proactive Monitoring: Regular monitoring of the model's performance and infrastructure utilization is essential to maintaining optimal operation and preventing potential issues before they impact performance.

- Comprehensive Logging: Detailed logging is crucial for troubleshooting, optimizing processes, and ensuring the model continues to perform as expected in real-world scenarios.

- Advanced Metric Tracking: Monitoring key metrics like CPU usage, memory, and network throughput helps in understanding the model's behavior and adjusting resources accordingly.

Lessons Learned:

- The Importance of MLOps: Implementing MLOps practices, such as continuous monitoring and maintenance, is critical for sustaining the long-term success of NLP models in production environments.

- The Need for Continuous Optimization: Regular updates and systematic refinements are necessary to ensure the model remains accurate and relevant as data evolves.

- Continuous Monitoring Is Essential: Postdeployment, continuous monitoring and maintenance are vital to keep the model performing at its best, ensuring it adapts to any changes in real-world data.

Responsible AI

We couldn't finish this book without addressing an important subject that, while referred to throughout the chapters from the perspective of cost-saving opportunities and strategies, has not yet been discussed explicitly. Some may overlook it, but it is a critical topic that deserves attention: the environmental impact of training and deploying NLP models, particularly Transformer-based models.

Carbon emissions and their relation to AI in general is an integral part of building "responsible AI." While the terms "responsible AI" and "AI ethics" are sometimes used interchangeably, they address different aspects of ensuring AI's impact on society is positive. Responsible AI focuses on the practical implementation of ethical principles, such as creating transparent, fair, and accountable systems. AI ethics, however, delves into the moral and philosophical implications of AI's development and use.

The environmental cost of AI, particularly with models like Transformers, has garnered significant attention in recent years. Headlines have highlighted that training a single AI model can produce as much CO_2 as five cars do over their lifetime. However, this is only sometimes true and depends on several key factors. The primary determinant of a model's carbon footprint is the type of energy used during training. If renewable energy sources such as solar, wind, or hydroelectric power are utilized, the carbon emissions are negligible. In contrast, training on nonrenewable energy sources like coal significantly increases the carbon footprint due to high greenhouse gas emissions.

Another critical factor is the duration of the training process. The longer a model is trained, the more energy it consumes, directly correlating to higher carbon emissions. This cumulative effect becomes particularly concerning when training large models over extended periods—days, weeks, or even months. Furthermore, the hardware used during training plays an important role. Some GPUs are more energy-efficient than others, meaning that using high-efficiency GPUs and maximizing their utilization can substantially reduce energy consumption and, by extension, the carbon footprint.

In addition to energy sources, training time, and hardware efficiency, other aspects such as I/O operations and data management also contribute to the overall carbon emissions. However, the three factors mentioned—energy type, training duration, and hardware efficiency—are the most significant and should be the primary focus when aiming to minimize the environmental impact of AI development.

Understanding the implications of energy sources and carbon intensity is vital. For example, the carbon footprint of a cloud computing instance can vary dramatically depending on its geographic location. For instance in Mumbai, India, may emit 920 grams of CO_2 per kilowatt-hour, while in Montreal, Canada, the emissions could be as low as 20 grams of CO_2 per kilowatt-hour. This vast disparity—nearly 40 times more carbon emissions in Mumbai than in Montreal—can accumulate quickly, especially during prolonged training sessions. Thus, selecting a computing instance in a low-carbon region is one of the most impactful decisions you can make to reduce emissions.

Additionally, leveraging pretrained models is akin to recycling in the realm of machine learning. When you use pretrained models, you avoid the carbon emissions associated with training from scratch, as no additional training is required. This approach, combined with fine-tuning existing models rather than training entirely new ones, can significantly cut down on energy use and emissions. For instance, if you find a model that nearly meets your needs, fine-tuning its final layers to align with your specific goals is much more energy-efficient than training a large Transformer model from the ground up.

Several tools are available to help estimate and track the carbon emissions generated during AI model development. The "Machine Learning Submissions Calculator" is one such tool, allowing users to manually input details like hardware specifications, usage hours, and geographic location to estimate the resulting CO_2 emissions. Another tool, CodeCarbon, offers a programmatic solution. Once installed via pip, CodeCarbon runs alongside your code, tracking energy usage throughout the training process and providing a detailed CSV report of the emissions generated. This allows for easy comparisons and a better understanding of the environmental impact. CodeCarbon also features a visual interface that compares emissions to everyday activities, such as driving a car or watching television, providing further context to the results.

For our case study, building a performant yet cost-effective NLP model was a priority due to budget constraints imposed by MedTALN Inc. on this particular NLP initiative. However, beyond the financial considerations, the strategies we outlined in this book are economically beneficial and critical for reducing carbon emissions associated with developing Transformer-based models. As the global demand for AI-powered solutions continues to rise, so does the environmental impact of training these large-scale models. Therefore, it is essential to consider the carbon footprint of AI development alongside cost and performance.

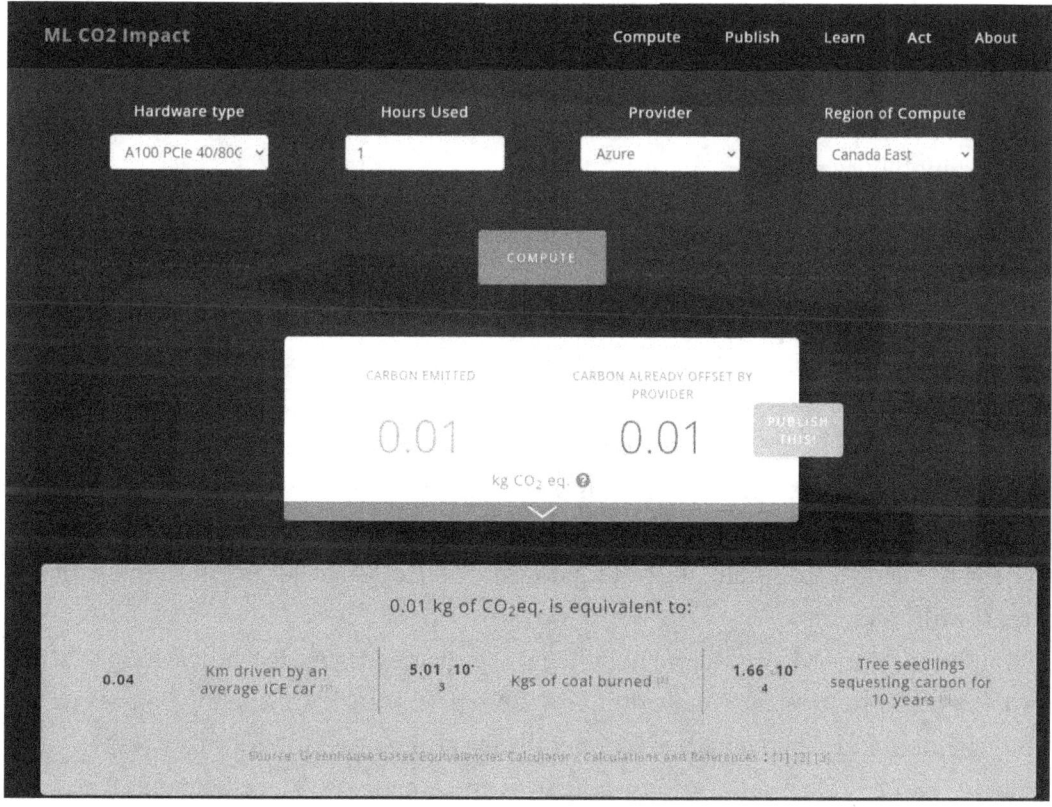

Figure 8-15. Estimate for our case study CO_2 emission[2]

---

[2] The tool Machine Learning CO2 Impact Calculator can be found at https://mlco2.github.io/impact/. Since OCI is not included in the list of providers, we chose Azure to be able to select the Canada East region, which is important in the calculation due to its reliance on clean energy.

CHAPTER 8   MLOPS AND CONCLUSION

Throughout the book, we've explored approaches like using prelabeled datasets, using transfer learning where we fine-tune pretrained models rather than training from scratch, leveraging GPU only for training time, etc. These strategies not only help reduce costs in terms of computational resources but also have a broader environmental impact by cutting down on energy consumption and carbon emissions. While cost saving has been a focus, these same methods contribute to more sustainable NLP practices.

Additionally, Oracle Cloud Infrastructure offers several features that support responsible AI development and reduced carbon emissions. Notably, 100% of OCI data centers in Europe run on renewable energy, with a goal to achieve this globally by 2025 (Oracle, n.d.). Additionally, OCI utilizes energy-efficient processors specifically designed for machine learning tasks and employs optimized cloud infrastructure that allows for better energy management and location optimization to minimize the carbon footprint.

As show in Figure 8-15, by implementing the cost-saving strategies discussed in this book, it was possible to build high-performance NLP models that is not only cost-effective but also environmentally responsible, aligning with both budgetary constraints and sustainability goals.

Summary

As we conclude our journey through NLP on Oracle Cloud Infrastructure, let's take a moment to reflect on the path we've traveled together. Through the fictional yet expertly crafted narrative of John Doe and MedTALN Inc., we've explored the foundations of NLP, tackled the complexities of real-world implementations, and uncovered the powerful potential of OCI for advanced NLP solutions. This book has been more than just a guide; it's been a practical roadmap, illustrating how the concepts we've covered can be applied in real-world scenarios.

Although centered around the healthcare sector, the lessons learned throughout this book extend far beyond any single domain. From preparing high-quality datasets to fine-tuning models and deploying them efficiently, the strategies and best practices we've discussed can be adapted to various industries and languages. The challenges faced in healthcare NLP—such as data scarcity, language specificity, and domain relevance—are universal hurdles that NLP practitioners in any field will encounter.

The methods we employed are grounded in real-world experiences from actual projects at typica.ai, and though the MedTALN Inc. narrative is fictional, the solutions we provided are practical and tested. This book serves as a toolkit for NLP professionals, offering actionable insights to navigate the intricate processes of dataset preparation, model training, and deployment with confidence.

Our case study approach has shown how to tackle these challenges effectively, but we also understand that the journey continues beyond technical implementation. Critical considerations like cost-effectiveness, environmental impact, and the ethical dimensions of NLP have come to the forefront of modern machine learning practices. As we look toward the future, we've touched upon emerging trends in NLP and the evolving landscape of cloud computing, preparing you to stay ahead in this rapidly changing field.

Ultimately, knowledge is only as powerful as its application. Now, as this book comes to a close, the real challenge begins: how you will apply these insights in your own work. This final chapter is your launchpad—designed to empower you to take the lessons learned and bring them into your projects, balancing technical expertise with strategic business decisions. While our journey together may be ending, your own adventure in NLP on OCI is just beginning. We hope this book has equipped you with the tools, the confidence, and the curiosity to continue exploring, innovating, and making a lasting impact in the world of NLP.

Reference

Oracle. (n.d.). *Oracle Cloud sustainability*. Retrieved from Oracle: `https://www.oracle.com/ca-en/sustainability/green-cloud/`

Index

A

A/B testing, 363
Accelerated Data Science (ADS) library, 323
AI, *see* Artificial intelligence (AI)
Artificial intelligence (AI), 3, 52
ASR, *see* Automatic speech recognition (ASR)
Automated machine learning (AutoML), 57
Automatic speech recognition (ASR), 55
AutoML, *see* Automated machine learning (AutoML)
AutoTokenizer, 270

B

BERT-large, 255
Blue-green deployments, 363

C

CAM, *see* Customer Account Management (CAM)
Cloud computing, 395
Customer Account Management (CAM), 245

D, E

Data Labeling Service (DLS), 66, 195, 245

Data Science Environment setup
   notebook sessions
      costs, 142
      CPU-based compute shape, 144–146, 148, 150–155, 157–162
      GPU-based notebook session, 162–169
      JupyterLab file, 143
      local directories, 142
      teams, 143
   project, 139, 141
Dataset labeling
   annotations, 227, 229
   dataset creation, 234–238
   dataset import, 216–221, 223–227
   dataset list, 228
   details page, 229
   DLS, 215, 216
   DLS UI, import dataset, 239–244
   instructions window, 230
   QA, 231–233
   record count limit, 245, 246
Dataset life cycle
   candidate healthcare NER dataset, 187–191
   dataset selection, 182
   Hugging Face Hub, 183–187
   training dataset preparation, 180, 181
Dataset preparation techniques
   cost comparative analysis, 176–179
   cost saving, 174

INDEX

Dataset preparation techniques (*cont.*)
    labeled datasets, 172, 173
    OCI ML Services, 171
    off-the-shelf datasets, 174, 175
    prelabeled datasets, 172
Datasets, 10
Dedicated Virtual Machine
        Hosts (DVH), 46
Deep learning (DL), 10, 13
DL, *see* Deep learning (DL)
DLS, *see* Data Labeling Service (DLS)
DVH, *see* Dedicated Virtual Machine
        Hosts (DVH)

F

Fine-tuning pretrained models, 260

G

GenAI, 69
General-purpose language models, 257
Generative Language Model (GLM), 253
GLM, *see* Generative Language
        Model (GLM)
GPT-3, 255

H

Healthcare NER
    architecture
        methodology, 90
        optimal approach, selection, 96, 97
        preselection, candidate solution
            options, 90–95, 97
Healthcare NER model evaluation
    GPU-based notebook, 300
    initialization, 301–303
    load best checkpoints, 304, 305
    model best checkpoints, 305–308,
        310, 311
    save best model, 312
    test best model, 313–317
Healthcare NER model fine-tuning
    automated checkpoint
        selection, 296–299
    strating training, 292, 293
    training and evaluating, 276
    training dataset, creating notebook,
        276, 277, 279–282
    training notebook
        GPU-based notebook, 284, 285
        high-level data flow interactions,
            283, 284
        initialize training, 287–291
        loading training dataset, 285, 287
    visual analysis, 293, 294
Healthcare NLP models, 73
HF, *see* Hugging Face (HF)
Hugging Face (HF), 262
    ecosystem, 334
    libraries, 249
    transformers library, 271
Human-in-the-loop (HITL) validation
    method, 231

I

IaaS, *see* Infrastructure as a Service (IaaS)
IAM, *see* Identity and Access
        Management (IAM)
Identity and Access Management (IAM),
    38, 51
Infrastructure as a Service (IaaS), 35

J, K

JupyterLab notebooks, 66

INDEX

L

Language models (LMs), 11
    acronyms, 253, 254
    cost saving strategies, training phase, 259, 260
    domain specific model, healthcare, 257–259
    evolution
        LLMs, 252
        neural language models, 250
        PLMs, 252
        transformers, 252
        word embeddings, 251
    pretrained MLM selection, 262, 263, 265–272, 274, 275
    taxonomy, PLMs, 254–257
    transfer learning-basedfine tuning, 260–262

Large language models (LLMs), 11, 55, 56, 252

LLMs, *see* Large language models (LLMs)

LMs, *see* Language models (LMs)

load_dataset() function, 196

M

Machine learning (ML), 10, 13, 56, 57

Masked Language Models (MLM), 24, 253, 261, 262, 367

Massachusetts Institute of Technology (MIT), 4

MedicalNER_Fr, 191

MedTALN Inc.
    advanced statistical algorithms, 74
    components, 75
    healthcare analytics solution, 75
    healthcare NER initiative, 79–83

    healthcare NLP, 76, 77
    inception phase, NER
        assembling team, 88, 89
        NLP experts, 85
        requirements, 86, 87
        scope, 85, 86
    solution
        development, 105
        high level approach, 100, 101
        high level architecture, 97–99
        project preparation, 101–104

MIT, *see* Massachusetts Institute of Technology (MIT)

ML, *see* Machine learning (ML)

MLM, *see* Masked Language Models (MLM)

MLOps
    data science pipelines, 365
    NER model life cycle
        data preparation, 380–383
        model deployment/monitoring, 386–391
        model training/evaluation, 383–385
        responsible AI, 391, 392, 394
    OCI data science pipelines
        DevOps principles, 365
        example, 366, 367
        pipeline creation, 367, 368, 370–373, 375–380

Model deployment
    authenticate
        call mode summary, 346
        generated model artifacts, verify, 347
        initialize hugging face pipeline, 342
        manually correct score.py, 344
        prepare model artifact, 342–344

INDEX

Model deployment (*cont.*)
    resource principals, 341
    run introspection test, 344, 345
  deploy/invoke, 356
  high-level deployment process, 335
  libraries, 334
  notebook
    deploy/invoke, 353–357
    model to the model
      catalog, 347–353
    resources, 340, 341
  OCI Data Science, 336–338
  Oracle ADS HuggingFacePipeline
    Model, 338–340
  Oracle Data Science Model Catalog,
    335, 336
Model inference
  cost saving strategies, 326–328
  preparing environment
    publish custom condav env.,
      333, 334
    setting up logging, 329–333
    setting up policies, 328, 329
  vs. training, 324, 325
Monitoring and maintenance
  inference calls, NLP model, 358
  logs, 360
  metrics, 361, 362
  testing notebook, 358, 359
MultiCoNER v2, 189
MultiNERD, 189

N

Named Entity Recognition (NER), 6, 24,
    79, 81, 84, 88, 109, 171, 172, 179,
    181, 182, 275, 317, 324, 365, 383

Natural Language Processing (NLP),
    74, 174, 179, 250, 317
  adoption
    data, 32
    models, 29–31
    team, 32, 33
  advancement, 5
  approaches and workflow, 9
  challenges, 14
  components, 10–12
  computer science and linguistics, 3
  history, 4
  tasks, 6–8
  tools, 12–14
  transformers, 15
    architecture, 15–20
    hugging face ecosystem, 25–27
    taxonomy, 20, 22, 23
    transfer learning, 23–25
NER, *see* Named Entity Recognition (NER)
Network security groups (NSGs), 43
Neural language models, 250
NLP, *see* Natural Language
    Processing (NLP)
NSGs, *see* Network security groups (NSGs)

O

OCI, *see* Oracle Cloud Infrastructure (OCI)
OCI Data Labeling service (DLS), 215
OCI Data Science Notebook Sessions, 246
Oracle Cloud Infrastructure (OCI), 74, 86,
    88, 101, 323, 362, 366, 386, 388
  AI
    infrastructure, 57, 58
    ML services, 56, 57
    services, 54–56

stack, 53, 54
strategy, 52
history, 35–37
NLP
    aI samples, 66
    data labeling, 64–66
    data science, 62–64
    hih level flow, 68, 69
    implementations, 59
    language, 59–61
    solutions, 70
regions/realms, 39
resources, 43, 45
    compute instances, 43–46
    IAM, 51
    networking, 41–43
    storage, 47–50
services, 37, 38
tenancy/compartment, 40, 41
Oracle's Accelerated Data Science (ADS), 336, 338, 362
Oracle Services Network (OSN), 43
OSN, *see* Oracle Services Network (OSN)

P

Part-of-Speech (POS), 79
Personally Identifiable Information (PII), 59, 61, 78
PHI, *see* Protected Health Information (PHI)
PII, *see* Personally Identifiable Information (PII)
PLMs, *see* Pretrained language models (PLMs)
"plot_distributions" function, 202
plot_radar() function, 310
POS, *see* Part-of-Speech (POS)

.prepare() method, 339
Pretrained language models (PLMs), 249, 252, 253
Protected Health Information (PHI), 61

Q

QA, *see* Quality assurance (QA)
Quality assurance (QA), 231

R

RAG, *see* Retrieval-augmented generation (RAG)
Random undersampling approach, 204
Retrieval-augmented generation (RAG), 52, 55, 56
Rule-based systems, 5

S

SaaS, *see* Software as a Service (SaaS)
.save() method, 339
Security list, 42
Self-attention mechanism, 252
SLM, *see* Small language model (SLM)
Small language model (SLM), 254
Software as a Service (SaaS), 35
.summary_status() method, 346
Support vector machines (SVM), 12
SVM, *see* Support vector machines (SVM)

T

Tenancy preparation
    compartment creation, 111, 112, 114
    data labelers, IAM setup, 134–137, 139
    IAM setup, data scientists architecture, 125

INDEX

Tenancy preparation (*cont.*)
    dynamic groups, 129, 130
    policies, 131, 133
    users/groups, 126–128
  identity/security, 124
  network configuration, 114, 116–118
  setup activities, MedTALN, 111
  storage, 119–124
  strategies, 110
Term Frequency-Inverse Document Frequency (TF-IDF), 12
Text-to-speech (TTS) models, 55
TF-IDF, *see* Term Frequency-Inverse Document Frequency (TF-IDF)
Tokenization, 10
Tokens, 10
Training dataset preparation
  data wrangling process, 194
  high-level architecture, 192, 193
  hugging face dataset, 194
  Notebook
    loading, 196, 197
    wrangling steps, 198–203, 205–211, 213–215
  workflow, 192

"Train_ner_models" notebook, 301
Transfer learning, 23
Transformer-based models, 252
"Transformers" library, 267
TTS models, *see* Text-to-speech (TTS) models

U

"undersample_label" function, 204

V

VCNs, *see* Virtual Cloud Networks (VCNs)
.verify() method, 339
Virtual Cloud Networks (VCNs), 38, 42, 111
Virtual machine (VM), 38
VM, *see* Virtual machine (VM)
VM.GPU.GU1, 163

W, X, Y, Z

Wikiann, 188
WikiNER, 188

GPSR Compliance

The European Union's (EU) General Product Safety Regulation (GPSR) is a set of rules that requires consumer products to be safe and our obligations to ensure this.

If you have any concerns about our products, you can contact us on

ProductSafety@springernature.com

In case Publisher is established outside the EU, the EU authorized representative is:

Springer Nature Customer Service Center GmbH
Europaplatz 3
69115 Heidelberg, Germany

www.ingramcontent.com/pod-product-compliance
Lightning Source LLC
LaVergne TN
LVHW080310260326
834688LV00038B/1034